WAI PASIFIKA

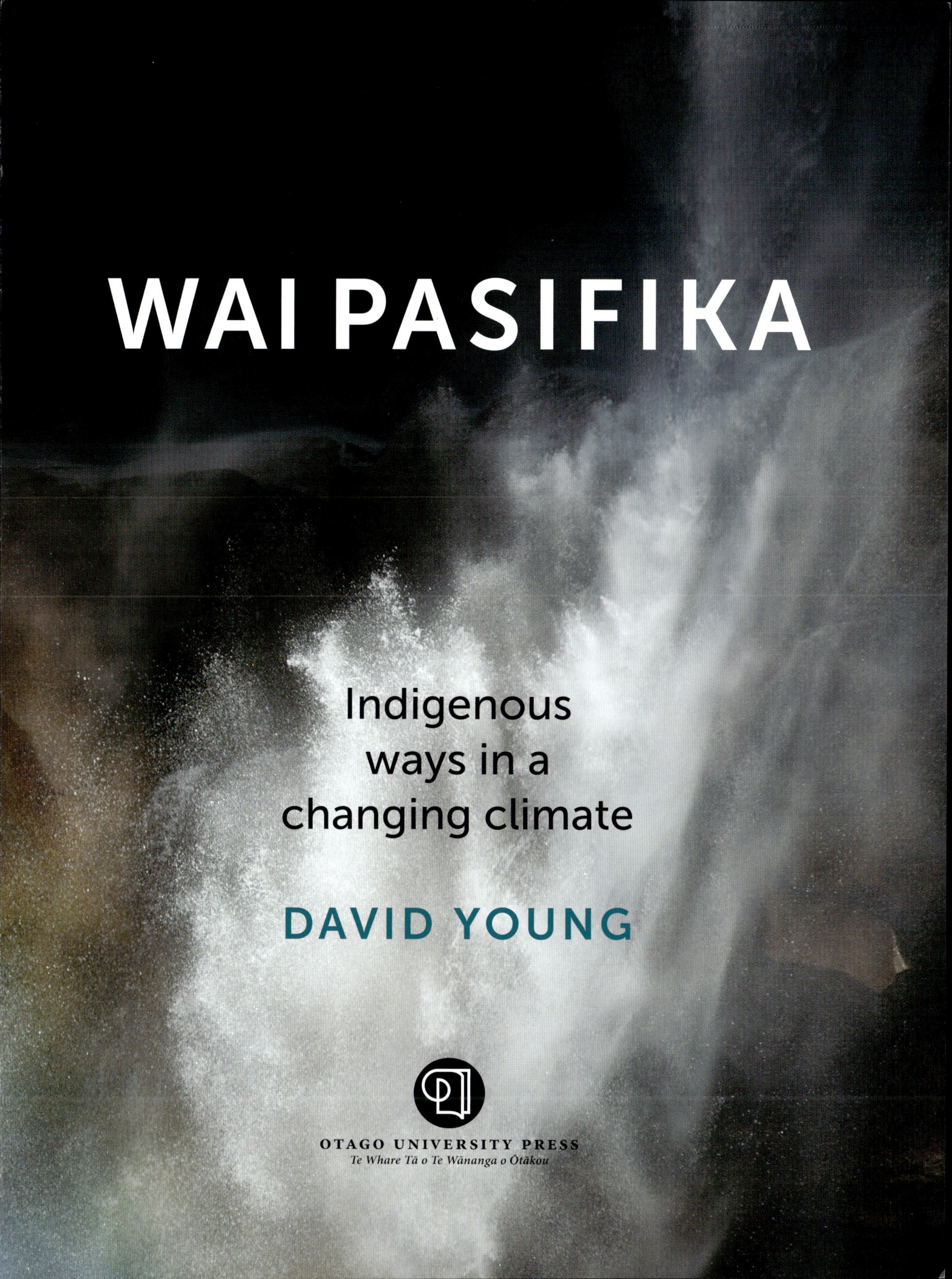

WAI PASIFIKA

Indigenous ways in a changing climate

DAVID YOUNG

OTAGO UNIVERSITY PRESS
Te Whare Tā o Te Wānanga o Ōtākou

Published by Otago University Press
Te Whare Tā o Te Wānanga o Ōtākou
533 Castle Street
Dunedin, New Zealand
university.press@otago.ac.nz
www.otago.ac.nz/press

First published 2021

ISBN 978-1-99-004807-4

A catalogue record for this book is available from the National Library of New Zealand.

Published with the assistance of Creative New Zealand

creative nz
Arts Council of New Zealand Toi Aotearoa

Editors: Gillian Tewsley and Imogen Coxhead
Design and layout: Fiona Moffat
Index: Lee Slater
Cover and title page photographs: Aliscia Young
Internal photographs by David Young unless otherwise credited

Printed in China through Asia Pacific Offset

FSC
www.fsc.org
MIX
Paper from responsible sources
FSC® C136333

Contents

Lake Wānaka from Roys Peak: new snow is one of the elemental foundations of fresh water.
Photo Richard Sidey

PROLOGUE

Pulsing wisdom

Water's timeless passage marked by smoothed river stones, alpine river, Aotearoa. *Photo Richard Sidey*

The inception of this project was over 30 years ago when a tohunga told me a story from the Māori sacred realm of wisdom that differentiated waters from the mountains intrinsically, each according to its source.[1]

The distinctions are similarly asserted in the gift of Tāne Pūkohurangi, god of mist, ice, spring and snow: 'I a ia te ua, te kohu, te hukapapa, te hukarere. He is the author of rain, mist, ice and snow.'[2] When I mentioned this to eminent Judaeo-Christian scholar Don Akensen, he grinned. 'That's very rabbinic,' he said. It was one of those occasions when entirely separate pathways of ancient knowledge appear to accord, not just with each other but, as it turns out, with modern scientific understanding – such as in Dale Stokes and Doc White's *The Fish in the Forest*:

> Over long periods of time, the slight differences in heavy/light isotype balance in the environment can create distinctive isotopic signatures. For instance, the processes of evaporation and condensation alter the isotopic composition of hydrogen molecules in water and create different isotopic signatures for water falling as rain, evaporating from the surface of the sea, or existing in a glacier high on a mountainside.[3]

As we shall see, for Māori, the workings of Tāne Pūkohurangi are always fundamental to the waters of life. According to Ngārangi Rangiuia of Aitanga a Hauiti, an authoritative voice on Māoritanga, these qualities are covered in old Māori knowledge by the term kū a makere, or makere whatu. 'That's a reference to water that's up in space. So that water, comes down – "chi-chi-chi", trickles down.' However, Ngārangi believes that makere whatu (literally 'a stone that's moving – through the heavens, like somebody tossed it and it's got a trail it leaves behind') has today come to mean simply 'hail'.

There is also more than a hint here that the old knowledge of makere whatu might gesture to the relatively recent science that informs us that much of our green planet's quantum of water arrived from outer space, deep-frozen onto giant asteroids. If this was known to Polynesians, originally one of eight indigenous tribes in Taiwan, how did such fine distinctions of water knowledge serve Māori? One might speculate that, in the sometimes extreme conditions of the Central Plateau in the North Island, for example, it may have given an edge that enabled in-stream patterns for harvesting fish. Indeed, science has found that the kōaro, once abundant in many parts of Aotearoa New Zealand, including Lake Rotoaira on the Volcanic Plateau, is commonly 'associated with spring-fed tributary lakes'. The kōaro is quite temperature sensitive, 'preferring waters around 13°C' – the usual temperature of such lakes.[4] Once again, such explanations bring me up short: the reach and depth of indigenous knowledge, even in its modern outlines, often seems to lie, as it were, in an excavation offering shards and jaspers.

This was a reminder that science can be as wondrous as the explanations that come out of earth-rooted cultures. Polynesians, like the inhabitants of Oceania generally, have always lived in a state of heightened awareness of the profundity and subtlety of nature's moods and interrelationships. Theirs was a holistic view of the world and their place in it. As David Abram argues, these 'oral cultures' were oral not simply in the transmission of their knowledge, but in ways that embodied every aspect of their encounter with the world, the cosmos, that enfolded them. In this light, science may be seen simply as the step-wise advance of scaffolding on the face of the deep unknown.

From the 1950s, the recognition of traditional cultures generally and their retention and integration with the modern world was understood as an 'alternative development theory'.[5] It is often stated that we of the so-called developed world have much to learn from indigenous culture – far more, I believe, than we can ever imagine.[6] When simultaneously two thinkers from backgrounds as disparate as that of prominent sustainability advocate and former Green leader Sir Jonathon Porritt and veteran American social and political analyst Noam Chomsky each ask the world to listen carefully to the voices of the indigenous, then it probably is time to do just that. 'It's a kind of incredible irony,' Chomsky said in an interview, 'that all over the world the leading forces in trying to prevent a race to disaster are the Indigenous communities.'[7]

For focus on indigenous voices in the Pacific, my lens is freshwater. I have few illusions about this; when it comes to water, the idea of 'focus' is something of a paradox. As Seamus Heaney has written, moving freshwater is ever in 'the continuous present' – which is part of its attraction and mystery to humans.[8] Albert Wendt's beautiful line, 'Or the compulsive search by water always for its apt shape' speaks to its quicksilver quality of presence and elusiveness.[9]

The challenge is that in this context, elaboration on water themes quickly becomes a consideration of life, not least as it pertains to plants, trees and humans. And as Herman Melville proclaimed, perhaps more with the sea in mind, 'Yes, as everyone knows, meditation and water are wedded for ever.'[10]

In 'The Waking', American poet Theodore Roethke wrote:

> I came where the river
> Ran over stones:
> My ears knew
> An early joy.

Water droplets shaken from a whio's wings catch the light, Kahurangi National Park, Aotearoa.
Photo Richard Sidey

And all the water
Of all the streams
Sang in my veins
That summer day.[11]

From early childhood I have a number of memories like this: of pure joy, of water singing in the veins. As an adult, I felt this on the Manganuioteao, a pitching, cleaving, heaving thread of life, a river that binds the snows of Ruapehu to the upper Whanganui River. This is a riverscape I have found myself returning to often – to address threats to build hydroelectric dams; its eventual protection under a Water Conservation Order; its status as a sanctuary for New Zealand's whio (blue duck). Once to be found up and down streams the length of the country, this territorial duck and bioindicator of river health has been gravely diminished in numbers from its ubiquity of 150 years ago – notwithstanding the efforts of a more recent excellent recovery programme.

In the multi-laminate history of Māori, this river is also a traditional pathway into the volcanic interior, and it was the centre of power for one of the most powerful chiefs in nineteenth-century Aotearoa, Te Pēhi Tūroa. For his descendants, knowledge of their history anchors them not only into place, but into a timeless universe.

For a short pitch of water, the Manganuioteao has a distinctive energy. One summer afternoon I stood in stillness on a hill above the valley, the river snaking below me a kilometre away along a far cliff edge. As I watched, through my ears came the river's song – perhaps its mauri (life force) or maybe an auditory trick, who knows? – a persistent susurration, pulsing through my skull. All that was joy, contentment and a sense of moment. The Manganuioteao debouches into the Whanganui River through twin headlands. It was on the Whanganui where, urged on by two notable elders, Titi Tihu and Hikaia Amohia, both now deceased, my cultural exploration of freshwater began.[12] That river still stands as genius loci across time for the many strands of this book – although certainly not without references to other tribes, waterways and practices in Aotearoa.

Similarly affecting was an experience in January 2018, up one of the rivers leading out of an intricate cave system through a marble mountain in northwest Nelson. Exploration of these caves over many decades has been lovingly documented in words, photography and design in a book that is epic in subject matter and execution.[13] As we came into the widest part of what is in many places a river gorge, it was Richard, a photographer in this book, who noticed that he was picking up a tingling in his fingers and hands. I had been distracted by whio, but when he mentioned it, it was there, immediately obvious. Indeed, the tingling was so strong, it ran up our arms, almost to the elbows. If there isn't a scientific explanation yet for this – and it may be to do with a transition from a substrate of slate to limestone – I am sure one day there will be. However, I shall continue to think of it as mauri.[14]

When this book was almost completed I was beginning another large writing project. One day, under the stillness of the sequoia in our garden, I heard the thrum and quiver of claws enclosing the tensioned tynes of the branches. This vibration was repeated as the bird made a purposeful descent, criss-crossing from one suitable branch to the next; the sound from where I sat was almost marimba. It was a tūī, a bird so vocal it comes with a dual voicebox. I watched it descend until it reached the last branch above me. That was its station. From there it had to look down to see me, and to do this it had to cock its head to one side so as to cast its eye in my direction. As our eyes met the tūī broke into energetic bird language. The visit lasted some minutes, during which it continued to talk in bursts. I knew not what it spoke of, but the feeling was of reassurance.

The point is that nature – and not least water – is full of surprises, especially when we are open to them. And further, that the gift of indigenous people to their colonisers is that they offer systematic and different concepts of being in and experiencing nature.

At the heart of it all are shared insights rooted in millennia of intimate engagement with the natural world. As Anne Salmond writes: 'Māori ways of living and thinking are rarely thought of as being on the same footing as modernity and Western "civilisation", governance and science.'[15] For Polynesians this is about a place to stand, summed up in the term tūrangawaewae. In writing this book I realised that tūrangawaewae is not simply a place of grounded belonging and kinship, it is also the space that is created by

Fast flowing stream, Aotearoa. *Photo Richard Sidey*

a culturally evolved way of seeing the world. Tūrangawaewae is a way of being, a way of perception, of being enfolded by a culture with its beliefs, ethics and lore.[16] This place and space of being, with its particular focus on freshwater, is the thrust of it.

It is not about becoming Māori, or of any other indigenous culture. When these older, locally informed world views are not discounted, they can anchor us into a deeper appreciation of place. With that can come a sense both of belonging and of responsibility and the abandonment of entitlement or, as Ross Gibson puts it, 'presumption and violence'[17] towards the earth's creation. I wrote my first journalism articles about the environment in the 1970s, just as the Māori rights movement was getting a little media traction. By the 1980s I began, with the help of some learned Māori leaders, to understand that when it came to the environment, the Māori world view – while it was not the same – had far more to share than some environmentalists at the time were willing to allow.

Not that engagement with Māori is without its challenges, and nor should it be. Filmmaker Vincent Ward likes to say that living – consciously – in Aotearoa's cultural relations is 'like living on a faultline'. Those who choose to do so, on what is already shaky ground geologically, live in a kind of double jeopardy. But given this opportunity to engage, who would pass it up? Always, however – and this is the thrust of the book – it all rests on developing the discipline of deep respect for place, for planet and for life in its myriad forms.

Where freshwater meets the ocean: Stirling Falls in Milford Sound, Aotearoa. *Photo Richard Sidey*

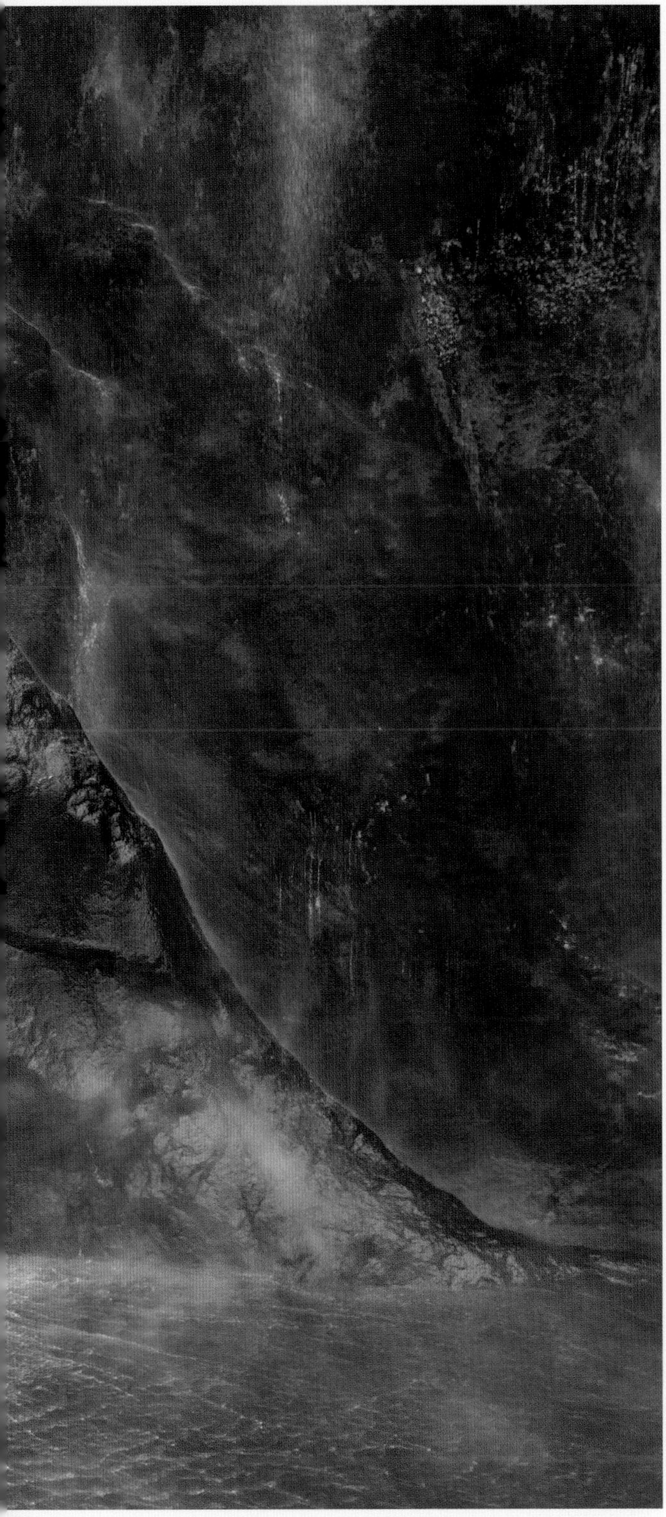

Introduction

In Polynesia, wai, or freshwater, is an invitation to many possibilities. Underlying the invitation are Pacific indigenous perceptions that offer glimpses of cultures whose focus on nature challenges the very basis of Western assumptions.

The movement of water – from cloud traffic to the multiple capacities of the ocean to both absorb and release in the water cycle, to the patterns of rivers' interstices across the land – is captured in myths and stories and 'god energies'. Generally speaking, several centuries ago, what is now the materially developed world ceased to consider in any indigenous way water's proffered layers, depths and possibilities.

This book is, then, an exploration of the indigenous reflections and uses of water from the perspective of an Aotearoa New Zealand Pākehā. It has an islander's outward-looking point of view, into the wider Pacific. It grew into an attempt to convey the cultural sacredness of water to a world that has, for the most part – and to its peril – turned its back on such thinking. In all chapters, what is to be found is a revival of *relationship* between communities of the land and a sense of what they have come to understand about water in mythical, moral and practical terms.

Such an exploration is concerned with valuing that which the trajectory of Western progress has gradually and largely discounted. 'Enlightenment' structures require us to create a dispassionate distance and separation from what we perceive. That has granted Westerners many advantages, each with an inherent cost to the planet. Indigenous communities and societies, by contrast, create a weave of experience that embraces the protean energies of the universe with, fundamentally, a living earth, inviting profound engagement with all our senses. This book shows how the indigenous position rests on so much more than survival – on respect for place, for planet, life in its many forms and interdependencies.

Despite hopeful predictions, it remains uncertain that even the Covid-19 virus today stalking the planet will necessarily shift the West's world view significantly. It has certainly succeeded in challenging it and applying a handbrake to the breakneck pace of modern living. One of the qualities about being deliberately in the presence of still or flowing water is that it slows us down psychologically. It draws us towards being 'in the moment', invites reassessment if not contemplation, and therefore the possibility for reconsideration. Not everyone, even in countries such as Aotearoa New Zealand or Hawai`i, has personal access to the contexts for these experiences that were refined over millennia by probably all indigenous cultures.

As chapters unfold, the knowledge of, particularly, Polynesian place and world view is juxtaposed with personal experience at the interface of water–nature. A pattern is set up that moves between the old ways and the author's encounters. It ranges between myth and legend and hydrological and ecological science, between anthropology and archaeology and the ritual, custom and beliefs from out of the mouths or records, where possible, of indigenous people.

Chapter 1 is rooted – mind, body and spirit – in one place, personal to the author, that inspires meditation of the connections of water and nature back through time. The strangeness to New Zealand of the exotic sequoia's presence in a largely indigenous-oriented discussion acknowledges another symbol of the colonisation of people and plants across the Pacific.

The design of the remaining chapters picks up on that span of the Pacific geographically and across time. Chapter 2, located largely in Aotearoa, discusses the one Polynesian-differentiated water aspect to have its own chapter, mist – linking this to another indigenous approach to fog in the water-short Atacama Desert, Chile. The science of these perspectives is also considered. The book moves next to the oldest continuous culture in the world, living often with water's absence or elusive presence hard at the western edge of the Pacific: Australia's First Nation people. After at least 60,000 years of being part of the Red Continent, such was the elegance and sophistication of their customary ways and their integration into that land, that their relationship was largely unseen, underestimated and dismissed by their colonisers. Their subjection, extirpation or manipulation for most of the past 250 years may just be on the turn. This chapter is written out of my respect for that most deliberative culture, and a reluctance to presume too much further without a far greater depth of knowledge.

Having started with ancient Australia's First Nation's take on water on one side of the Pacific, the book ends on the 'far' side of that ocean, alighting on the efforts of Portland, Oregon, a progressive, growing city with decades of application to water management.

Between them is the heart of the book. There are four chapters plus a shorter essay on the Hawai`i–Aotearoa connection. This is seen through the cultural lens of freshwater as it is expressed on these two, somewhat different archipelagos. Water's lack of universal availability in one and its plentiful aspect in the other shaped differently the forms of both within their shared cosmology, myths, values and customs. Today, as old knowledge is recalled and re-established, the regeneration of culture and nature in community and production is taking hold in both. The Polynesian societies in Sāmoa and Rapanui Easter Island are examined through big themes that are also around water and life.

For Sāmoa, the role of tropical cyclones in challenging sustainability – natural and cultural – is the theme. For Rapanui, the importance of controlling pests for the sustaining of forests and freshwater is explored. The most urgent topic of discussion is undoubtedly in the Micronesian chapter, with Kiribati's many low-lying atolls – Tarawa, Tamana and Banaba especially – in the frontline of climate change. Again, the local, traditional ways and responses are considered.

The final chapter discusses what represents, for New Zealanders, 'the other side' of the Pacific – Oregon's freshwater life, ancient and modern. Here the framing of old ways with respect to water has somewhat more emphasis on the ecological than the indigenous. Salmon rivers figure strongly, in a context of broken habitat and rapidly diminishing wild salmon stock. Modernism and Portland city's efforts over the past 50 years to seriously address those many urgent 'calls of nature' are explored. This awakening took place as the Club of Rome's uncannily accurate global predictions, the 1972 Limits to Growth, were published. Over the last 50 years, this stake in the ground has become a measure of the West's inability to take charge of its runaway exploitation of nature. There are unexpected and sometimes startling parallels between Oregon State and the nation of New Zealand. But Portland's response to the kind of challenges described in this publication have been impressive in both its planning and water management. In contrast, Aotearoa has advanced its Treaty relations and settlements in ways that are now likely to improve management of freshwater.

Another, equally important reason for writing this must also be mentioned. For some years now, where two or more ecologists or conservationists are gathered, the conversation turns rapidly to the task of winning over hearts and minds. Nothing necessary, it seems, will take place without a change in consciousness – a change that can only occur voluntarily with all members of the gestalt that is humanity. The politics of environmental awareness and science seems to get us only so far. So often, what is missing is a sense of history – of what has worked, what has not and why. This is true of colonial history, as well as the indigenous. In policy formulation and in our lives, there has to be a stance that takes the time to hear nature and the voices for nature. The entire paradigm on which our increasingly uncivil civilisation teeters ignores the one word that indigenous people return to again and again: respect for life and life-giving systems on earth. Nowhere is this more important than for the co-inhabitants of places where the indigenous evolved their understanding. For them, but hopefully for others too, this book is offered as one way into the discussion.

'The Senate' cluster (left) and 'The General Sherman' (above, centre), Sequoia National Park, California. *Photos Aliscia Young*

CHAPTER 1

Sequoia

LOCATION IN MEDITATION

Once it chanced that I stood in the very abutment of a rainbow's arch, which filled the lower stratum of the atmosphere, tinging the grass and leaves around, and dazzling me as if I looked through colored crystal. It was a lake of rainbow light, in which, for a short while, I lived like a dolphin.

—*David Henry Thoreau*[1]

To brush beneath the skirts of the Californian coastal redwood (*Sequoia sempervirens*), a tree established in surprising corners of Aotearoa,[2] is to enter an inviting, private room. Its roots have pushed up a deep cushion of humus, needles and strips of bark shed around the base of the trunk. This affords a magnetic place to be seated, inside the halo of one of the earth's great species, born out of deep time. These are giant beings that can live for several thousand years. Despite the sequoia's soaring solidity, its bark is strangely soft and yielding and, since I sit here often, the back of my head and my trunk have left their impress on it. Still a young 'stripling' of about 55 years, this living tower pushes steadfastly skyward, already over 30 metres tall and climbing year on year. Despite its busyness, its radiant serenity offers a perfect place for contemplation.

As I settle, a thrush darts in. Astonished to find me there, she is off again even faster, bleeping out her shrill alarm call. Over in the Douglas fir I can hear a pair of endemic kererū (native wood pigeon) cooing to each other. For some years these sparks of life from a once prolific species have made their nests in this close of exotic trees – high enough to keep their progeny safe from other, bandit exotics such as rats, possums and stoats. From there they must also avoid the attention of the raptors, the kāhu and kārearea, forever shadowing their nests. The four-legged pests also compete for the kererū's food.

Like so much of lowland New Zealand, the native birds and their habitat forests have a rarity value that continues to inspire community restoration projects, especially in recent decades. Some of that regeneration is a modest work in progress on our property – but given room enough for a couple of California sequoia, planted in the mid-1970s, long before we came to live here, who could gainsay their presence, even though – like me – the sequoia is not indigenous?

Looking up I can see the scrumple of an old kererū nest that the wind must have taken, caught like an abandoned scarf in the sequoia's fine needles. Curiously, the soft mating calls of the kererū are almost the only vocal sounds they make – like us humans, they are audience to our dawn chorus, not contributors. They are big birds, and for some 60 million years their consumption of native drupes and seeds and their potential to fly in a 35-km range daily has meant they have been a vector in the continuing life of much of New Zealand's forest: they are a keystone species. In the mating season they declare courtship on their exuberant flightways, strafing the driveway, lifting steeply on cliffside thermals to stall point, then diving into freefall before soaring into another loop. Aerial tai chi. Within a few minutes of my settling into the sequoia realm, the pair are perched aloft directly above me.

Broken light plays through foliage bobbing gently on the ends of spar-like branches. Is it these appendages, with their feathery tips, that allow sequoia to grow without our rough trade winds leaving a mark or a bend in their essential form? The sequoia's efficiency in converting sunlight, water and CO_2 into new tree cells – and generating fresh oxygen – ranks with that of other great trees. Science tells us that. There are other ways to address the tree. Contemporary landscape writer Robert Macfarlane described American transcendentalist John Muir's interest as: 'worshipping in the crypt-light of the sequoia groves'.[3] The world owes much to Muir for the advancement of national parks.

A single sequoia, even among other lovely exotics, is not 'wildness' in any New Zealand sense, but the tree's grandeur, energy and shelter beckon. And nature worship en plein air like this allows nature to admonish any possibility of altar-boy piety: one warm evening a blackbird, I suspect, let fly from far up in the tree with bombardier precision. Forehead and spectacles took the brunt of it. As summer warmth draws forth the cicadas from the ground, their collective beat sets up the rhythm of the mantra. Always, here, has been a feeling of old familiarity, of primeval nurture – a place of magic where, no matter my condition, my energy invariably lifts. And seated here charged with negative ions, the atmosphere is like that of being behind a waterfall. And that's exactly what it is – a delicate, upside-down waterfall, the tree's capillaries silently drawing water and minerals out of Papatūānuku and raising it up in a miracle of nature, back and out towards the seminal light of late afternoon sun.

Others have been here thinking before me. Callum Coates, disciple of the traditional Austrian forester and radical thinker about water, Viktor Schauberger, has observed that 'trees were once columns of water'.[4] As Colin Tudge writes elegantly: 'It would be wonderful with X-ray eyes to see a forest without timber. It would be a colony of ghosts, each tree a spectral sheath of rising water.'[5] Richard Preston, who documented the exploration of the crowns of Californian redwood and sequoia, put it this way:

> Redwoods, it appears, may be 'aquifers aloft', with one hectare of forest containing somewhere between 25,000 and 50,000 litres of water in the mats of vegetation and epiphytes that hang out in them. But more to the point, water taken in by the roots of a giant redwood is on an upward journey through the sap wood, beneath the bark and enclosing the heart wood, to its top taking at least two weeks.[6]

The kererū, Aotearoa's native woodpigeon, is a critical disperser of seeds in the forest.
Photo Richard Sidey

The medium for transmission is the cambium, which is why, if a tree is ring-barked, it dies. When the going gets tough and groundwater supplies dry up, the sequoia can switch from earth to sky and absorb fog-moisture through its needles.[7] Such powers of adaptation help explain its enormous timespan as an inhabitant of the planet.

Natural forest in Aotearoa, not least kauri forest, contains several rainforest ecosystems that are rich in epiphytes. Ancient cultures revered and often worshipped trees; and the Japanese we now know like to engage in 'forest bathing'.[8] It's hardly surprising that, given a chance, humanity is still supplicant to and adorer of forests.

One of New Zealand's distinguished conservation leaders, Kevin Prime, a Ngāti

Hine kaumātua (elder), has led his people for decades in the restoration of kūkupa (kererū) numbers through intensive predator control. Prime sums up being around his big trees in a different way. He describes mauri as the energy of living things, including ecosystems. Here he speaks particularly of the kauri forests, which contain Aotearoa's largest trees; some of the most venerable are antipodean close matches in size for redwoods: 'A kauri tree has stood in the forest for hundreds of years, absorbing sunlight, and has stored that energy as its individual life force. With lots of trees in the forest their combined energy is something you can feel and absorb.'[9]

Right now, entire forests in many parts of the globe are under siege on many fronts, from the crazed policies to clear forests for ranching in Brazil to the ongoing destruction of forests in West Papua. In northern Aotearoa the ancient kauri forests are threatened by kauri dieback disease – tiny spores that begin on the roots, invade the tree and slay these giants, threatening entire remnant forests of Northland. This is but one iconic species struggling to survive in the new millenium. In 2017 we were alerted to the risk of myrtle rust. This disease, possibly carried here by high winds from Australia, has already infected some of Aotearoa's most beloved myrtles, among them pōhutukawa, rātā and mānuka.[10]

Even with the sequoia's built-in fire resistance, California's forests, for other reasons, are at risk. So, too, are Australia's. The effects of the 2018 El Niño event brought drought and concomitant fires to the northwest coast of America, destroying 100 million trees. The area destroyed by California's then worst historical 'fire complex' was more than 766,439ha: close to 90 people lost their lives, and the cost of the fires was around US$3.5 billion. About a million dead trees, both climate-induced and of natural mortality, provided the fuel for most of these fires.[11] 'Heat,' warned Callum Coates, 'is the destroyer of the natural world.'[12]

There is a kind of progressive rampancy here – with less forest equalling more CO_2 released into the atmosphere, leading to overall rising temperatures and the risk of more and increasingly intense forest fires, and the possibility of more flattened forest from intense storms. This is a deadly circularity right out of a linear economy. It seems astonishing that, in the face of such evidence, any resistance to a cap on warming the planet even by 2°C should continue, especially since a 1.5°C cap would give a far greater survivability potential for most life on earth. But the latter cap, after many words and insufficient action, now requires rapid, far-reaching changes in all aspects of society.[13]

What happened in Australia and California and in the different forests of Oregon and British Columbia in 2017–20, and is likely to happen for summers to come, reminds us of the relatively narrow margins of temperature survivability in which humankind and its supporting biological services have been able to prosper. Just as the giant adult redwoods are reliant on the summer heat of the mountains to burst their cones for release of their pollen and seed-throw, their taller cousins, the sequoia, rely on the fog that blows in from the Pacific Ocean for vital moisture.[14] And, not always apparent to the naked eye, sea fog rolls in off Cook Strait across Tasman Bay at the northern end of the South Island. Once it moistened the leaves of great forests; today it can slightly ease the heat of a summer's day.

Much of our discussion so far is about energy and processes, usually summed up as photosynthesis. Richard Preston, with his intimate understanding of California's big trees, explains it thus:

Tane Moana, the largest remaining coastal kauri, Northland, where entire forests are threatened by a fungus-like organism. *Photo Aliscia Young*

> The sapwood transports water through the tree, whereas the heartwood is entirely dead and doesn't transport water. All it does is support the tree, providing a strong superstructure. Underneath the bark lies a layer of actively dividing cells called the vascular cambium. The vascular cambium is entirely alive.[15]

The genesis of this process, back in the conception of original life on the planet, is even more amazing. It begins at the thick end of a billion years ago with an ancestral bacterium called a cyanobacterium and an amoeba, which find each other in water. Although the amoeba engulfs the bacterium, a deal is struck. The amoeba is a eukaryotic cell, whose membrane contains a nucleus and DNA. The card the amoeba holds is that it can convert sugars into energy. The unwitting resident bacterium, with its host, now assumes a much larger role in Darwin's dance. The new binary unit evolves into what is known scientifically as a chloroplast, the crucible wherein the miracle of modern photosynthesis takes place.

This coming together was one of those extraordinary moments, a colossal tipping point in the evolution of life on earth. Out of the symbiosis of these two simple organisms and their utilisation of sunlight and water, photosynthesis began. From this came forth all land-dwelling plants. Thus the entirety of the (vascular) plant kingdom, from lowly pond-edge grasses to the giants of rainforest and all that depend on and sustain them, can trace their origins back to this ancestral alchemy.

The redwoods figure in this evolution from about 280 million years back. The ancestors of the great forest trees of New Zealand, such as tōtara and rimu among the pines, hark back some 230 million years, nudging like redwood back into the Permian era. A little later in geological time they are followed by the country's tallest, much-loved, oft-mourned denizen of our diminished swamps, kahikatea. The kauri is a little more recent – potentially even bulkier than the kahikatea and suddenly also at risk.[16] And the binary union these organisms 'invented' remains axiomatic to the way so many forest species continue to propagate, grow and render the planet habitable.

So much depends on how the trees are able to manage water.

Science credits this as the first successful union between animal and vegetable kingdoms on the planet. As such, it reminds us, as descendants of that process, that when Polynesians include in their whakapapa the recitation of ancestors that are rivers, plants, animals and even insects, these are more than poetic inventions. The lineages acknowledge what science has, for some time now, amply confirmed. Equally significant, each recitation frames the speaker not only in time and in their landscape; it also reminds them and their audience that there is a continuous thread of life between humans and nature. We are nature, each and all defined by fate, by the throw of the double helix, and by acts of evolutionary 'necessity'. This, in turn, informs our relationship with nature.

These connections are elaborated at length in the Polynesian whakapapa/origins chant the Hawaiians call the *Kumulipo*. It begins with the rise of Pleiades, and the stirrings in the long darkness out of which came 'the slime'. This discussion continues in chapter 4.

The chloroplasts' main role, then, is to perform photosynthesis, which, thanks to science, we are able to describe. In trees, the leaf pigment captures energy from sunlight and converts it in storage molecules, while at the same time freeing the oxygen from water. The sunlight splits off the water atom, fixing its carbon from atmospheric carbon dioxide into the body of the tree – as carbohydrate – while the plant releases the freed oxygen back into the atmosphere. This takes place at the heart of the leaf, in the spongy mesophyll, where a gas exchange occurs; transpiration occurs as water evaporates, but only at the edge of the leaf. It is from this process that arises the popular understanding that 'our forests are the lungs of the earth'. Yes, but more than this, the efficacy of the entire water cycle, the biosphere and hydrosphere depend upon it. The obvious has to be restated: water and plants are inseparable; and, together with other life, humans are bound into a shared history and destiny.[17]

Through aeons, as vascular forests and plants evolved, trees found ways to capitalise on this basic mechanism, so that the taller ones can carry water higher and higher through capillary action, induced by evaporation. This is thanks to specialised kinds of water-carrying cells – the xylem. Being a reinforced cell designed to withstand collapse from overload, the xylem can overcome the difficulties that water tension presents to an organism. It is these cells that provide the conduit that takes water up the plant. Without them, no plant would stand taller than a moss, as primitive plant forms did.[18]

Trees are most actively engaged in their 'sun salutes' in the morning, when the leaf is fully hydrated and before it becomes too warm; and sometimes also in the late afternoon. The leaf holds stomata, each with two specialised cells

The ancient exchange: photosynthesis under way in the sequoia's glade, Nelson. *Photo Aliscia Young*

called guard cells. These can shrink and shut off the movement of air from the inside to the outside of the leaf. For photosynthesis to be achieved, the stomata have to be open. That's the lock through which trees take in carbon dioxide gas. Some big trees, like sequoia and redwood and most of the giants in Aotearoa's native forests, have evolved small leaves that allow trees to survive drought and cold. In much of the northern hemisphere, deciduous trees shed their leaves to shut down their water uptake through winter, since in many locations temperatures would freeze them into scalpels of ice that would fatally lacerate the tree internally. Where appropriate, the wood design has even evolved a natural anti-freeze.

Indeed it is the sapwood wherein the life processes are contained. This is on the trunk rim, just under the bark where, on the annual rings, the capillaries and cells for water transportation build outwards. The xylem cells housed within it are designed to allow the tree to withstand desiccation and enable it to grow and climb competitively for light. It is that competitiveness that brings forth enhanced resilience.[19]

So much depends on how the trees are able to manage water. Indeed, the sapwood is the site of active upwards transport to the very top of the tree, while the heart timber is the wood that persists inside the tree, impregnated with resins; where the cross-sectional rings bear witness to its history and age. Together with the ocean, collectively, the wood of the world houses what we in the Anthropocene (the age in which humans have become the potential regulator of both the hydrosphere and the biosphere) call the carbon sink.[20]

But with such dizzying imbalances now so evident, the great truth articulated by the proponent of holism, Alexander von Humboldt, that 'the wooded region acts in a threefold manner in diminishing the temperature by cooling shade, by evaporation, and by radiation', may no longer have the certainty it had during the Enlightenment.[21]

That's why trees matter, why water matters to trees, and why trees' survival en masse matters to the survival of human beings. Without the trees, and their long-evolved relationship to water, sun and atmosphere, we simply cannot exist.

It may be that those with the privilege of living beside indigenous people have reached a point where the pathway to urgent but considered action for collective life on the planet is through their profound understanding of nature. Unlike the metropolitan, 'developed' view of the world – essentially that of the coloniser – the indigenous ancestors knew fully the meaning of the word 'consequences'.

PETRIFIED FOREST

These exposed petrified deposits of Metasequoia trees (a family of sequoia on the remote coast of the Alaska Peninsula) are 25 million years old, from the Miocene-Oligocene period. The living forest was engulfed by a lahar from a volcanic eruption. These fossils indicate the history and range of the ancient forests of earlier times, extant when warm-blooded dinosaurs ranged the Alaskan North Slope. Over time, the remnants were exposed to water: these subtle colours were set by mineral wash-through replacing the original organic material. The 2600-hectare site is part of the much larger Alaska Maritime National Wildlife Refuge.

'hotos
liscia Young

Valley fog below the Pisa Conservation Area, Otago.
Photo Richard Sidey

CHAPTER 2

Both Sides Now

FOG CAPTURE IN THE PACIFIC

'Mike Mohi has made several visits and what is interesting is that when he came back the last time, that hill out there, Kapowai, was covered in fog, which is unusual here,' says Sonny George, chair of the Hoori Whanau Trust.

'It looked like a white cloak and I said to him, "Our tūpuna [ancestors] are thanking you for what you did, they are paying you their dues for looking after them. They are saying you have put a rāhui on this land and we are happy with what you have done and appreciate it because if we hadn't done it then things would have gone way back."'[1]

Fog. It rolls in from the sea, sometimes as mist, clagging the mountain passes. Bedraggled, it will cling for days like snared plastic bags to the windward hills of New Zealand's capital, Wellington, in the strongest of northerlies. Equally, on a clear, still day in Tasman Bay, the sea mist's presence is invisible, detectable only as pricks of moisture on the skin of a bare forearm.[2]

The clamminess of mist is recognised in Māori culture in the acknowledgement of Tāne, god of the forest in old Aotearoa. 'If you're walking through the mist and the water gathers and collects on you, that's Tāne Tokorangi,' explains Ngārangi Rangiuia, a holder of the old knowledge. 'Tāne Tokorangi also refers to how the trees get it, the leaves get it, the bark gets it. The toko is what is collected and what comes together.' It also subtly acknowledges, he says, the tremor of leaves in a mist.[3] In a particular example, the unusual but timely appearance of fog was seen as an acknowledgement by the forest to those present – a thanks for the rāhui (protection) they have put in place.[4]

In modern thinking, fog figures hardly at all in our comprehension of the water cycle. We the public have tended to overlook it. Scientific understanding of the relationship between fog, snow tussock and water uptake has been relatively recent. Fog is integral to the evolution and persistence of the Californian redwoods, whose ancient and equally massive forebears were set in fossils millions of years ago in Alaska. Metasequoias – pioneers but by no means the earliest of those trees now seen in northern California – grew 45 million years ago at 82 degrees north and into Siberia and Alaska during the Eocene epoch.[5] A sequoia can take up to 30 percent of its moisture needs from fog.[6]

Fog is a vital but largely unheralded contributor to the life of special plants like Aotearoa's snow tussock. As individuals and as networks these grasses can take up the life force of water expressed as fog and store it in considerable quantities for release as part of its contribution to the natural health of the high country.

What has been learnt by indigenous people about fog capture in entirely different circumstances, on opposite sides of the Pacific, is still available to us today.

The descendants of Waitaha tell how their early forebears, who inhabited Te Waipounamu (the South Island of New Zealand), engaged in systematic replanting of snow tussock following the great fires that left nothing but, as a song of theirs goes, 'Puehu e, puehu e, puehu e [dust, grit and grime] …' The details, which acknowledge the snow tussock's role in 'capturing the dew and the fog', seem internally consistent and are clearly based on old, orally transmitted knowledge. Intriguing though it is, the account, as far as can be ascertained, still awaits further supporting evidence.[7]

Across the Pacific, indigenous inhabitants of Chile's Atacama Desert – a 1000-km narrow shelf wedged between the Andes and the ocean – have had to pit their ingenuity in the world's driest desert. They have been part of the region for at least 6000 years, and have adapted to long-interval changing climate – some groups surviving by coaxing water from fog.[8]

Over the slow time of Aotearoa's geological evolution, some ecosystems have evolved above the treeline in the mountain ranges – notably the tall tussock grasslands that grow in several parts of the world, 'particularly [in] temperate Oceanic conditions, as may be found in New Zealand'.[9] Some are endemic snow tussocks that originated during the Miocene but, responding to climate change, 'came into their own' during the following epochs in glacial periods, as George Gibbs explains. Their position on the mountains in relation to the treeline depended on whether cold favoured the tussocks or not.[10] When they were able, they spread 'tribally' across available open spaces, extending above and sometimes below the treeline, the species differing according to location.[11] These tussocks contribute to the retention of water and, therefore and more obviously, soil conservation in steep, sometimes unstable country. However, in the globally competitive league tables of 'environmentally at-risk rankings', temperate grasslands are listed as 'the world's most beleaguered biome'.[12]

In modern thinking, fog figures hardly at all in our comprehension of the water cycle.

Between 1300 and 1450 AD in the drier, eastern region of Te Waipounamu, Māori fired vast tracts of the forest and scrub of the eastern dry 'rain shadow' region, effectively sparing only the Catlins, Banks Peninsula and some inland ranges such as the Kaikōura mountains.[13] In this way, assisted by millennia of natural fires, 'narrow-leaved and slim snow tussocks extended their ranges below the treeline on the drier eastern side of the mountains'.[14] In the 1840s the lower-dwelling short tussocks made up an estimated 44 percent of the country's total grasslands; and the snow tussocks 13 percent.[15] Then, after an interval of about 500 years, it was European settlers who touched the flame to country by then dominated by these native grasslands.[16]

As described by the incendiary Mary Jane Barker, in the 1860s European runholders of the South Island high country blazed the native tussock (*Chionochloa spp.*) with a naïve enthusiasm that became smoked into their culture. Not all runholders were so feckless.[17] Nevertheless, for the second time, the high country was to be radically transformed by humans, this time in the service largely of sheep farming. The pastoralist's cycle was to fire, then sow grass into the cooling ash before turning sheep onto the new green shoots – and potash.[18] Initially selected on a haphazard basis, 'more suitable' grass seed could then be broadcast for the grazing of sheep. For 100 years sheep's meat and wool were New Zealand's number one export items. But this combined sequence of fire, partial recovery, then browsing affected nutrients in ways that inhibited the recovery of tussock foliage.[19]

Regularly fired tussock enabled the recent settlers to keep an eye on their flocks, otherwise lost in the fenceless swathes of tall grasses.[20] Deployment of exotic seed in what was officially dubbed 'waste land' seemed the perfect adaptation – akin in its presumption to the Australian runholder's exploitation of 'terra nullius' ('nobody's land'), the Australian interior.[21] As early as 1910 a Liberal Government commission observed that the penetration of invasive weeds into the Mackenzie Country was likely caused by firing by farmers, dating from the 1870s. Historian Roberta McIntyre wrote, 'The Commission concluded that nothing had contributed more to the widespread destruction of pastoral high country than the indiscriminate burning

Alan Mark, defender of the high country with his grandson, Henri Gordon, at Mt Burns, Fiordland.

Photo supplied by Alan Mark

of pastoral high country.'[22] If pastoral life appeared to prosper from firing, it was a century before this received wisdom was up-ended by the dedicated work of Alan Mark and his co-scientists.

The endemic snow tussock is distinctive. With its clumps of long, tapering narrow leaves flying in the wind from their tethers, it is found in the lower alpine reaches of Te Waipounamu. *Chionochloa rigida* is in the southwestern South Island, but is more widespread east of the Main Divide from about 500m above the snowline. This is a golden, narrow-leaved species standing over 1.5m in the glory of its long flower heads. While *C. macra* (slim snow tussock) can withstand drier climate runs as far north as Marlborough, it tends to replace *C. rigida* at higher altitude south of the Rakaia River.[23] Resistant to wind and snow, the snow tussock has the capacity to retain most of the moisture it absorbs. These tussocks are, of course, perfectly engineered to place and purpose. Their stomata (the openings through which water vapour and gas are exchanged) are wrapped in 'the deeply incised furrows' within the furl of the leaves.[24]

Otago University emeritus professor Sir Alan Mark has spent a lifetime of systematic botanising in the high country that rises up and along the spine of New Zealand's alpine uplift. With some 200 papers to his name, his has been a life of research and conservation activism. He became nationally prominent for his role in the Save Manapouri campaign, the nation's portal into the modern environmental era, from the late 1960s. His commitment to protection of high-country plants and ecosystems continues undiminished. With science still their watchword, the Wise Response group he chairs has been an influential force among those who finally saw years of research and campaigning realised in the enactment of the Zero Carbon Bill in New Zealand in 2018.

Tussock high country, Otago. *Photo Richard Sidey*

Mark was curious as to the role and capacity of snow tussock to collect and retain water on often steep country, and the effect of fire on these slow-growth plants. Indeed, their life history was such that his early findings challenged high-country farming interests, as well as the prevailing orthodoxy of the New Zealand Forest Service. These two production sectors viewed both the native conifers and some of the drier-region tussock grasslands as being in a senescent phase. They argued that, under the effects of epochal climate change in an era that preceded the evidence of global warming, both had failed to adapt.[25] As a result of this hypothesis, forestry practice in the mid-twentieth century also significantly reduced native forests, replacing them with massive man-made exotic pine forests. Turning back these practices in various places occupied the environmental movement in New Zealand for many decades until 1999.

Contrary, however, to this view of tussock in its death throes, even Mark's early work showed that these ancient plants were not only vital and viable, but were so keenly adapted to their mountainside habitat that in the seasons when they did flower they displayed synchronous or mast flowering: even genetically different

The setting sun picks out a fenceline through the snow tussock in the Oteake Conservation Park.
Photo Richard Sidey

tussocks on the same mountainside tuned in to an orchestration of florescence.[26]

Mark knew the firing of tussock could give the normally slow-growing, long-lived tussock a burst of new growth. That was the point of firing: like the slow-developing sequoia across the Pacific, they could readily withstand fire. But these new shoots proved most attractive to sheep, and this made the plants more vulnerable to degradation by the effects of over-grazing than to the original fire. Mark concluded that by his reckoning, without such interference and their capacity for recovery from natural disasters, including fire, these plants were 'potentially immortal'. At the very least, they could survive in this habitat for as long as New Zealand's longest-lived trees, the kauri of the north[27] – that is, for perhaps 2000 years. For that reason, he said, it was 'important to care for them'.[28] Indeed, Mark challenges anyone to find a dead tussock in healthy, unmodified grasslands.[29]

But it was the capacity of snow tussock to absorb and retain water from whatever source – with its implications for water retention and erosion prevention – that took these studies to another level.[30] After completing his PhD in the United States, Mark's research with the regional catchment board in an old stamping ground of his, the Old Man Range in Central Otago, revealed gradients that were far more marked in temperature and moisture from north to south along the ridgelines than previously understood.[31]

In 1966, after a visit to the Cairngorms in Scotland's rugged uplands, Mark began to apply the techniques of Dr F.W.H. Green, who measured the water yield of these moorlands using modified 44-gallon drums – cut down, sealed with tar, dug snug into the ground and fitted with a filter and outlet piping water to a reservoir. Mark and his team monitored mature snow tussock of the region in groups, with some as control studies along different contours. The idea under test was that the larger the tussock, the more the moisture yielded.

The high-country, narrow-leaved snow tussock grows naturally at three to the square metre. Each tank took up a similar space to a snow tussock, and contained one plant. Other groups of plants provided the experimental controls: one group was burnt, one 'severely clipped', and another group provided different control measures.

The initial studies, carried out with PhD student Jennifer Rowley, were 1000m up in the Rock and Pillar Range in Otago, a conservation area that takes its name from its distinctive resistant schist tors. It is a place rich in diversity: the snow tussocks grow on its lower slopes among native grasses and herbs. Mark and Rowley's research demonstrated with some precision the volume of water a tussock would collect that was 'surplus to evapo-transpiration', and the losses in tanks abutting tussock 'represented the water yield from each'. The annual yield from the unmodified snow tussocks produced a phenomenal '63 percent water surplus of the measured precipitation (i.e. all forms of water) of 1350mm'.[32] In some parts of the Lammerlaw Range, 80–86 percent of measured rainfall was yielded by large tussocks.[33]

The data collected also allowed Mark to explain how perfectly snow tussocks as a species were evolved to the high country. They could cope surprisingly well with burning, but heavy grazing in their recovery phase could be damaging – if not lethal – to these otherwise long-lived plants. Mark showed how, contrary to conventional wisdom, in country where fog is common, 'the long fine foliage of snow tussocks is very efficient at intercepting or straining water from wind-driven fog'.[34] Despite events such as tussocks being held under winter snow, 'We estimated that the water gained through the tussock interception of fog may increase water yield up to 25 percent from the most fog-prone sites on the Otago uplands.'[35] He explained, 'It was that narrow, inwardly curved leaf design that enabled the leaves to resist evapo-transpiration back into the atmosphere, at the same time releasing moisture gradually into the soil.'[36]

The ability of the tussocks to confound human assumptions continued. On Mt Cargill – Dunedin's antipodean answer to Edinburgh's Arthur's Seat – they caught up to half a metre an hour from tussock in conditions where no measurable rainfall was determined. As Mark reflected, summarising this work, 'The bigger the tussock, the better the catch.'

they caught up to half a metre an hour from tussock in conditions where no measurable rainfall was determined

A paired catchment study followed in Glendhu Bay on Lake Wānaka, comparing water uptake of exotic pine plantations and snow tussocks. By year six, the water yield in the catchment planted in pine started to reduce, and by year 22 there was a 46 percent reduction in water yield.[37] Other paired catchment studies conducted over 22 years – initially by the Forest Service and continued by Landcare Research – showed that exotic forests (usually *Pinus radiata*) planted in snow tussock country steadily reduced the water yielded over that time, by a similar amount. This was relative to indigenous grassland in an adjacent catchment.[38] Yields from snow tussock were even greater, they found, in the higher regions of fog.

There have been periodic challenges to the science from vested interests, including some runholders – one of whom even covertly sponsored an alternative line of research. Like Mike Joy, scientist and critic of chaotic government freshwater policies in this century, Mark in his 'tussock days' was, however, able to withstand the hostility of vested interests through the integrity of his work – and a steadfast university administration.

But Mark was far from finished. Neil Ingraham, a Californian geohydrologist at the University of Auckland, contacted him to say he could differentiate water derived as fog from rainwater and groundwater, using stable isotopes. Condensation of saturated air produces fog first, enriched in stable isotopes of oxygen and hydrogen; rain comes later and shows a reduced level of these isotopes. Groundwater is a mix of the two.[39] Such a differentiation between types of water, depending on its source from mountain regions, has been long held in old Māori wisdom.

Using these isotopes, Mark and Ingraham were able to further demonstrate the elegance of these high-country ecosystems thriving in their climatically often severe environments, with the tussock trapping and drawing down fog with efficiency. In both the Rock and Pillar and Lammerlaw ranges, they found significantly high percentages of wind-blown moisture from fog – and rain – trapped in the tufted clumps, picked up by the plant's long, slender leaves billowing in the wind.[40] At full stretch, single snow tussocks were capable of taking up to half a litre of water an hour 'from dense fog during periods with negligible recorded rain.'[41]

The work in the Lammerlaws confirmed again the importance of fog. Most fogs in the area, whether light or heavy, are associated

ATACAMA DESERT: TRAPPING FOG AND MINING LITHIUM

In Chile's Atacama Desert, some of its long-established indigenous people learnt centuries ago how they could survive by literally deriving water from fog. In some places, no rainfall has ever been recorded, so nothing at all, plant or animal, grows there. In others, unprepossessing, shrubby, prostrate plants dominate, performing their own special survival routines. At a seminar in Honolulu, Professor Javier Núñez of the University of Santiago described the coastal fog or camanchaca in Chile and Peru, and how it has been been traditionally trapped for water for human consumption: 'The desert is close by the ocean, so the fog is very localised and captured by the high mountains.'[42] The fog rises out of the cold Humboldt Current mingling with the warmer air of northern Chile.[43]

It was during a severe drought in 1956 that retired mathematician Carlos Espinosa Arancibia turned his mind to the idea of fog harvesting. Since then, and especially in this century, improving technology in the form of huge fog-trapping nets has brought water in livable quantities to inhabitants of the virtually uninhabitable.[44] Today nets are raised in the Chilean Atacama Desert to catch fog. Traditionally fog was collected for water on assembled piles of stones or 'in arrangements of leaves', Núñez explained. For fog to be trapped, it needs to be moving in banks.

Above: Indigenous Atacamenos, as the Spanish called them, in a Pachamama ceremony, honouring the Andean fertility god, overseer of planting and harvesting, who also embodies mountains and earthquakes.
Photo: Aliscia Young

Above left: The Atacama Desert, Antofogasta, Chile.
Photo: Aliscia Young

Paradoxically, in this place where rainfall amounts to about 25cm per year, the salt flats of Salar de Atacama, site of up to one third of the world's lithium, is being mined using 'huge amounts of water'. The water is trapped beneath the salt pans containing the rare earths. The photographer who graphically brought this mining to the world's attention, David Maisel, has said that in our use of electronic devices we are all complicit in this business. Not least of these are electric vehicle batteries, on which the world is rapidly becoming so dependent. After some years of freely granting water permits in the area, the Chilean government appears to be on the point of introducing regulation to this 21st-century bonanza.[45]

with cold fronts from the southwest. Heavy fogs contributed 154mm of deposited water over a 306-day period; and, including the light fogs, total water derived from the deposition of fog over that period was 277mm – an impressive 22 percent addition to the mean precipitation recorded.[46]

As a prominent conservation advocate, Mark's research set him up for a career in protecting high-country ecological systems, breaking open popular characterisations of plants like snow tussocks into a fuller appreciation of their fundamental ecological function as well as their beauty, in an era of mass tourism and relentless photography. His research was pioneering and highly influential. His collaborative studies in the northern Andes gained similar results on Ecuador's upland páramo tall bunchgrass ecosystems. He found that, like the snow tussocks in New Zealand, the tall bunchgrasses in the páramo ecosystem were important for their maintenance of soil moisture.[47]

In an ecological approach to such complex problems as climate change mitigation, Alan Mark and co-researcher Katharine Dickinson took as a precept that unprecedented population growth makes the accessibility of quality freshwater even more of a global essential. They outlined this in a contribution to *Frontiers in Ecology and Environment* in 2008. With projected marked shifts in both rainfall and temperature, however, 'Resolution of system-wide problems involves a combination of behavioural, institutional and technical factors,' they wrote. Trade-offs will almost certainly be part of such an integrated approach. The draw-off of water, for example for domestic supply as opposed to hydroelectricity or irrigation, is 'often with associated displacement of people and downstream effects on biological conservation and other ecosystem services'.[48]

It is estimated almost half the world's population depends in one way or another on the water derived from mountain catchments. In their paper, Mark and Dickinson foresaw a tension that is assuming greater importance in Aotearoa today – between protection of indigenous tall tussock grasslands and the planting of exotic forests for both wood production and sequestration of carbon in upland watersheds. They argue that the latter cannot be at the expense of quality water production: 'There are important issues to be addressed in balancing afforestation/re-afforestation with maintenance of a vegetation cover that maximises water yield.'[49] In spite of the generally widespread availability of freshwater in New Zealand, Jan Wright, parliamentary commissioner for the environment, listed the protection of freshwater at the top of her concerns in a 2009 report on high-country tenure review.[50]

It follows that the determination of mutual benefits, whereby competing goals may be reconciled, will demand diverse science – including social science – and superb governance. Achieving this will require a step up from the interest group-driven decision-making that has been a characteristic of nations like New Zealand, locked in three-year election cycles of unwavering pragmatism. Water management issues are notoriously 'wicked problems', and ecosystem services – a management term for natural processes such as nutrient cycling – are tied in to that problem. Mark and Dickinson remind us that climate change issues only compound this complexity. They argue for a kind of thinking that can bring plans together for the integration of all these elements, as well as ensuring socioeconomic benefits, as essential to the way we humans govern ourselves.[51]

Today the yield of water from healthy snow tussock grassland in the high country is widely accepted scientifically (though not necessarily politically) as a vital ecosystem service. Tussocks are also a highly recognisable cultural feature of our conservation estate. Almost all the high country of the South Island was held in legal title by the Crown as commons, most of it leased to sheep runholders. However, a tenure review process that has been running since 1992 delivered about a third of a million hectares of the high country into official conservation land.

A similar proportion of the attached 'low country', perhaps more suited to farming, has been sold willy-nilly into private hands.[52] What once had been considered a smart idea was challenged. Criticism was levelled at the resultant divisiveness and lack of strategic, or even landscape vision in the review process. The sharp-edged patchworking of land resulting from onselling and subdivision can also be ecologically disruptive;[53] not to mention the controversial sales to foreign-based buyers and excessive profit-taking by opportunists.[54] But more to the point, rather than solving the problems it was set up for,[55] this highly segmented approach to landscape added yet another layer of complexity to potential management solutions. Hence, the process of tenure review ended in February 2019.[56]

Water will always be imperative to landscape management, not least because it is water that will ultimately bear the consequences of whatever decisions are made. In the high country, the fog tussock still provides as good a focus as any for the study of the high country's continuing viability.

The Red Continent at dusk. *Photo Richard Sidey*

CHAPTER 3

Fire and Water, Rock and Fish

ABORIGINAL MANAGEMENT OF AUSTRALIA

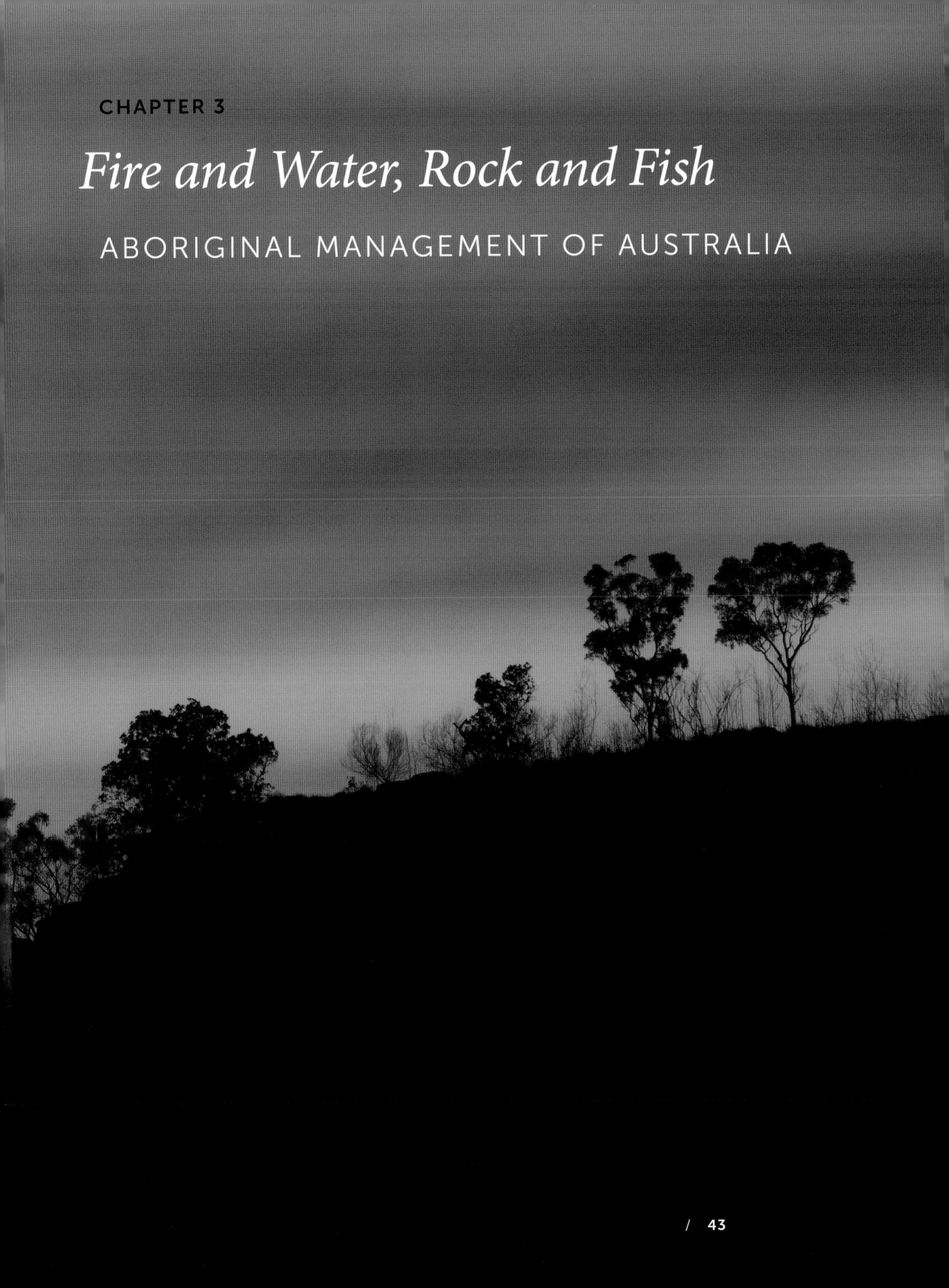

In Warrane (Sydney) Cove, a freshwater stream (Tank Stream) flowed from a spring down a series of waterfalls to the harbour at today's Circular Quay. Sydney from the Western Side of the Cove, c. 1803, *by George William Evan, Mitchell Library, State Library of New South Wales*

Bennelong Point, where the 'wings' of the Sydney Opera House dart high above the harbour, is as recognisable a landmark as any, certainly in the southern hemisphere. What is less well known is that it is named for Woollarawarre Bennelong, a warrior of the Wangal, from the south shore of the Parramatta River.

Initially kidnapped, Bennelong was placed under 'house arrest' by Governor Phillip in his home in old Port Jackson, where the governor sought to learn Aboriginal ways and language from him. Later, Bennelong occupied a hut at the Point, built on the orders of Phillip. The governor, by no means consistently well disposed to the Aboriginal race, came to hold him in high regard; he found Bennelong to be an engaging, adaptable and intelligent intermediary in the establishment of the new colony.[1]

Yet it was not until Eleanor Dark's original research from the Mitchell Library for her 1941 trilogy *The Timeless Land* that a dawning comprehension as to the nature of those whose lands they had occupied was held out to the wider Australian public. Despite the efforts of many since then, the sun of that dawning still sits low in the eastern sky. But Phillip's recognition of Bennelong's humanity – if not that of all Aboriginal people – had been signposted 160 years before.[2]

Bennelong is often written about in the same breath as his younger companion and kinsman

Woollarawarre Bennelong, a senior man of the Eora, an Aboriginal Koori people of the Port Jackson area, was born around 1764. He was respected by Governor Phillip and visited Europe. Watercolour pre 1806 by William Waterhouse. *Mitchell Library, State Library of New South Wales*

Yemmerrawanne. Both men accompanied Phillip on a six-month voyage to England from December 1792.[3] As the first of their race to visit, they were curiosities and, to the open-minded, diplomats who cut a path for a number of Aboriginal people of standing, as well as Māori, who were to follow. Too often, though, these pioneers out of the Pacific lacked the immunity of their hosts to Western pathogens. After a long illness Yemmerrawanne died, aged about 19, of a lung infection. He is buried near London.[4]

Just along the waterfront and east behind Bennelong Point is modern Sydney's CBD, enclosing the approximately 60ha catchment of Tank Stream. This name described a series of reservoirs cut into the sandstone, a technique brought by the British who had been colonists in India.[5] It was this tributary running out of a swamp near today's Hyde Park, at the head of Sydney Cove – 'the run of fresh water that stole silently along through a very thick wood' – that decided Phillip's location of his new settlement at Sydney Cove. As Phillip's soldiers and convicts knew (and as artefacts have since confirmed), its water had made the cove a gathering place through thousands of years of Aboriginal settlement.[6]

Nothing, however, not even the 'very thick wood', was as it seemed. This was an Aboriginal landscape, planted along the stream with 'fire-welcoming and fire-sensitive plants [that] grew together', the entirety regularly burnt to manage sustainably the relationship between plants, humans and animals.[7] It is now widely understood that, apart from keeping the more flammable understorey in check, judicious Aboriginal firing enabled 'the older animals to be taken more easily and provided a new crop of sweeter grasses for the rising generation of the forest'.[8] No wonder it seemed to Phillip such an ideal site for a settlement.

But the newcomers failed to ask any of the right questions. The cooperative approach to early settlement at Sydney Cove from the local Eora people 'rapidly evaporated with European tree-clearing,' suggested Michael Cathcart. The arrivals' set-up was 'a devastating and unprecedented act of trespass'.[9] By January 1891, within just a few years of European occupation, 'most of the streams or runs of water about the cove dried up'.[10] This may have had many causes, among them tree-clearing and the island continent's perennial climatic aberrations. But by 1795, the new colonists had fouled their nest:

> Having been permitted to build themselves huts on each side of and near the stream of water which supplied the town of Sydney, they had, for the convenience of procuring water, opened the paling, and made paths from each hut; by which, in rainy weather, a great quantity of filth ran into the stream, polluting the water of which everyone drank. It therefore became an object of police; and the governor prohibited removing the paling, or keeping hogs in the neighbourhood of the stream, under penalty to the offender that his house should be pulled down.[11]

A year later people were disobeying the governor's orders and 'not only still opened the paling, but took with dirty vessels the water which they wanted above the tanks, thereby disturbing and polluting the whole stream below'.[12]

Today, Tank Stream is encased entirely in concrete. Its consignment to modern engineering requirements is literally 9000 kilometres from the relationship that evolved over millennia between the original people of Australia and water.

Bennelong's kinsman Yemmerrawanne had been given a second name, Kebbarrah, which signified that he had had a tooth removed as an initiation into manhood.[13] The custom was widespread among Australian Aboriginal peoples. Author and artist Jacques Arago, who accompanied the French explorer Louis de Freycinet on a scientific voyage around the world from 1817 to 1820, wrote and drew extensively on his travels and later published his drawings as lithographs. Arago observed a double incisor extraction south of Jackson Bay. The occasion was not an initiation into manhood: it was performed on a woman, perhaps as a sign of marriage. The male (possibly the groom) executing the operation applied a thin piece of hardwood to the tooth and struck it with a polished stone: he removed two incisors in this way.[14]

Marriage is but one of what seem to be several, often interrelated reasons for such a ritual. For many Aboriginal people, observance of weather and water is identified as central to this practice. But anthropologists W. Baldwin Spencer and Frances Gillen, who spent from 1875 until 1912 with the Arunta 'confederation' in the heart of Australia and observed the 'knocking out of teeth' ritual, did not find it so.[15] More widely practised among the Arunta to the north and northeast of Alice Springs, 'amongst the Arunta and other Central Australian natives … it has nothing to do with initiation, and is in fact practised by men and women, the rite having no sacred significance of any kind … [In] the central area, it has lost its old meaning.'[16] However, Spencer and Gillen were granted access to a 48-hour ceremony of the Purula and Kumara, a 'sub-division' of the Arunta, located about 85km east of Alice Springs, in a region known as Kartwia Quatcha or 'rain country'. Here the knocking out of teeth was the most important of ceremonies. A web of carefully observed protocols for initiation from the totem group preceded the 'rain-dance' itself, led by 'a celebrated rain-maker'. Ornate costumes, body paint, accoutrements and actual words used in the ceremony are set down in their report.[17] For the Purula and Kumara a similar tooth extraction ceremony was explained as being associated with:

> the rain totem … the object of the rite is to produce in the face a resemblance to what they call *Alailinga*, which is the name applied to certain clouds, dark with a light margin, which are of peculiar appearance and are said to portend the coming of rain. There evidently was … a time when the ceremony had a much deeper meaning than it has at the present day.[18]

We might speculate as to how the original meaning of the 'knocking out of teeth' ritual had been lost. Climate changes affecting availability of drinking water during the late Holocene in the pre-European period (700–1500 AD) may have played a part in this.[19] The Arunta will, of course, have their own explanations. But deep respect for water seems to have underlain this tradition of personal sacrifice for some groups, at least, in the interior.

Whatever the availability of water, particularly for inland nomadic clan groups, the vagaries of its supply and that of associated food would demand exceptional skills of Aboriginal people. Procurement, associated with customs and rituals such as those described, was a strategic issue. After 60,000 years of life in Australia they knew how to survive on a day-to-day basis; their cultural adaptation to the land was so profound it seems likely that there was no separation between long-term and quotidian thinking. Once European exploration of the interior began, it was Aboriginal memory of sites and ingenuity that, on occasions too many to count, saved the strangers from certain death in what appeared to them to be a pitiless and waterless land. And time and again, the gift of Australia's native people to the newcomers was repaid in anything but kind.[20]

So familiar was 'country' to them, Aboriginal Australians had an array of choices that were not within the imagination of their white 'explorer' dependants. Just as Polynesian voyagers noted the appearance of certain birds to show them land, Aboriginal people used birds, including the zebra finch and, in the Western and Simpson deserts, budgerigars, to guide them to standing water.[21] Again like Māori elders, who still use a tokotoko or talking stick as a memory prompt on the marae, desert Aboriginal people had notched sticks that were mnemonics to assist in recalling the location of tens of waterholes.[22]

Ian Bayly, a New Zealander who became an authority on the biology of salt lakes in Australia, surveyed the extraordinary ways in which they took water from extreme and unlikely landscapes (as detailed by white explorers). Dew droplets were collected from leaves using a ball of grass as a sponge; and frogs that they found in a clay-pan that was 'quite dry and baked hard by the sun' contained enough freshwater to slake the thirst of a desperate white man and his Aboriginal guide: 'Subsequent digging disclosed more frogs, from all of which so great a supply of water was squeezed that not only he and his boy, but the horses also were saved from a terrible death!'[23]

Other examples showed how the local Aboriginal people could read different landscapes and what wells or 'soaks' might lie hidden within them. Sometimes these might be in swales between sand ridges, the sides secured with timber and brushwood. R.L. Jack found that the Indulkana Springs in South Australia were 'better described as holes in a saturated sponge of alluvial than as true springs'.[24] Gnammas – natural holes that hold water in hard, impermeable rock – were known, and sometimes covered with a rock or branches to prevent evaporation. Bayly also described 'impoundments' – clay dug out of waterholes with wooden shovels to form a dam, and semi-permanent waterholes that formed from dry, ephemeral streams, as well as 'mound springs' fed by carbonated water under hydrostatic pressure. Tree-trunk hollows and tree roots were other sources of water for the Aboriginal people[25] – and, out in the desert, eucalypts and mallee. Water-mallee roots lie 5–25cm below the surface, running out from the plant for up to 20 metres. European observers described how the indigenous people would 'excavate the root with a wood shovel or spear-point. The root would be pulled up and broken off near the trunk, then broken into lengths of 50–100cm. These were drained into a coolamor or marsupial-skin bag.'[26] Explorer Thomas Mitchell described them sweetening water in a muddy billabong by filtering it in scooped sand and sucking the clear water through a grass straw. 'In time, the explorers would learn to rely on Aborigines to show the way to their traditional watering places – a kindness which, in the early contact between the two cultures, the Aborigines often seemed happy to perform.'[27]

But how could their knowledge be otherwise? This was a culture entire unto itself. Locally, regionally and, connected by the mysteries of the songlines, across the huge continent, this was Aboriginal country. They had fashioned it that way. We learn from Babbage's momentous scholarship on Aboriginal 'landscaping', ingenuously documented by early explorers, that the entire continent was an artefact of the indigenous inhabitants' unerring grasp of country. Nothing was overlooked – and most certainly not soil management.[28] Their almost uncanny understanding of sustainable relationships in matters of climate, fire control, flows and cycles of water, with their implications for soils, insects, plants, fish and animals, were of an order that the best discipline of Western land usage – 'landscape architecture' for instance – cannot match. Not yet.

Academic Tyson Yunkaporta, from the Apalech Clan in far north Queensland, suggests in his *Sand Talk: How Indigenous thinking can save the world* that 'Aboriginal knowledge' is no less accessible than the science of genetic sequencing; it 'just requires a bit of discernment, humility and awareness'. Armed with these qualities, he maintains, there is a capacity for adaptation, innovation and learning that enables you 'to realise your true status as a single node in a cooperative network'. From this position you can appreciate diversity, retaining your autonomy 'without clustering into narcissistic flash mobs'. The rules for effective participation in interactions for sustainability are, Yunkaporta reminds us, to 'connect, diversify, interact and adapt'.[29] These ideas resurface throughout this present book. In the decades since the Mabo decision[30] that nullified the culturally annihilating 'terra nullius' doctrine and, after 200 years, recognised Aboriginal land rights in common law, the traditional water space has been largely subject to state decisions under the National Water Initiative. Progress is slow.[31]

In these kinds of discussions, the term 'country' is often used to describe the land, in its manifold associations, to which an Aboriginal group is attached: for Aboriginal people it is defined as both the life and the

Sandy Creek, south of Albury in Victoria, Australia: natural waterholes can be muddy, often sandy, water sources. *State Library of Victoria*

landscape of belonging and 'its seasons, stories, and creation spirits ... a place of belonging and a way of being'.[32] Unsurprisingly, it is architects and landscape architects along with planners who have declared their professional interest in working with the people of country. The educators' guide developed in 2018 for these professional groups is based on Aboriginal rights to and knowledge of country. Country is seen as particularly relevant to these disciplines 'as it relates to the practice of statutory and strategic land planning, management, spatial knowledge exchange and landscape design'. The guide acknowledges the circular and reciprocal nature of indigenous systems of learning and knowledge acquisition, while Western systems are linear and non-reciprocal – another theme that recurs in this book. The guide emphasises relationship, respect for people, for country and for ancestry. It recognises the ties between people and country – in the words of Peter Berg: 'People and *Country* take care of each other.' As Ngarrindjeri elder Tom Trevorrow says: 'Our traditional management plan was don't be greedy, don't take any more than you need and respect everything around you. That's the management plan – it's such a simple management plan, but so hard for people to carry out.' A number of principles are set out, recognising self-determination, rights to the intangible heritage, and respecting and upholding traditional knowledge, practices and innovations. As with reconciliation across the globe, development of these principles is recognised as being an interactive, ongoing process.[33]

One such initiative, in December 2019, was at the Maroondah Federation Estate Gallery in Ringwood, Melbourne, with finalists in the R. & M. McGivern Art Contest exhibiting interpretive work on the theme 'Anthropocene'.[34] A seminar, part of a wider programme by the Maroondah City Council in the Ringwood–Croydon district, was run in conjunction with the festival.[35] The presentations at the seminar were from Aboriginal thinkers. Having watched the stuttering process of Aboriginal reconciliation over a span of some 40 years, I realised that this event, where white Australians became attentive listeners, was part of a gradual move towards fuller (if belated) recognition and interest in indigenous knowledge nationally.

With a population of more than five million, Melbourne is arguably Australia's largest centre of economic power and people. The autonomous hub of Maroondah, northeast of Melbourne CBD, serves a community of more than 110,000. Its city council is working across sectors to build a culture that aims to be more inclusive and attentive both to its deep past and to the challenges now pressing the biosphere's capabilities. Part of the process is to reach, through the creative arts, to the sensibility, mind maps and reciprocal cultural contours of country through its Aboriginal community to the wider population. The council's partnership with the locally based Mullum Mullum Indigenous Gathering Place is becoming a significant cultural growth point.

At the seminar Tyson argued that, rather than the Anthropocene, we were living in what he called the 'complexity of the Octocene', referring to the (loosely) eight rich men who control the world, its water, energy use and waste generation. He encouraged his audience to 'Embrace it. Listen, it's still there – remember what it is to be a human being. Do not bury yourself in introspection, it's a thing that has got into the Green Movement, don't be too hard on yourself, we've got to take out the Octo Thing.'

Another indigenous speaker, Janet Turpie-Johnstone (known as Aunty Janet), is living her talk. Aunty Janet, an elder who grew up in the Portland area, is an artist and is researching a PhD at Australian National University, Canberra, on indigenous knowing of the land and colonial attitudes. Along with several of the other speakers, she is a board member of the Mullum Mullum Indigenous Gathering Place. Acknowledging what has gone before, 'the thousands of generations – not just people, but animals and trees, waterways', she referred to 'the colonial invasion and all our people who stood strong in the face of that', and identified the ongoing consequences as 'arrogance based on insecurity'. Together with massive population increases, the cumulative effects have left an imprint so 'indelible we are changing the way the biosphere works,' she said. These effects include climate change, increasing invasive species, technofossil production (including plastics), anthroturbation (mining deep into the earth), terraforming (a super-effficient way of building), and redirection of water – not least the Yarra River, which has been subdued and diminished by flood-control measures. These changes are being exacerbated by runaway increases in population. 'We are now all operating in a single feedback loop,' she said, reflecting an idea Tyson had expressed about embracing complexity and eschewing the conventional.

She referred to the impacts since the early 'explorers' of Victoria, John Batman and John Pascoe Fawkner, with their terra nullius views, had come with some seven Aboriginal men from the Sydney area. At that time no settlers had yet arrived. 'By 1855, 300,000 people lived here and the vast swamps were extinguished,' she said. It was a remarkable increase in population in such a short timespan.

Palaeontological probing of a few remaining billabongs is now beginning to provide indications of what went on before this influx. Along with the destruction of these waterholes went massive extinction of species, including koala and kangaroo. The small coastal settlement of Portland – long-time place of the Gunditjmara people, where Aunty Janet grew up – had 90cm of annual rainfall; today it averages 82cm.[36]

At the time of the seminar, multiple fires burning in New South Wales since October 2019 were flickering in and out of megafire status, generating a smoke haze across greater Sydney that, with a following wind, had reached Melbourne. Then South Australia's wine-growing hills were ablaze; parts of Western Australia also started to burn, and Queensland had endured unusually early summer fires. Pleas to federal government – from senior firefighters for a commitment to meaningful climate change policies, and from skilled Aboriginal back-firers – had still not been acted on by early 2020. The seminar coincided with the latest round of UN Climate Change Conference discussions. These talks foundered on the intransigence of four of the planet's largest contributors to CO_2 levels, including, on a pro-rata basis, Australia – the unrepentant coal-exporting 'lucky country'.

Aunty Janet suggested that an Aboriginal sense of time and of planetary life were both generational and relational. She sees their role as being to uphold 'a sense of keeping everything alive'. Belonging gave her 'a sense of obligation even to the land, to the animals'. She saw this relationship as a circular, or spiral process. 'Yet this country at the moment is defaulting to no-memory,' she said. Fundamentally, hers is a philosophy based on a sense of creative engagement with the earth; of being able to feel the land and respond with an embodiment of that feeling in appropriate action. She recalled how her people's relationship with rivers, including the nearby Yarra, actually worked as a legal system, and of how 'the laws were about our relationship to water and the land'.

Stone fish traps of unparalleled antiquity

At the upper reaches of the Murray River in northern New South Wales is the village of Brewarrina. It stands beside what is possibly human history's oldest stone structure[37] – the Ngunnhu stone fish traps. The fish traps were laid out according to the ways of the river and its fish and the needs of local people, the Ngemba. Half a kilometre long and estimated to be 50,000 years old, they were built to be of the river, withstanding its natural alterations – to endure – after the ancestral spirit of that special place, Baiame, threw his net out into the river and created the design for the fish traps. With some huge rocks already in position, the ancestors rolled and placed others beside a natural weir of desert sandstone, making a maze of winding, labyrinthine proportions that directed and gradually narrowed the fishes' passage.[38] Each division of the tribe had an allotted pen or trap passed on from Baiame, each with a name. It was, according to Michael Cathcart, 'once a swirling maze of graceful stone walls that snaked its way through the water … It was the largest network of stone traps in Australia and is a triumph of human creativity.'[39]

In 1901, one or two tribal elders living on a government sheep station nearby, who retained traditional knowledge of this place,[40] passed this knowledge on to visiting anthropologist Robert Mathews, who documented it. His visit took place, however, only after a path had been cleared for riverboats, allowing them to steam to the limits of navigability up the mighty

Brewarrina: 50,000-year-old Aboriginal rock fish traps on the Upper Darling River, before a channel was cleared through them for riverboats in the late 19th century. This happened around the time that eel traps were removed for riverboats on the Whanganui River in New Zealand. *State Library Victoria*

Murray into the Darling River. Some restoration work was undertaken on the fish traps from the mid-1980s.[41] At the time of Mathews' visit, there had been a considerable period of drought, so that the river ceased to flow, enabling 'a careful survey'. The maze was a series of dykes of neatly fitted stones, as in drystone walling, about half a metre in height. These were interlocked in such a way that they withstood flooding events. As Bunurong man Bruce Pascoe writes, 'More information about that technique would allow us greater insights into the engineering skills of the Ngemba.'[42] Mathews found that the fish pens were maintained 'in good working order, in anticipation of the spawning season and also of freshets in the river'. Each division of the tribe had an allotted pen or trap passed on from Baiame, each with a name, and Mathews recorded that 'new pens were sometimes added'.[43]

'The traps also allowed … breeding stock to pass through so that upstream fisheries could gain a share,' writes Pascoe.[44] In traditional times of higher flows, fish moved upstream 'in immense numbers'. They swam into the narrowing stone traps, herded by the Ngemba, and once sufficient had entered their escape was closed off with rocks. The 'big-fellow fish' were then speared or clubbed to death, the smaller ones caught and strung together by their gills on a cord.[45] They took mainly Murray cod, yellowbelly and black bream – their favourite. At least one species they fished, now extinct, grew – like the critically endangered Murray cod – to at least a metre long.[46]

Over time the area became an annual gathering place for some 20 nations, 'including the Morowari, Paarkinji, Weilwan, Barabinja, Ualarai and Kamilaroi people', for a working congress.[47] 'Like most waterholes, Nguunhu is a place bathed in spiritual power,' writes Cathcart; 'it was fashioned by the great dreamtime ancestor Baiame, who first spoke its many names and taught the people how they should use it.' Baiame was the creator of their rivers, forests and mountains and source of their customs, laws and songs and folkways.[48]

As historian Marcia Langton explains, 'The various land and waters were a living drama of history.' Waters flowed with sacred meanings and memories, and with the social obligations by which each individual defined him or herself as a person of that place. Langton, who is a descendant of the Yiman and Bidjara nations, tells how the elders of one Cape York community sometimes 'baptise' newcomers with local water. By participating in this ritual, strangers make themselves known to the land, and signal to the ancestors that they will treat the land with respect.[49]

Since its swift recovery from the global financial crisis of 2008, the new millennium has not been kind to the 'lucky country'. As a new colony, it was founded on, grew up with, and prospered on the extractive and industrial processes of the carbon economy. The continued and cumulative effects of this culture now threaten, simultaneously, the very existence of its farming interior, the brush and eucalyptus suburban edges of its great cities and the political stability of federal governance. In the past decade five prime ministers – four of whom sought to take serious measures to curb carbon emissions – have lost their job, either electorally or directly at the hands of their colleagues, largely on this issue.

Malcolm Turnbull, who led the Liberal Party in Opposition and later as prime minister in the Liberal–National coalition government until August 2018, well understands the effects of climate change. He lost both roles over time, mainly over divisions within his party on this issue. So, too, did Labour prime ministers Kevin Rudd (twice) and Julia Gillard, as well as the climate-equivocating Liberal prime minister Tony Abbott, for his dalliance with greenhouse gas emission reductions.

Climate change delivers some brutal ironies. Australia still exports 37 percent of global traded coal, with much more still in prospect. But by 2016, 30 percent of the priceless Great Barrier Reef, over which Australia holds stewardship, had succumbed to the heat-bleaching effects of CO_2-warmed seas. It was its worst year on record.[50] In the heartland of 'big hat Australian' myth and National voter support in regions such as the New South Wales central west, farming is in crisis. In some of the worst-affected places, a 10-year drought lasting into 2020 exacted an unforgiving toll of dead stock, suicide and farm abandonment.[51] For those who remain in such places, an intergenerational way of life is in ecological and existential crisis, surviving only through state and federal support. Ironically, it was not until the early 1980s that mining overtook pastoralism as Australia's main export.[52]

Other parts of the Red Continent are being similarly impacted. The El Niño cycle predicted for 2020 brought further drought.[53] The ensuing much-feared fires delivered death and destruction, at times with an unimaginable intensity that sometimes turned firefighters into observer-managers-at-a-distance. As ecological scientist Tim Flannery predicted a quarter-century ago, Australia has little margin for adjustment. With a population now at 25 million, it continues to deny its ecological footprint as much as its colonial past. Flannery suggested long-term a far more sustainable population was somewhere between 12 and 19 million.[54] He is one of the few voices of leadership who has consistently tied climate change and growing population together in his reasoning.

The implications of climate change for Australia carry all the usual risks of debilitating temperature rises, agricultural losses, sea-level rise (250,000 houses at risk this century), but also – and notwithstanding an increased frequency of deluges – water shortages that will affect both food production and viability of human settlement. A major concomitant of droughts is the risk of fire to life, property and biodiversity; and excess demand for bore water brings the risk of ruinous salinisation.

Above: Cyclone manifest: ancient Aboriginal rock art in the Kimberley, Western Australia. *Photo Richard Sidey*

Left: King George Falls, the Kimberley, Western Australia, home of the Wandjina, the 'cloud and rain spirits' in local Aboriginal tradition. *Photo Richard Sidey*

Aboriginal people have so many challenges of their own, no one would expect them to be asked to play a part in solving the problems of the society that in so many respects has engulfed their ancient ways. At the same time, their deep attachment to and respect for the lands they roamed and managed moves their leadership to continue to offer their skills in land, water and fire management. If there was ever a time to listen, it must be now.

At the time of writing this chapter I watched *Gurrumul*, a deeply affecting film about the Aboriginal music and singing genius Geoffrey Gurrumul Yunupingu.[55] One of his colleagues likens Gurrumul to one of the idiosyncratically curved tiles on the roof of the Sydney Opera House, which need constant maintenance. In speaking of Gurrumul's early death, his colleague said that every day, somewhere in Australia, another piece of Aboriginal knowledge is lost to the greater detriment of humankind. Each death is like another tile falling off the roof of the Sydney Opera House.

Theatre curtain clouds draped over the sea at sunrise. *Photo Richard Sidey*

CHAPTER 4

Hawai`i — Aotearoa

THE SHAPE OF WATER

The great frigatebird or `iwa (*Fregata minor*) is spoken of in Hawaiian mythology. *iStock.com/randimal*

The river is me: A life of balance

Before dawn we headed out to the lookout at Makapu`u to watch the sunrise. The drive takes about 30 minutes from the sleepy boulevard of Sans Souci at the eastern end of Waikīkī. Out along the coast SUVs hulking like gridiron players are parked up in the oyster light that quietens everything before autumn's dawn. We walk away from the road and upward, along a sealed track. A fire has been through recently, but I can hear the calls of at least four different kinds of birds. Strangely, no one is there when we arrive and no one stumps up the pathway until well after the actual sunrise.

There is a pink smudge on the horizon where clouds are draped like theatre curtains, waiting for something to go up. A few minutes before 6.20am they turn to distant fire, ember-light rather than blaze. Scattered across the ramparts of the vista the light reveals concrete bunkers from when the threat of Japanese expansionism spawned these structures along Pacific coasts, from Hawai`i to Aotearoa.

We have a few minutes to chat and look out along the curving beach of Makapu`u northwards and out to the offshore islands. A full half of the moon is still in the sky. It remains there as the sun comes up. Beyond the moon, a fragment of rainbow breaks through from a sketch of clouds, just for a moment. This is perfection.

A large bird on the hover appears above the descending path. My friend is excited: 'It's a frigatebird – we have them in Rotuma, these birds have always been in my life since I was a child. Often you will see them in the morning going out for food – usually they return with something for their mate.'

In Hawai`i, this is the mythical bird `iwa which, together with the halulu and the kīwa`a, 'appear in stories as bearers overseas or to the heavens'; the kīwa`a is 'said to be the pilot bird which conducts the navigator in to the canoe shed at the landing place'.[1]

This beautiful sea creature of the tropics maintains a geostationary stance above us before catching a small thermal where, in profile to us, it performs a ritual of repeatedly tilting its neck and dropping it. Right now is fishing time for them but this creature dallies aloft. We take its presence as a gift. 'It's huge – the biggest frigatebird I have ever seen,' my friend says.

I reflect on a similar bird experience I had as I was researching *Woven by Water: Histories from the Whanganui River*. A ruru (morepork) showed itself one evening. Larger than life, it sat unblinking on the bough of a pōhutukawa in the front of our place in Wellington. Years later, I learnt that the presence of the ruru is regarded as a messenger; or a kēhua (ghost), as author Keri Hulme explained to me.

I never saw that ruru again, but once, in the presence of tangata whenua at the sacred whenua of Tawata on the upper Whanganui, I heard a flock of ruru, birds of the night, calling out in broad daylight. The cry was quite uncharacteristic: shrill and chattering. The woman with me, a descendant of the prophet leader at Tawatā, Te Kere Ngātaierua, hushed them as if they were restless tūpuna, or even mokopuna (ancestors or descendants).[2]

That southeastern coast of O`ahu is a place of antique and modern associations of great significance. Behind Makapu`u is the site of a mighty battleground. In sight of Makapu`u is Kāne`ohe Bay, laminated in associations right back to voyaging Tahitian chiefs and even the fabled adventurer of the Polynesian dawn, Laka (Rātā). In honour of those traditions, the revival of Polynesian navigation began there with the launch of the famous waka kaulua *Hōkūle`a*.[3]

In Honolulu my morning meditation begins with a sound, ebbing fast like a recent dream, of an ancient chant that resonates into my body and up into my skull. The chant is carried by a rich baritone and, to my ears, is from another dimension. It is disembodied, clearly articulated, with strange rhythms and cadences – as though the chanter has transcended emotion, as would a spirit. As if carried down a lava tube – coursing like water through time and space, and with a little context, it needs no translation. Then wakefulness releases me.

I rerun the sound in my mind. The baritone commands status, upholding a mana steeped in a sense of balance, deliberation and reflection. It does not exhort but its notes are cautionary. Clearly, the live performance I had witnessed the night before had not left me alone in the night.

These great chants endured because they are about living, about being in balance, 'the life of righteousness' as King Kamehameha III put it: 'Ua mau ke ea o ka `āina i ka pono', 'The life of the land is perpetuated in righteousness.'[4] This statement – in language that carries a missionary watermark and will not reflect the full meaning of the words in Hawaiian – now stands as the motto of the state of Hawai`i. The Waitangi Tribunal in New Zealand speaks in a similar vein in this passage:

> Kaitiakitanga is the obligation, arising from the kin relationship, to nurture or care for a person or thing ... In the human realm, those who have *mana* (or, to use Treaty terminology, *rangatiratanga*) must exercise it in accordance with the values of kaitiakitanga – to act unselfishly, with right mind and heart, and with proper procedure. Mana and kaitiakitanga go together as right and responsibility, and that kaitiakitanga responsibility can be understood not only as a cultural principle but as a system of law.[5]

These values may be seen as inseparable elements of a complex of relational values that include rāhui (prohibition of entry or of harvest; or explicit conservation), utu (maintenance of balance and harmony), manaaki (hospitality) and tuku (transfer). 'These concepts resulted in a legal system embedded in values, rather than rules, and conveyed in stories of creation, songs, dance and artwork, rather than statute books.'[6] This considered approach may also be found in modern discourse on sustainability issues.[7]

Before anything else, then, there is the inescapable discussion of values. These were shared, and were as foundational to the voyages to new destinations as the waka that carried Polynesians to their last three islands of settlement, defining the further reaches of their oceanic territory – Hawai`i, Aotearoa and Rapanui Easter Island.[8]

All three waka groups originated in Eastern Polynesia, in Tahiti and the Marquesas. Each kin group began their voyaging and settled into their far-flung island homes within a few hundred years of each other. They carefully packed such taonga (treasures) as ornaments and mauri stones and weapons, as well as

Flocks of brown noddies fly over the Phoenix Islands, Kiribati. *Photo Richard Sidey*

special plants and animals, and freshwater.[9] Sailing north, south and east, they bore common genes, language and cosmology, expressed in shared myths, chants, songs and stories. Memory of their whakapapa was paramount: it informed and saturated their world view. An exquisite capability for navigation was essential to their survival. By night they read the stars. By day, in their island-finding, it was wind, weather and the many subtle indications of land, from the behaviour of currents such as vegetation drift, animal pathways and phosphorescence.[10] Some of these elements also betoken the Pacific as an ecosystem, the largest on the planet, whose edges, or sections, may be joined by the once prolific birds whose much smaller numbers today still make seasonal flights across large expanses of water. These include some of the birds described by the crew of the Tuia 250 voyages undertaken in 2019 to commemorate 250 years since James Cook's first encounters with Māori and Pākehā: the bartailed godwit, the long-tailed cuckoo and the shining cuckoo.[11] These are 'out-flown' by the tītī (muttonbirds) that travel between Aotearoa and Oregon.[12]

Underpinning it, as Rakato Te Rangiita of Ngāti Tūwharetoa explained, was 'the principle of Te Ika a Māui as held by [tohunga] Ngātoroirangi, Māui and other great original explorers … that of the moving wave, the point of acceleration, of a people who are always headed for somewhere else'. It continues today.[13] Having arrived, the greatest challenge for these later voyaging Polynesians was to apply this principle to the demands of these new and last lands they were to make their own: to adapt to the ocean conditions and to the new islands that connected back to their place of origin, Hawaiki, like fishhooks on their lines back across the Pacific. And as they worked to this end, the land was working on them.

Cascading starlight over the South Pacific. *Photo Richard Sidey*

The big adjustment for those who became the Māori, especially those who migrated south from the Muriwhenua district in the north and Tairāwhiti in the east, was to the temperature. In winter it was cold – and colder the further south they went: the arc of rock, forests and waterways they occupied lay directly in the path of the roaring forties. Now they were in temperate, not subtropical or tropical climes. It would have been something of a shock.

Furthermore, not since their passage out of Asia thousands of years before had they, or their ancestors, lived in such an abundance of freshwater. Here was a paradox embedded in one branch of the Polynesian migration. Notwithstanding its extraordinary availability in Aotearoa, the honouring of water has continued to this day, some 30 generations after its scarcity might have been experienced on any scale.

As we shall see, it was different for Hawaiians, and their gratitude for freshwater was as physical as it was metaphysical. E.S. Craighill Handy explained the Hawaiian position from the 'most frequently' invoked of the 'gods', the highly venerated supreme god Kāne, 'the procreator: Kane-i-ka-wai-ola [ora]'. Kāne was invoked when planting taro in streams 'for the planter he was, as embodied in fresh water for irrigation, ka-wai-ola-a-Kane'.[14]

'Gods' has been a widely-used European term for the whakapapa-based entities whose embodiment in the natural world encompasses so much more of the subtleties of Polynesian spirituality than personification, or energies out of the cosmological and natural world, than a single English word can ever describe.

The landscapes encountered by each of the three voyaging groups to the three far corners of Polynesia – Hawai`i, Aotearoa and Rapanui – differed from each other.[15] Aotearoa was large by comparison with any other island in the Pacific. Apart from its wider temperature range, its islands were notably more variable and diverse in latitude, altitude and, of course, biota.[16] Hawai`i, though it was considerably smaller, has far more endemism – some 15,000 species, including nearly all of its native plants and invertebrates and 25 percent of reef fishes, are found nowhere else. And Rapanui is tiny – less than the area of the seventh smallest island in the Hawaiian chain, Ni`ihau.[17] But more important yet was the single element that defined Polynesian life – freshwater, a fundamental driver in their voyages from and to the periphery of the Pacific.[18]

But freshwater is only of value to human life when it is accessible. Most of the places Polynesians made their home owe their existence to volcanoes. Hawai`i's wild landscape remains a visual, even visceral delight, but its geology may be taunting to a water-questor. Its mountain ranges rear up unexpectedly, like startled horses; its peaks are craggy, and its volcanic edges, some rising more than 9000m from the ocean floor, can be steep and impenetrable, faced vertically with flange-like dykes. Valleys are often wooded, deep and mysterious, their heads marked in favourably disposed places by trailing, often paired waterfalls, sometimes dropping in tiers. Out on the bluffs and salt-blasted headlands, red-footed boobies teeter in to land, white-tailed tropicbirds soar, and shearwaters and the `iwa or frigatebirds ride the air over winter seas tumbling in on basalt. Here, too, one can sense the sound of the unblown conch shell – the call from a culture never losing its commitment to kin and land, staggering now to stand again but doing it anyway. So much of what was denied Hawaiians, systematically and presumptively, with colonisation, had to do with control over water.[19]

In the past 10 years especially, having long admired the continuing surge of Māori reassertion of rights and culture, Hawaiians have made spectacular headway, especially in what can be summed up as biocultural regeneration. This is unfolding within a Polynesian culture again rich in association and in the primacy of connectedness in all dealings, imaginings and storying of its landscapes and natural life – a vivid corner of the Polynesian tapa mat that we refer to as the Pacific.

A plastic bag billows as it is picked up by the wind: it holds the illusion of fullness, carrying the shadow of death out onto the ocean. The antithesis of the conch shell, it summons the age of disconnection – all wind, its only vibration the crackle of carelessness. Its harvest is choked fish, dead marine mammals and contamination right up the food chain.[20] Before Cook's arrival, this was a concentrated, seemingly healthy population of humans, of astounding productivity, inhabiting an elaborate social order. Within little more than a century, however, the Native Hawaiian population of at least 600,000 had been decimated.[21]

The landscape carries an auditory undertow from the sea to the mountains, tugging the bell notes of slack-key guitars and the music of the hula danced under the big tourist skies of Waikīkī. Though the undertow is about loss, everyone on the beach is staring out at the waves. Beyond is a current that even today pulls back deep into the land. Only history, geological and human, can now divine this.

One explanation lies in the intricate nature of Hawai`i's makeup, where freshwater may be found in the most unlikely places – or not. On all of these volcanic islands are springs, sometimes colossal, of often inaccessible provenance. Where water surfaced, there was productivity. James Macrae wrote of the Pearl River, O`ahu, in the 1820s:

> we saw … a ravine stocked with bananas, taro and healthy breadfruit trees just forming their fruit. Here we met with an old Englishman, who told us there was on the opposite side of the ravine a large river coming out under the ground. We went to the place and found that what he had told us was correct, and stood admiring the subterranean stream of fine, cool water. Its source was rapid, forming a cascade nearly 20 feet in height, having ferns and mosses on its sides.[22]

From such places float chants of enlightenment, of wonder and connection that carry their culture forward from the seeds of human time.

In Hawai`i the lodestone of knowledge is the *Kumulipo*, its 2077 lines divided in two – a first part, of the spirit world, and a second, of humans:[23]

O ke au i kahuli wela ka honua
O ke au i kahuli lole ka lani
O ke au i kuka`iaka ka la
E ho`omalamalama i ka malama
O ke au o Makali`i ka po
O ka walewale ho`okumu honua ia
O ke kumu o ka lipo, i lipo ai
O ke kumu o ka Po, i po ai
O ka lipolipo, o ka lipolipo
O ka lipo o ka la, o ka lipo o ka po
Po wale ho—`i[24]

When space turned round, the earth heated
When space turned over, the sky reversed
When the sun appeared standing in the shadows
To cause light to make bright the moon,
When the Pleiades are small eyes in the night.

From the source in the slime was the earth formed
From the source in the dark was darkness formed
From the source in the night was night formed
From the depths of the darkness, darkness so deep
Darkness of day, darkness of night
Of night alone[25]

Unsurprisingly, this chant is not so different in Aotearoa. This was summed up in 1855 by the Rev. Richard Taylor:

> In fact, their traditions of creation go back far beyond even the gods themselves. They begin with nothing, which produced something, and that brought forth something more, and generated a power of increasing. Spirit, being more subtle than matter, arose before it, and thought, being supposed to be more so than spirit, the commencement comes with its birth. …
>
> The first period may be styled the epoch of thought –
>
> 'From the conception the increase
> From the increase the thought,
> From the thought the remembrance,
> From the remembrance the consciousness,
> From the consciousness the desire.'
>
> The second period is that of night –
>
> '… The great night, the long night,
> The lowest night, the loftiest night …'

The third period is that of light –

> 'From the nothing the begetting,
> From the nothing the increase,
> From the nothing the abundance,
> The power of increasing,
> The living breath;
> It dwelt with the empty space, and produced
> the atmosphere which is above us,
> The atmosphere which floats above the
> earth;
> The great firmament above us, dwelt with
> the early dawn ...'[26]

For Hawaiians, traditional life centred on pono, or 'fullness' and, as Carlos Andrade put it, 'balance on all levels';[27] Kanalu Young defines it as 'proper, good, true'.[28] One aspect of this awareness was that every plant on land had its correspondent in the ocean – from the first coral polyp for a reef, to the bats and birds who made their way there from another place. It is almost as if, were a species removed from the ocean, then something living, say, up a river had to disappear too. In geological terms, as the Hawaiian island chain continues to build itself up and out into the ocean over its fiery, ever-moving vent; as islands rise – expressions of Pele's volcanism – in Hawaiian perception, they create new balances. Watching live volcanism on Hawai`i (Big Island)'s southern shore, the knowledge that Pele, goddess of fire, has a brother, Kamohoali`i, the shark god, carries its own internal logic. In this construction, ocean and fire are dramatically and relationally fused.

Starting with Kaua`i at the northwest end of the chain, each island appeared, fully declaring itself as a volcano before settling slowly into extinction. Never still, Pele's eternal cycle of creation and destruction has continued east and southward to form new land for the past three million years. As 2018 reminded the televised world, the beloved Pele dances, Shiva-like, at the still-forming edge of the newest island, Hawai`i. Her being is a statement of a

Above: Ngārangi Rangiuia

Opposite: The tide comes in and the tide goes out. Kiribati's low-lying atolls are at the front line of climate change. *Photo Richard Sidey*

geocultural heritage in its poetics, of a science in its own right, upheld by Polynesian social and political structures and customary law.

Balance between male and female dualities, writes Jonathan Kay Kamakawiwo`ole Osorio, leads to a life that is 'wise, judicious, proper, good', bringing balance between light and darkness. It is easy to see how the reciprocity dynamic fits with this.[29]

Pono, meaning 'balance', is alive in both Hawaiian and Māori societies, and an emphasis on reciprocity endures strongly for Māori too. The idea of utu, an equivalent of pono, was a guiding principle for Māori, be it in matters of aroha (love) or vengeance. However, keeper of traditional knowledge Ngārangi Rangiuia suggests that behind today's 'hard' notion of utu tradition lies a 'soft', more labile principle that underpinned relationships in traditional society: 'I like to sing, "Tai tumu, tai pari, tai ope, tai oha" – The tide comes in and the tide goes out. The idea of balance, of ebb and flow. It's about the gentle weaving together of people, of meeting and of greeting, the treating of people with kindness, a strong flow-on effect – whereas utu and aroha have a point at which they stop.' This knowledge is from a society that aligned its thinking deeply with nature. But there were times, in the pre-musket era, of ritual human sacrifice: a society in which hierarchical power and sacrifice were bound together.[30]

One can almost imagine that, from the Hawaiian perspective, their model of balance works so powerfully that there could be no random island appearing off Hawai`i without some similar event in Aotearoa; indeed, no Hawai`i without Aotearoa and no Hawaiians without Māori who, as mentioned earlier, began their thirteenth-century journeys almost simultaneously in opposite directions out of the Marquesas group.[31]

In Hawai`i, observation of nature took symmetry to a deeper level. As with every plant on land, the Polynesian-introduced taro in its various forms had its counterpart growing in the ocean. The examples are reeled out in the rhythms of the *Kumulipo* chant: 'coralline seaweed living in the sea' and 'the bird's nest fern living on land'; 'the Manauea seaweed living in the sea' and 'the Manauea taro living on land'. The following extract from that text casts man and woman in their assigned and merging aspects of fresh and saltwater in the same kind of reciprocity:

> Born was the nerita shellfish, the sand-burrowing shellfish his child came forth
> Born was the freshwater shellfish, his child the small freshwater shellfish came forth
> Born was man for the narrow stream, the woman for the broad stream.

The village of Hanavave on the island of Fatu Hiva, Marquesas Islands. *Photo Richard Sidey*

Born was the Ekaha moss living in the sea
Guarded by the Ekahakaha fern living on land
Darkness slips into light
Earth and water are the food of the plant
The god enters, man can not enter

Man for the narrow stream, woman for the broad stream
Born was the tough seagrass living in the sea
Guarded by the tough landgrass living on land.[32]

Martha Beckwith has noted strong similarities between this, the first part of the chant and that part of the chant contained in the whare wānanga (house of learning) of the East Coast, Aotearoa. The similarities are found, too, in chants of the 'coming into being' in Tahiti, the Marquesas and the Tuamotu Islands. Even distant Rēkohu's foundational myth carries some similarities.[33]

Extracting the essence of the land

Thrusting miraculously out of the sea under fire and smoke, the Hawai`i islands are formed from pyroclastic flows; they are riddled with chambers and tubes formed by trapped air and the capricious twists and turns of lava memory. Such architecture produces extraordinary opportunities for the presence of freshwater in unlikely places. Early human arrivals, though, in many parts found its availability tantalising – close, just out of reach, submerged, but not available.

In the 1890s when he first sailed in to Honolulu, Robert Louis Stevenson rapidly apprised himself of the lie of the land, and seized on the importance of water in Hawai`i. In the late 1880s he wrote of the old steamship landing place of Ho`okena, Big Island:

Hawai`i's old volcanic mountains covered in mist at dawn, Haleakalā, Maui. *Photo Richard Sidey*

Rainbow Falls at Hilo Big Island, Hawai`i. *Photo supplied*

The Limahuli Valley descends from the Makana Mountain ridge, on the Hawaiian island of Kaua`i.

> Streams of lava have followed and submerged each other down this slope, and overflowed into the sea. These cooled and shrank, and were buried under fresh inundations, or dislocated by fresh tremors of the mountain. A multiplicity of caves is the result. The mouths of caves are everywhere; the lava is tunnelled with corridors and halls; under houses high on the mountain, the sea can be heard throbbing in the bowels of the land, and there is one gallery of miles, which has been used by armies as a pass. Streams are thus unknown. The rain falls continually in the highlands: an isle that rises nearly fourteen thousand feet sheer from the sea could never fail of rain; but the treasure is squandered on a sieve; and by sunless conduits returns unseen into the ocean. Corrugated slopes of lava, bristling lava cliffs, sprouts of metallic clinkers, miles of coast without a well or rivulet, scarce anywhere a beach, nowhere a harbour: here seems a singular land to be contended for in battle as a seat for courts and princes.[34]

Stevenson was right – water was abundant but not always available. As their population increased over time, Hawaiians could only survive by bending to the islands' ways; and as some sources of naturally available food declined, the organisation of work became imperative.

The word for water, wai, is at the base of the compound word for law: kānāwai – literally 'of the water' or 'that which pertains to regulation of waters' – which serves to remind us 'that the fundamental concept of law is associated with water rights'. And the Hawaiian word for 'wealth' is the word for freshwater, wai, doubled for emphasis to give waiwai – 'abundance of water'.[35] Defined property rights, communally based, became part of Hawaiian culture. In both the Hawai`i and Aotearoa archipelagos survival meant hard work, but in Hawai`i the difference was that often their organisation was around not only work, but the reliability of freshwater supply. Indeed, there was no conception of property pertaining to land or water, only to water use.[36]

The most expressive feature of life shaped in the Hawai`i landscape was the ahupua`a, a term for radial land division, and for an integrated method of farming that emerged from this: the original 'mountains-to-sea', or 'ridge-to-reef' tiered horticulture that took the best of island architecture and climate, turning it into a highly organised production unit that fed several thousand people by sustainable means.

The ahupua`a system of farming ran from the chisel-edged mountain tops, place of the gods and the birds, drawing from springs and streams to make irrigated terraces for growing precious taro.[37] Freshwater fish fed on stream plants and contributed nutrients; birds consumed fish. Below, in the bay, penned between the reefs, sea fish grew fat on the discharged nutrients from the gardens. Not all islands and by no means all districts were suitable for this method: 'Only on Kaua`i and O`ahu, together with western Maui and ancient Moloka`i's southern coast, can ahupua`a exist in their entirety, mountains-to-sea.'[38]

In order to channel water, labour was organised, sometimes on an epic scale – shaping, hefting and then placing rock into position. Conducting water flows through a sloping valley was often challenging: Handy described how Hawaiians would engineer its course around a jutting cliff with lava blocks cut, jointed and interlocked. He cited a tunnel 'used to bring water from a sunken stream bed to an adjoining valley'. Sometimes stems of large bamboo, `ohe (*Schizostachyum glaucifolium*) – a cargo plant of migrating Hawaiians, and the time-honoured water plumbing of the tropics – were used to draw water out of streams and `auwai (irrigation channels). On southern O`ahu, Kāne`ohe Bay, the site of an ancient ahupua`a with fishpond, combines the names of Kāne the god and `ohe, bamboo.[39]

In Honolulu, fishponds (loko i`a) were formed by rolling blocks of lava into the ocean to create a partial barrier to retain sea fish.[40] The work required to build Moloka`i's extensive ancient fishponds (just one component of an ahupua`a) is estimated to have required 10,000 workers. As one of Hawai`i's earliest settlements, Moloka`i was also the last location for a fishpond built in those islands, in 1829. Since the 1990s a number of these have been restored on this rugged island.[41]

Aotearoa's abundant water

There is evidence of volcanism in many parts of Aotearoa, notably in the North Island. Central Auckland, where its most recent eruption occurred just 350 years ago, sits over several hundred lava caves.[42] Quiescent Mt Taranaki is by no means extinct: when Cook sailed past 250 years ago it was smoking. Volcanism has shaped many parts of the country, including in Northland, obviously the central North Island, the Nelson region, Banks Peninsula and Dunedin.[43]

The difference between the southernmost islands of Polynesia and Hawai`i is that in Aotearoa freshwater was available almost everywhere: if not as mist or rain, then below the mountain peaks, either running in streams or moderate-sized rivers, while lakes and wetlands abounded. Springs, too, were often plentiful.

Many of the rivers of Aotearoa are far broader and longer than any in Hawai`i, and only a few waterfalls of the Hawai`i group have even a hint of the grandeur of the water systems of their southern counterparts. The longest river in Hawai`i is the Wailuku (which flows between the lava flows of two big mountains), at 45km. The longest river on Kaua`i is the Wailua at 32km.[44] Only five rivers in the entire island group are at all navigable. Aoteoroa's largest rivers are the Waikato (425km) and the full-bodied Clutha (322km) – which is actually larger in volume. The longest navigable river is the Whanganui at 290km – navigable for about 225km from Taumarunui to Whanganui township.[45]

Aotearoa also holds massive artesian reservoirs, many of which are known to Māori. One such aquifer, spilling from still active Mt Ruapehu on the Central Plateau, emerges in springs off the southwest coast of Te Ika a Māui (the North Island) and the top

Water reappears at the Riuwaka Resurgence from deep within the limestone hills.
Photo Aliscia Young

of Te Waipounamu, has long been known and described by Māori. They believe that the Riuwaka Resurgence, which spouts into the Riuwaka River out of a tree-fringed portal in the side of Tākaka marble mountain, is connected into this system. The guardian spirit here is Huriawa, whose influence is over the entire system, through which she is said to be able to move. The name Riuwaka, long misspelt as Riwaka, was corrected following a Waitangi Tribunal settlement in 2014. Speaking of Polynesians' fine-grained understanding of water, riuwaka means 'the bilgewater in an ocean-going waka'; it was named by the great voyager Hui Te Rangiora as he sat contemplating the river's riffles. He was 'one of the most renowned Polynesian navigators', a grandfather of Māui. Ngārangi Rangiuia says that his exploration was so extensive he sailed even beyond the Pacific – to Antarctica, which he named Te Anu Matou, 'to know cold'.[46]

One way to appreciate the level of detail of Māori knowledge of water is in the number of names they had for it. For them, discovery of a landmass of such size and opportunity in which water scarcity was rarely an issue was a new phenomenon. Water came down out of the sky; it lay extensively upon the earth in lakes and wetlands; it blocked their path, whether they were roaming the glaciated and pluvial wilds of Fiordland or picking their way through the great rolling forests of the central North Island.

For the Western mentality, rarity is so often a precondition of respect and of value, a principle well understood in Western conservation. Polynesia's sacred regard for water, born of frequent deprivation, might have declined given exposure over a period of 700 years to generally plentiful water. But the contrary applies. Even as the arrivals found water's ubiquity across the lowlands to be frequently a barrier and often a threat, Māori gratitude for and awe of water was both profound and enduring. The root word for a spring of water in both Hawaiian and Aotearoan cultures is the axiomatic puna, and – given the holistic Polynesian view there is no other way to put it – everything flows from this.

Of course there were also times, especially in the east, when water was not so readily available and accessing it demanded ingenuity. Māori, when travelling, were known to adze a rātā vine and drink their fill from the cut.[47] The response of Māori, as they evolved, was to go beyond what their parent culture prescribed. In Aotearoa, over time, they were to attach themselves in name, in whakapapa and mythology and in deed, to their river. Today, in any formal introduction or mihimihi, Māori characteristically begin by identifying with the trinity of their maunga (mountain), their awa, roto or moana (river, lake or sea) and, through their ariki (chief), with their kinship group. All Polynesia reveres land. It is the interweaving of whakapapa and respect for tūpuna and mokopuna that brings home the Polynesian power of connection with water and place. Place is rendered poetically as identity – mana whenua. If relationship is everything in Polynesia, for Māori, water can be as binding as blood. As an ancestor, the river is provider, healer and mauri. Its essence is held in the elegant pepeha of the Whanganui iwi: 'Ko au te awa, ko te awa ko au', 'I am the river and the river is me.'

For other tribes, their identification with a specific body of water can be just as explicit. The pepeha of the Warahoe, who preceded Ngāti Awa in the district of the lower Rangitaiki River in the Bay of Plenty, for example, was: 'Ko Warahoe te awa, ko Warahoe te tangata', 'The people of Warahoe are the river, the river is the people.' The numerous Tainui have a similar expression; for them, their Waikato River carries ancestral and, therefore, profound cultural embodiment.[48] For all iwi, it is integral to their trinity of belonging.

Land works its mysteries upon us in a tangle of interactions, culturally mediated. Patrick Vinton Kirch, an archaeologist and haole (non-Native Hawaiian), argues that the cultural evolution in Hawaiian society up until Cook's arrival may be expressed in the evolution of two words: kainanga and kāinga. Kainanga was an ancient Proto-Polynesian term 'that in old Hawaiki had referred to a group of people descended from a common ancestor', who travelled to Hawai`i via the Marquesas as matakāinanga. In Hawai`i they became maka`āinana, a relatively new concept defined as a 'collectivity of commoners' who worked the land. Kirch holds, controversially, that while the land was ancestral and shared, their genealogies were lost, suborned to the chiefs, their masters.[49] Hawaiians – Kawika Winter among them – take issue with this interpretation. 'There is a saying,' says Kawika, '"I ali`i ke kānaka", which means that the ali`i [chief] is no one without his people. If the ali`i didn't provide good leadership and fairness, they killed him.'[50] Again, this is an assertion of the balancing principle of pono as an agency of correction.

A second word, kāinga, referred to a group of people living together in a house, sharing lands and gardens. The word has changed little in Māori, where it embraces those concepts, but it usually refers to a village and, by implication, a kinship group. In Hawai`i, the same social–political change that altered kāinanga also changed the meaning of kāinga – planing it down to the word `āina, meaning land, as in land to which people are attached. For Māori, their evolution took quite a different turn. It brought their association with tribal/hapū land so close that the word for the umbilicus, whenua, also served them as the term for land, and those who belonged became tangata whenua.[51] After the fifteenth century, Kirch argues, there was little room in the social organisation of Hawaiians for such a concept – certainly not publicly. It was, he writes, 'the radical transformation of ancient kāinanga to make maka`āinana and of kāinga to `āina [that] underscores the deep changes that took place in Hawaiian society in the 16th century'.[52]

The change was one by which Hawaiian society became more highly organised and chiefs and priests more powerful. What Māori still hold to as whakapapa (ancestral lineage, literally 'connecting to the earth') was no longer the gift of Hawaiians, but was, argues Kirch, the sole possession of leadership. It was they, the Hawaiian royalty, who connected their lines to gods and power. With junior royalty, known as konohiki, deployed to run the agricultural allotments and the `āina, tribute was paid under threat of severe sanctions, in taro, pigs and other items such as mats and barkcloth. Ali`i in turn offered stability and protection to the kānaka (ordinary people) as the divine right of kings. In Kirch's estimation this created a 'radically new social idea' in which the chiefly or 'feathered' kings headed a social organisation that became a state. Rivalry between kings then drove innovation in the construction and development of agriculture, be it dryland or wetland farming. Over time these competing and sometimes warring states became essentially island states.[53] Some scholars have critiqued significant details of Kirch's work around the relationship of resources to chiefly power. Paul D'Arcy, for example, puts the emphasis back on pono: 'The expectation of reciprocity underlying maka`āinana obedience to sacred leaders, and the limited coercive power of those leaders relative to subordinate ali`i, combined to restrain the concentration of power by mō`ī [sacred leaders].'[54]

By contrast, it was not until the Kīngitanga movement began in mid-nineteenth-century

New Zealand that a large number of North Island tribes, particularly, took such a step in the face of a common enemy – a land-grabbing Crown. By that stage in Hawai`i, kingship over the entirety of islands, consolidated under Kamehameha I in the eighteenth century, was in decline. With it went the systems of traditional land use. Indeed, the correlation between the decline of the monarchy, especially from 1874 under King David Kalākaua, and the rise of the water-hungry sugar barons, characterises the power shift.[55] As Jonathan Osorio details, the majority of native Hawaiians 'overwhelmingly, but not unanimously' opposed the 1887 constitution under which the wealthy, including some native Hawaiians, were enfranchised to vote for representation into both houses – the House of Representatives and the State Senate. But the law, which Osorio sees as the issue for his people, and the government were already stacked against them. The law meant a loss of sovereignty for native Hawaiians as they were 'manipulated into a position of actual inferiority by haole settlers'. Collusion between 'Americans in the US State Department and American businessmen in Hawai`i' marked the end of an expansive century of American acquisition of territory.[56]

So much emerges from these fundamentals. Many suitable Hawaiian landscapes, beginning in early times on the old island of Moloka`i until the incursion of armed American opportunists, had been subsumed to the business of taro growing. As Hawaiians lost their land and their traditional rights and status – their cultural landscapes – their capacity for sustainability was replaced by systems that, while they were highly profitable for a few well-known American families and companies, had little regard for those dispossessed.

In Aotearoa, as populations demanded and where flows allowed, its generally much larger rivers were marked by pā tuna – ingenious large, wooden weirs that supported the hīnaki (eel pots).[57] These were raised with such tapu and respect for the ancestral river in all matters, including its larder – its manaaki – that its builders never ate while they were engaged in its construction. In times of crisis, restraints and prohibitions might be defined by a rāhui, a declaration by a chief, usually to protect a crop or harvest from human or spiritual defilement, the latter induced by an event such as a drowning or a violent death. However, traditionally rāhui were a relatively rare event in Aotearoa.[58] In Hawai`i, the ahupua`a, the use of kapu (tapu), upheld the system of management rights of the konihi (the managers of the land) on behalf of the ali`i. But this sanction seems to have been applied more stringently in the coastal/fisheries aspect of this highly disciplined system. Rules were upheld by severe sanctions that ensured sustainable harvest. They also included the taking of species such as Pacific threadfin, reserved for chiefs, and ulua (jack), for other men of rank. The kumu (endemic goatfish) and honu (turtles), for example, embodied the gods and their consumption was restricted to women.[59]

But other, similar measures were also available. For instance, on the Whanganui in 1864, tensions leading to the Battle of Moutoa prompted lower-river elders to set up a declared 'no go' waystation. This was a signal by way of an aukati – a barrier of woven supplejack across the river – to the forces pressing from upriver.[60] Today, when tragedy occurs, say on a river, rāhui are more often invoked than was previously the case.[61] The need may be greater now. The current outbreak of the species-threatening kauri dieback disease in Auckland's Waitākere Ranges prompted a rāhui by Te Kawerau ā Maki, mana whenua of the region, in order to reduce dispersal of disease spores by human agency.[62]

A tuna (eel) or piharau (lamprey) weir at Pungarehu, on the banks of the Whanganui River. The piharau travel in shoals along the bottom of the river. When they suddenly hit fast-rippling water generated by the trap, they turn into it. The uprights are secured together by woven mānuka brush and fern is compressed between the poles. *1/1-000482-G, photographed by William James Harding between 1856 and 1889, Whanganui Regional Museum*

Outliers of old Hawai`i where some threads to the past were never broken today provide ways back to models for a renewed application of rāhui practice. This is being applied as part of the reintegration of traditional ecological knowledge, for example in the strategies for inshore and offshore fisheries management. This combines traditional spiritual and subsistence practices of each particular island, according to its history and current resources. While kapu is regularly invoked for a restriction or prohibition of a defined fishery, the term rāhui itself is not always invoked.[63]

The revival may be found in many far-flung parts of the Pacific. In the French-held Tuamotu island of Fakarava, rāhui 'has taken on a new, ecological direction. New priorities have emerged since the atoll was declared a biosphere reserve.' The Fisheries Service 'now invokes the word rāhui for a period of legal and regulated operation', with a complete ban in the tapu phase.[64] 'In today's world,' Ngārangi suggests, 'rāhui suits the purpose it is being used for. It shows an ability to use the language in the modern world, because we're keeping people safe – it's just that the concept has changed meaning.'[65]

Just as population pressures challenged Hawaiians to measures of greater ingenuity and industry in their use of water, Māori, when pressed, undertook comparable feats of water engineering. In a few particular regions, iwi Māori cut flow channels. In Marlborough, close to the oldest of studied sites of human habitation in the entire country, massive earthworks were executed in the estuarine wetlands of the Wairau River. First recorded in 1902, this considerable array of excavated canals from coastal wetlands at the mouth of the Wairau River, Cloudy Bay, Marlborough, remained navigable by a small canoe. Surveyor and foundation member of the Polynesian Society, W.H. Skinner, appraised the network as having been constructed largely to enhance the capture of waterfowl, eels and other tidal fish. This entailed linking existing lagoons by excavating the waterways. Spreading in total almost 20km, it was sometimes more than three metres wide, and one canal extended 6.5km at an average depth of two metres. An estimated 55,000 tonnes of soil was excavated by the Rangitāne/Ngāti Māmoe people, using the simple kō for digging and stretchers made of mānuka and woven flax for soil removal. The techniques employed would have been refined, Skinner suggested, in their construction of maioro, or fortified earthworks.[66]

Oral history sources in the district placed this usage at about nine generations earlier – around the time of Cook's southern voyages. As is the case at Limahuli Gardens, on the Hawaiian island of Kaua`i, a few surviving traditions at hand suggest that this construction was the work of earlier Polynesians. Certainly, Ngāi Tahu traditions hold that their ancestors and predecessors, Ngāti Māmoe, 'were drawn south by an abundance of manu (birds), tuna (eels) and ika (fish) resources of the Wairau estuaries and lagoons'.[67]

The planning and physical effort that went into these constructions, which are continuously subject to erosion from weather and tidal surges, betoken two things: that the potential harvest of fish and wildlife enabled by the works justified this level of effort; and that the people who constructed them were fit, healthy and largely disease-free – and numerous, before the influx of foreign contagion and muskets that devastated their race.[68] As in Hawai`i, then, the demands of a growing population appear to have motivated the undertaking of hydraulic works.

Each body of water in the area was named: Mataroa 'Big Lagoon' (about 809ha), Ohineanaumate 'Upper Lagoon', Te Awaaroiti 'Chandlers' and Wāhangaatangaroa 'Main Channel' (283ha).[69] At intervals, buttresses were erected to narrow the channel; it was here that eel traps and fish nets were set up in season. Close by were special sand pits 'into which the traps and nets were emptied'. The fish were killed, and carefully preserved by sprinkling them with fine dry grit or sand. Pūtangitangi or paradise shelduck were easily caught here, too, during the moulting season when they were flightless, when they were known as maumi, or flappers; as were pārera or grey duck. The birds were preserved: 'potted down in their own fat and stored in *ipus*, calabashes, or vessels, made from the bark of the totara tree, and also from the giant sea kelp, and put away in the *whata* or village storehouse for future use'.[70]

Some of Skinner's report is questionable. The main channel, Ohineanaumate, is described on several survey maps as the Main or Maori Channel. It runs from the edge of modern Blenheim township out into Big Lagoon. 'It dates only from about the 1880s. It was present in 1899, but not in 1848,' says Bruce McFadgen; he thinks it was constructed to lower the water table for the town site. Today it is called the Riverlands Co-op Drain.[71]

Modern science poses further questions about just what it was Skinner was describing. The land is so low-lying that even within the period of European settlement, much has changed. 'The current geomorphological configuration of the site has existed for less than 170 years,' write McFadgen and Peter Adds. Earthquakes in the region in 1848 and 1855, along the highly active subduction zone of the Pacific and Australian fault, altered the lagoons and the old boulder bank: both were lowered by 'about 1.5 metres due to earthquake compaction and subsidence'. Further, 'at least three tsunamis have impacted the bar since human settlement, the most recent being in 1855 as a result of the Wairarapa earthquake'.[72] Even since 1901 – only 11 years before Skinner's paper – the shoreline had receded noticeably;[73] and now, with global climate change, it faces more storms and rising sea levels.

By contrast, on the historically eel-rich Whanganui River for example, tuna were kept alive in large korotete or woven baskets; and eel cuts have been identified in the once considerable harakeke wetland behind the river mouth.[74] Further south, Muaūpoko iwi's numerous eel cuts – made through the sand from their food basket, the Horowhenua dune lake to the sea – enabled both ecological management and harvest.[75]

The largest lake in Te Waipounamu, Te Waihora Lake Ellesmere, is actually a huge lagoon; it is now protected[76] but, despite strenuous rehabilitation efforts, is still a sink for nitrates and eutrophication. Ngāi Tahu retain a similar tradition here to that of Muaūpoko: they open this historically critical food basket to the sea annually, in a ritual of flows – cultural and ecological as well as hydrological.[77]

However, in parts of New Zealand – for example in Te Tai Tokerau, where summers are warm and conditions for cultivation were otherwise attractive – there are indications that irrigation made the difference. One site, Motutangi, north of Kaitāia, was a simple raised-bed system of irrigation that is not easily dismissed as mere drainage. According to Ian Barber, who has surveyed the evidence, 'Fed by fresh spring water, this system would have maintained a more consistent distribution of nutrients to the plants.'[78] Barber found the Ōruru and Kaitāia river valley ditch systems had adapted elements of the natural flow to pass them through tillage areas, to keep soils moist. In the absence of reliable rainfall, this is how the wetland taro must be grown.[79]

A large korotete, a woven corf for keeping eels in freshwater, Whanganui River. They were fed on boiled potatoes. *Donated to Whanganui Regional Museum by K.W. Puohotaua in 1968*

Captain St Jean Roux describes what he calls 'canals' in one early journal. The voyage of Marion du Fresne ended in tragedy in the Bay of Islands when he and his crew were killed by provoked and aggrieved Māori. His surviving junior, St Jean Roux, who exacted reprisals, described how he and his crew struggled to obtain potable water in the far northeast in April 1772. After wild equinoctial gales had delayed their landing, they came ashore at Takapaukura (Tom Bowling Bay), just east of Spirits Bay, where several waterways, including Waitangi Stream, flow into the bay:

CALABASH

Both the coconut and gourd have spread across great distances of the central Pacific. The calabash or bottle gourd (*Lagenaria siceraria*) originated in South America, and reached Eastern Polynesia possibly by 'natural dispersion'.[80] The shell of the calabash can be polished easily, and they are used in the American tropics as utensils and receptacles for liquids, such as basins.[81] Calabashes and gourds were highly valued 'kit' for venturing Polynesians and were carefully cultivated by their descendants. In most parts of the Pacific, water was carried in an ipuwai and food such as taro was prepared in a calabash. In Aotearoa, however, where suitable receptacles were not so readily grown, the language was less clear cut. Often people just scooped water directly from its source with their hands.[82]

Māia Poroaki, brother-in-law of the great chief Uenuku, is associated with seven gourds, large and small. Defying a tapu, he escaped death at the hands of Uenuku in an early waka, *Te Ikaroa-a-Rauru*, and arrived in Tūranganui a Kiwa (Poverty Bay), bringing with him knowledge of gourd-growing. One gourd or hue that was a container for storing birds in their fat is named Rauru. Other larger ones were used as dippers, while long-necked 'pilgrim gourds' were for travelling. Some still grow at Muriwai on the East Coast. The pūnotinoti and te kārure were the smallest, just fist-sized; these were cured for use as drinking cups. 'Māia also brought Māori the tikanga for drinking from gourds,' Ngārangi explained. 'Lips never touched the gourd, there was no guzzling; a better way was to sip slowly.'[83]

Kelp does not usually grow in the tropics as the seas are too warm. Here it is pictured at the Chatham Islands (far left) and Enderby Island (left).

Photos Richard Sidey

Tahā huahua, a food container made from a gourd bound in fine-woven fibre, supported by carved wooden supports.
Whanganui Regional Museum

Investigations by scientists, including Lisa Matisoo-Smith, an archaeologist who has been tracking the bones of the Lapita and the Polynesian chicken and kiore right across the Pacific, indicate that Polynesians did have contact with American pre-contact populations.[84] But it was they who made return journeys and not, as Thor Heyerdahl would have it, the other way round. According to numerous Hawaiian traditions, the gourd was carried as a prized plant on board voyaging waka. Other gourds, usually for ceremonial purposes, were carved from suitable wood and decorated with fibre.[85]

The preferred vessel for ocean-going water storage outbound from Aotearoa was the hollow crown of kelp. Kelp forests grow in ocean temperatures of less than 20°C, which limits its availability in the tropics. S. Percy Smith described how, on long voyages that presumably originated in Aotearoa, the kelp pack was towed at night from a big waka. This ensured that the water inside stayed cool for consumption and the constant external remoistening prevented it from evaporating.[86] In Aotearoa, kelp was also a preferred storage container for seasonal fatty foods, such as kererū.[87] In all of Oceania it was the Hawaiian who, it is said, made most use of the calabash.[88]

> The plain on which they flowed had a fertile appearance and is divided by several streams. It appeared to me to have been cultivated. Every ten paces or so there were little canals through which the water flowed. The herbage was so high as to give certain proof of the fertility of the soil. I saw but few shrubs whose names are known to us, but some of the plants are the same as ours. I noticed the gourd, the wild endive, the daisy, the reed, the water flag (or reed), a species of vacoua [vacoas], pandanus, and others. Of the latter the natives make very fine nets and lines.[89]

Cook's early encounter in New Zealand at Uawa (Tolaga Bay) offered some insights into water management. In a rare show of comic relief in the cross-cultural encounters at Uawa Bay, he gave his generous hosts a keg of gunpowder, a musket and a lump of lead. Being a practical people, they assumed the powder was turnip seed and planted it in their extensive gardens of gourds, yams and kūmara. These gardens were irrigated from a natural spring and waterfall.[90] 'They tapped it,' said Ngārangi, who is of this bay, 'so that the waterfall flowed down, it flowed through the whole garden. If you look down from the hills, you can see the lines where the irrigation flowed.' Other small-scale irrigation was used at Kaiawa, also on the East Coast. 'They had a hill where they used to grow kūmara. They used to block it off with a stone, to tap into the river, then unblock it again.' At Te Maatu, a forest on Motueka's river-gravel plain, Māori planted in circles around kahikatea for companion planting: 'around the bottom of the tree,' Ngārangi explained, 'the kahikatea holds so much moisture in its roots, there was no need to irrigate in that kind of land.'

In general, anthropologists and archaeologists have concluded that Māori did not need to use canal irrigation[91] as there was a reliable water supply in most places in Aotearoa. However, there are parts of the country that do need water, especially in high summer, and it may well be that simple, small-scale manipulation of water as described above was quite widespread: deforestation and erosion, as well as farm cultivation, may have obscured earthworks that existed in drought-prone regions from earlier times.[92]

The agricultural landscape at Uawa is described in the journals of Cook's *Endeavour*. There were taro plantings in circles at Anaura Bay.[93] These have been identified as *Colocasia esculenta.*[94] William Colenso knew of 30 different varieties of taro, most of which had specific uses; he named two.[95] Taro was cultivated in New Zealand by pre-European Māori for food. This was, however, mainly under dry-form cultivation, found in a number of places from the north of the North Island to the top of the South. The Bay of Plenty, Waikato, Hawke's Bay and, in the south, Golden Bay have been identified as having the occasional historical site for taro growing.[96] Ngāti Te Wehi and Ngāti Maniapoto had two sites where taro was grown near Hawaiki-iti, where the waka *Tainui* came ashore. These were planted by Whakaterangi, wife of the *Tainui* navigator Hoturoa. One patch, in a creek, still grows these taonga.[97]

Gods and guardianship

Since this book is dedicated to the idea of sustainable ways from the past, it is important in discussing irrigation and agriculture to acknowledge that the forests, while not always vanquished, were subject to serious reduction, usually in the name of agriculture. This is an important discussion point, especially in an exercise such as this that seeks out broad principles and values.

Wherever humans settle and enlarge their populations, forest is removed. The Polynesians were survivors and, as their numbers grew, forests were modified and declined in spread and variety. The original 80 percent of forest cover of the New Zealand archipelago had been reduced by around 25 percent. The remainder – often beech (*Nothofagus*) – was largely in the steep, uplifted spinal country.[98] The approximately one quarter of forests that Māori had removed over 700 years was to be joined by another quarter, fired and felled systematically by the newcomers in little over a century; and, within a similar timeframe, the country's wetland systems –'areas of marsh, fen, peatland or water' – were decimated.[99]

The loss of forests in Hawai`i follows a similar pattern. Kū is 'god of forest and of rain' and of forest growth, worshipped at the heiau by canoe-builders.[100] Moreover, as Kanalu Young puts it, he 'represented the acquisition of *mana* by the spear to establish the ridge post of governance'.[101]

Ku-moku-hali`i
(Ku spreading over the land)

Ku-pulupulu
(Ku of the undergrowth)

Ku-olono-wao
(Ku of the deep forest)

Ku-holoholo-pali
(Ku sliding down steeps)

Ku-pepeiao-loa and -poko
(Big- and small-eared Ku)

Ku-ai-ke`e
(Adzing out the canoe)

Ku-mauna
(Ku of the mountain)

Ku-ka-ohia-laka
(Ku of the ohia-lehua tree)

Ku-ka-ieie
(Ku of the wild pandanus vine)[102]

The chant references the divine ancestor Hina and her husband Kū, the 'great ancestral gods of heaven and earth who have general control over the fruitfulness of the earth and the generations of mankind'. Hina and Kū represent the female and male principles – 'female fecundity and the power of growth and production' and 'the male generating power'.[103]

When adzing out the canoe, it is Kū of the deep forest, Kū spreading over land, Kū of both wet and dry farming and Kū of the `ōhi`a lehua tree. The `ōhi`a lehua, then, is associated with both Kū and Kāne.[104] But it is the Kāne/Tāne association that appears to unite Hawai`i and Aotearoa in terms of their association with forest energy. In Hawai`i, Kū and Laka become joined in name, writes Martha Beckwith:

> Ku-ka-ohia-laka is worshipped by canoe-builders in the body of the ohia lehua, the principal hardwood tree of the upland forest. His image in the form of a feather god is also worshipped in the heiau with Ku-nui-akea, Lono, Kane, and Kanaloa. He is the male Laka worshiped in the hula dance. That is why the altar in the dance hall is not complete without a branch of red lehua blossoms. In Tahiti, Rarotonga, and New Zealand, Rātā [Laka] is the name of the ohia tree.[105]

But it is even better than that. Modern botany informs us that the rātā/laka/`ōhi`a tree has spread widely across the Pacific, gracing the shores of Melanesian Fiji, Vanuatu, Papua New Guinea and New Caledonia as well as Polynesian islands, including those of Lord Howe, French Polynesia, Rarotonga and Hawai`i. The Polynesian varieties of the genus *Metrosideros* are related, with their origins in the southern Pacific from an ancient, inchoate landform that was still-emerging Zealandia.[106] As impressive as the botanical colonisation story is in its own right, when exploratory waka from the centre of Polynesia reached Aotearoa, the narrative they brought back was of a coastal tree in flower. The explorers had arrived around early summer. Polynesian voyagers who followed to settle Aotearoa were to look for the distinctive red flowers of the coastal tree in Te Tai Tokerau.[107]

Southern rātā in flower, Fiordland World Heritage Park, Aotearoa. *Photo Richard Sidey*

So, from the ends of Polynesia comes the story of a human Rātā, Lata, Laka – the voyager. It is bound into forest and bold voyaging. Even though the rātā was from Aotearoa's temperate forests, it was first the mataī and then the tōtara, not the rātā, that became the preferred tree for Māori in Aotearoa from which to shape a waka.[108]

Rātā's tale of personal discovery is similar in Tahiti, Sāmoa and New Zealand. In order to avenge the disrespect and death of his father in another land, Rātā wanted to build an ocean-going waka from a suitable tree.

Pōhutukawa (*Metrosideros excelsa*), the hardy coastal dweller, has spread itself widely across the Pacific.

Photo Aliscia Young

However, before felling it, he needed to secure the services of a tohunga to offer up suitable karakia, as laid down in tribal lore. But Rātā was in a rush, and neglected to honour Tāne. Twice he felled the tree and twice the patupaiarehe (fair-skinned mythical people; in some accounts, they were birds) overnight restored the tree to standing.

Rātā had to learn, as humans do, the hard way. Finally, chastened, he made a ritual offering of food. Only then could he claim a waka from his chosen tree – an action that may also be described as an acknowledgment of mana. 'After felling, the waka was completed and brought to the beach, where a baptism ceremony preceding its launch, named it *Va`a-i-ama*.'[109] His story is told in various forms from Hawai`i to the Marquesas, Tuamotus, Mangareva, the Society and southern Cooks, Sāmoa, Aotearoa, Niue, Tokelau and Banaba (Kiribati).[110]

For the Tuamotuans of eastern French Polynesia, who reputedly navigated the most tempestuous of Oceania's seas, Rātā's story holds a different emphasis. They still detail memory of Rātā's lesson of learning the ritual of acknowledgement and the chants for the tree-felling ritual – even down to the correct way to sharpen the axe. (On the Tuamotu archipelago suitable trees for hewing were possibly always scarce. Tuamotu waka were expertly crafted from many pieces, usually from the tou, 'a

Under construction across a colonial faultline: an adzed waka lying in its wood chips, Upper Whanganui River, 1902. *Denton Collection, Whanganui Regional Museum*

medium-sized tree favoured for the building of voyaging waka'.)[111] Living in such deep and dangerous waters, the Tuamotuans venerate Rātā more as vagabond navigator than as the beloved blunderer whose wise use of the forest came from hard-won knowledge. Key elements of the story are retained, however.[112] Long regarded as 'the most daring and successful navigator in the eastern Pacific', the chants on Rātā, grandson of Tāwhaki (Ka`aki, Tahaki),[113] remind Tuamotuans of the ancestor who still inspires their sailing:

> Rata! There is none like Rata …
> Bravely beating heart! …
> Ever seeking the destined land.
> Let the beloved assume his place.
> Let him stand proudly erect in his ship![114]

Rātā's lessons on the need for respect for mana and the laws of utu, and the long karakia (prayers) before the felling of trees that were crafted into waka, have been recorded. Further karakia were required before launching into their journeys. These prayers are known to have been used for those waka heading south to Aotearoa. So embedded are these rituals in Polynesian culture that even the traditional axes are named and the sacredness of Polynesia's supreme god Io invoked as 'the foundation of the waters of life'.[115] So, finally, the story returns to the sanctity and sanctifying qualities of freshwater. Again, the linkages between water and forests, water and life, in both lore and deed, are for Polynesians both visible and indivisible, joined by story.

CHAPTER 5

Waikīkī

GHOSTS, GLITZ AND GOBIES

Above: An 1873 engraving of Diamond Head as the backdrop to Waikīkī. *iStock.com/traveler1116*

Opposite: Landmark and keeper of history, Lē`ahi, Diamond Head, Honolulu. *Photo Cosmin Serban*

Waikīkī is a byword in tourism chic, a long-burnished image of golden sands brushed by hula winds bearing tropical fragrances and Hawaiian chords, of glittering high-rise condos and hotels and stretch limos cruising city blocks. With its smoked windows and mirror-glassed towers, it is an enclave of smoke and mirrors not only because of what it represents, but also what it obscures. Unrecognisable from a century before, it is a place transmuted into the fantasies of its conquerors.[1] When the already ill Robert Louis Stevenson sailed in on his journeying of Oceania from the late 1880s, he would stay at Sans Souci Hotel. Today this is the site of a fine restaurant at the far eastern end of Waikīkī Beach, dominated by the extinct volcanic crag, Lē`ahi, Diamond Head, Honolulu's most enduring landmark.

Archaeology at Lē`ahi has opened a time capsule. This begins with the goddess Pele's ground-breaking activities and moves to being a heiau for human sacrifice; it has also been a battleground for Hawaiian sovereignty battling against American piracy and, since annexation, a battery site, military complex and aviation control centre.[2] Today tourists trudge its internal steel stairways. By contrast, at Sans Souci one may stretch out, it is claimed, beneath the same hau tree that Stevenson lay under. Certainly a grand old hau does thrive there. He wrote of the place's 'lovely scenery, quiet, pure air, clean sea water, good food, and heavenly sunsets hung out before [our] eyes over the Pacific and the distant hills of Waianae'.[3]

Today that scene is Queen Kapi`olani Reserve. But before the park, this was marsh and streamland, rich in birdlife, islets and Hawaiian fishponds. Beyond that, stretching for miles and curving into the Mānoa Valley, beneath the Mānoa Falls, were the lo`i kalo, the flooded taro fields that Hawaiians had cultivated for many hundreds of years. Three main streams, Makiki, Pālolo and Mānoa, rose in the highlands, flowed and eventually trickled out into the ocean through the wetlands, 'the lungs/kidneys of islands', as poet Haunani-Kay Trask calls them.

Little remains of this today. In the 1980s students at University of Hawai`i, Mānoa, found `auwai in the undergrowth beside the Mānoa Stream. Known today as the Kānewai Cultural Garden, with its own cultural centre, it was an ancient and valued taro growing site. It became a royal possession after the dynastic Kamehameha the Great's conquest of O`ahu. Activists, including kupuna Uncle Harry Kunihi Mitchell, re-established traditional `auwai for taro growing and māla`ai for dryland taro planting, together with a hālau, or thatched pavilion. This has become a place of intensive cultural regeneration, where planting and weeding activities occur, and where youngsters can rediscover their traditions.[4]

In the hills behind Mānoa, the University of Hawai`i arboretum contains the beginnings of a restored lo`i kalo, on a site where the old taro terraces rise. The stream, ponds and taro are tended by Hawaiians. Invading trees have been cut back, allowing the light to filter through. In the scale of things the site is a tiny gesture that has already involved hard work. But it provides food, community and heritage. It is attended by Liloa, who greets us: 'You need to take that path there,' he says as he watches us picking our way through the squelchy kikuyu grass after early winter rain. These cultural regeneration efforts, which express the irrepressible will of the Hawaiian people since the Bayonet Constitution was imposed on them in 1887, are often out of sight of tourists.[5]

Since 1887 when Bernice Pauahi Bishop bequeathed her estate from King Kamehameha III, the Kamehameha Schools have held one ninth of the total land area of Hawai`i in

The Mānoa Stream, O`ahu, rises in the highlands and flows through wetlands to the ocean.
iStock.com/Anabella Rivera

stewardship, with about 73,000ha each in agriculture and conservation and a further 2000ha in commercial property.[6] On the North Shore, 800ha in a fertile valley have been returned to the Kamehameha Schools. 'About a hundred years ago it was taken with the water rights by sugar interests,' Kā`eo Duarte, vice-president for engagement and resources at the schools, explains. 'They built a weir high in the valley that diverted a lot of water into pipes for irrigation.' The sugar has gone now and the Hawaiians are walking a line between restoration of their land and upholding their water rights without alienating the haole settlers, who also depend on water. Kā`eo ponders on those farmers affected by water decisions that the native Hawaiians will make, such as the new pipes that must be installed, and how costs will be shared.[7] 'We know that we cannot restore the ways that were,' he says. 'But we do want to restore the communal ways, even if they are not all Hawaiian.' Under the deeds of the original trust, where the lands realise returns, these are used for advancing the education interests of Hawaiians.

Ten years later Kā`eo is able to report encouraging progress 'transitioning those sugar plantation lands towards a more vibrant portfolio that is better aligned with who we are as an indigenous organisation'. The ancient Hawaiian fishpond Loko Ea has been restored by the community with a local not-for-profit group. So too have the Ukoa wetlands, which are 'very significant traditionally for native fish and waterbirds,' Kā`eo says. 'It has been overgrown and choked with non-native vegetation since the plantation days.' The organisation now owns a majority of

Anahulu Stream in Haleiwa, O`ahu, whose valley is a place of antiquity and renewal under the direction of the Kamehameha Schools programme. *iStock.com/pikappa*

the kuleana (land grants from Kingdom of Hawai`i) that were held by the Dole Company (known for its pineapples). This has enabled Hawaiians 'to consolidate majority control of Anahulu Valley on the North Shore, which is rich with traditional agriculture sites, sacred sites and temples, burial sites and house sites,' he says. 'This valley has sat for over a hundred years – so much work to be done here.' Already, notable heiau have been restored and solar and wind energy projects installed. The Hawaiians have also developed several vertically integrated farming projects that incorporate native food crops, including a cacao plantation.

To most people, the idea of a mountains-to-sea, or ridge-to-reef management system, descending from original upland forest to cultivations and coastal fishponds, is a big ask. In many parts of Hawai`i, however, it was a way of life, practised within walking distance of Waikīkī, the stream-fed irrigated terraces of the ahupua`a cultivation plots. Some of these stream-fed gardens existed in Waikīkī into the 1920s.

As the epic creation chant the *Kumulipo* states, after 'the intense darkness, the deep darkness / Darkness of the sun, darkness of the night / Nothing but night …' emerged the panopoly of gods within a natural order that pivoted between yin and yang energies. Inevitably, this philosophy embraced the idea of reciprocity as a balancing mechanism within nature and for the actions of its children.[8]

From their world view of stabilising

symmetries and holistic practices Polynesians uncovered their explanations of counterposed forms of freshwater. Water's male energy was powerful and pulsing, while the generally slower nursery life of the fecund marshlands, the wetlands and broad estuaries were female in character. Modern ecology reveals that the latter are among the most productive ecosystems on earth.[9]

These ancient understandings of connected function and purpose have not always been shared by contemporary conservationists. Until now, most of the balance provided by the dominant modern practices to the ancient orders has been as an antagonist that operates within the linear, unreciprocal impositions of the foreign interloper. This attitude builds Roman-style highways that cut and bite down into the uplifted landscape like blue plastic strapping racheted down on cargo.

From the *Kumulipo*, the great, multidimensional Hawaiian chant of being, emerges an entirely different and deeply informed view of landscape – in this case one shaped by water.

> It is a night gliding through the passage
> Of an opening; a stream of water is the food
> of plants
> It is the god who enters; not as a human does
> he enter
> Male for the narrow waters
> Female for the broad waters.[10]

Haunani-Kay Trask explores this idea in a poem on sensuous love:

> Submerged minerals
> rivers of night flowers
> female foam hanging
> from caverns of mud.[11]

A Hawaiian nationalist, Haunani-Kay has campaigned much of her life for restoration of water rights and water flows. Some of her fights have been for wetlands, like He`eia Kāne`ohe

Ridge-to-reef, a sustainable system in which cultivation is governed by nature. *Photo Cole Keister*

Bay, which is a traditional site of restored fishponds, not far from Makapu`u, where I encountered the frigatebird at dawn. It also has associations with Pele, and was likely a site where King Kamehameha I worked at this 'great fishpond'.[12] Clearly, her poetic inspiration is inseparable from the cultural life she has worked to reclaim.

This understanding of yin and yang energies may also be found in traditional Chinese thought.[13] In Polynesian thinking, freshwater in a confined place can be destructive; in an open place it manifests as stillness, reflectiveness (wai ata) and nurture. Darkness and coolness enhance water's vitality in places like Kāne's waters, not reachable by human hand.

Above: Mangaore Stream.

Right: Native kōkopu, in the still vital Mangaore Stream in the Ruahine foothills, Manawatu, Aotearoa. They have no genetic relationship to the Hawaiian `o`opu but their similarities in form and behaviour are recognised by Polynesians, who use similar names for both. New Zealand has 33 species of freshwater fish. *Photo Aliscia Young*

`O`opu and kōkopu

The `o`opu is on the same cognate as kōkopu: brought by the Polynesian voyagers from home islands and applied widely to freshwater fish in Aotearoa. The adult fishes to which it refers are the kōkopu species, so little and so lately respected by New Zealand's more recent colonisers. For freshwater species expert Robert McDowall, the giant bully (*Gobiomorphus gobioides*) probably best resembled those fish that were familiar to Hawaiians from their earlier island homes.[14] The five anadromous `o`opu that are endemic to Hawai`i all have traditional links with the gods. Kāne ko`a is acknowledged in fish altars along streams 'to increase the catch of `o`opu fish', and 'Kāne and Kanaloa are said to have been worshipped with `awa (kava) and aholehole (whitefish) on their arrival from Kahihi' – the Hawaiian place of the gods, a version of Hawaiki.[15]

Kava and kawakawa

No plant in the cosmology and life of Māori occupies a more revered place than the kawakawa (*Piper excelsum*), a plant associated traditionally with freshwater, explained Ngārangi Rangiuia. With its 'osteo-knobbled' branch joints, it is named for the sharp taste of its heart-shaped leaves, and attendant

Io's gift, the humble, efficacious kawakawa, has genetic and name associations into wider Polynesia. In Māori lore it is a purifier of water.

Photo Rob Suisted, Naturespic

kawakawa looper moths perforate its leaves. Kawakawa is moisture-loving, 'a water plant', he said. From the beginning of creation, during the upheaval when all the gods were warring, 'Io, the supreme god, sent his attendants Ruatau and Rehua to bring an end to all the chaos, and the gift that they brought was kawakawa.' This supreme gift was also 'to remind them that there is always a kawa, a way of doing things.'

> Kawakawa helps to purify water, that's why it's always planted near waterways. And we never drank water pure, we always dropped in a leaf of kawakawa. It has a slightly bitter flavour.
>
> Kawakawa is the only plant that doesn't have a whakapapa back to Tāne and to some of his wives. That's because it was given by Io, the supreme god. It is there to always remind us, being called kawakawa, that there is a right way to do everything. Whereas tikanga changes and adapts according to circumstances – that's why it's so important to us – the kawakawa is a reminder to us to ever be mindful of our actions.

Hence the use of kawakawa leaves as a headdress in times of mourning or welcome: after being dipped in pure water, its leafy stems were used to sprinkle water as a purifier over a person or whenua or even streams. There were many ceremonies. People would bind kawakawa leaves into a bundle, breathe their thoughts into it for, say, a good harvest, and disperse it onto water. 'This was a very common process,' Ngārangi explained. Used as an infusion or poultice, the plant's leaves and the erect spikes of its orange female inflorescence have many medicinal uses. However, unlike its kava relation in warmer climes, this member of the Piper family 'does not have the relaxing properties' of its distant relative.[16]

Ngārangi described the relationship between kawakawa and the tutu plant. Tutu, he says, was traditionally planted 'right down alongside the water; the kawakawa is standing back behind it. They have this relationship with the water. Tutu sucks up toxins and is used today by tangata whenua to absorb heavy metals and nitrates. The leaves, fruit and bark are all still used as medicine by Māori today.'[17]

Cultural landscapes lost to view

The more recent arrivals to this landscape of old wetlands – the auwai, lo`i kalo and loko i`a (fishponds) and forests – were of a different rhythm. Theirs was the insistent beat of unquestioning progress – nature and natives subsumed to notions foreign to this land. Perhaps of all the changes, the most dramatic physically was Honolulu's Ala Wai canal. Lucius E. Pinkham, president of the Territorial Board of Health and, later, governor of Hawai`i, decided these 'insanitary' wetlands needed to be drained. The Ala Wai ('water pathway') canal, constructed in the 1920s, channelled the region's three streams into a single flow direct to the ocean. Excavated soil was used to fill the ancient fishponds and cultivations of the area. Some 600ha of Waikīkī's traditional food-growing area went under the mechanical diggers.

Writers such as Houston Wood have thoroughly investigated the idea of displacement of indigenous people out of their cultural landscapes and of what was constructed in its place.[18] It has been explored with a specific cultural landscape perspective by both Gaye Chan and Andrea Feeser; and Dennis Kawaharada has researched its pre-European character.[19] The imposition is there in many ways – physically, culturally and socially; even as a military presence, especially on modern O`ahu, with jets out of Hickham streaking the sky, the military highways slashing straight across whatever terrain to connect airbases with naval bases, including Pearl Harbor. To Jonathan Osorio, this militarism is entirely at odds with Hawaiian traditional values of 'inclusion and aloha aina' (deep ties to the land) and 'interdependency' with all life: 'It defends a very particular definition of a people, and we Kānaka Maoli are focused on a much larger society than the American nation. Indeed, we have nurtured and will continue to uphold a community that is larger than humanity itself.'[20]

While tourism development on new land enabled millions of tourists to have a 'Hawaiian experience', ironically what they experienced was in an almost entirely reinvented landscape. Stormwater off the newly hardened surfaces, including high-rise tourism complexes, topped up the canal's slow flows and triggered massive displacement of water. Haunani-Kay Trask recalls the time before the high-rise, in the 1950s: 'They destroyed a place – when I was a child it was so beautiful. But the State spent so much on promoting tourism and not on what our people needed.'[21] (There are warnings here for New Zealand, with its unprecedented expansion into mass tourism into the new millenium that, post Covid-19, demands a far more considered approach. Many sites in Aotearoa have been subject to similar modification, including some cityscapes on the scale of Honolulu. By the 1890s much of the biota of Auckland isthmus had been modified almost beyond recognition; and, more dramatically, Wellington's reclaimed waterfront and the draining of Christchurch's vast wetlands – finally recognised in reconstruction planning following the 2011 earthquakes – come to mind.)[22]

Sluggish flows and low oxygen levels meant the Ala Wai canal soon became a no-go zone for life. Aquatic recreational activities were restricted, and fishing was not only prohibited, it was inconceivable. Notoriously, in 2006 big floods burst a sewer line and about 220 million litres of untreated sewage entered the canal. The alternative was to discharge it into the

In the early twentieth century drains were driven through the old Mānoa wetlands and lo`I kalo, now downtown Waikīkī.
Municipal Records Center, Department of Customer Services, City and County of Honolulu

waters at the heart of the tourism district itself: the mayor decided he had no choice.

A young man out on a bender and suffering from open wounds fell into the canal at Ala Wai Yacht Harbor in 2006. His subsequent illness and death a few days later – in which blood poisoning played a major part – parlayed him into local myth. As folklore, his misfortune serves to protect people from a canal built, allegedly for sanitary purposes, from fields where taro grew.[23] It is also a place where mosquitos will breed.

'Wetlands', as those who would save them call them, or 'marshes' and 'swamps' as they are often known by those who would 'reclaim' them, have been designated 'unsanitary' by colonisers the world over. Ironically, the occupiers' replacement – in this case the Ala Wai canal – may represent more of a health risk than the natural systems whose abolition it was supposed to improve on.

Hawai`i was mosquito-free until 1826: mosquitos arrived in the same decade as the first missionaries.[24] It wasn't just people who were affected – the mosquito is believed to have been responsible for the extinction of a number of Hawaiian bird species. Displacement heaped on displacement. Channelisation of earth and water on the scale of the Ala Wai will almost inevitably produce its own physical and social responses, especially on a plain that sits barely above sea level.

But there are other unintended consequences that such massive redesign can have on a landscape and on a culture. One involves the location of the University of Hawai`i at Mānoa, Honolulu; the second is strictly cultural in nature. The physical impact is easily described and readily understood.

On All Hallows' Eve 2004, heavy rains flooded the Mānoa Stream, site of an old ahupua`a system. Pent-up water in the stream above the university broke through, finding an old course over the campus to the basement of the Hamilton Library, where 160,000 of the university's old maps were stored. These were precious and revered documents – among them unique collections from the Pacific. A class was in progress in the basement. Floodwaters washed around the windows above the students, and they were effectively trapped until a student broke a ground-level window with a chair, enabling them to escape. The floodwaters eventually broke through the fire doors and invaded the storage facilities, filling the basement with mud, African snails and debris. Sodden, torn maps, photographs and files littered the campus.

Displacement is an issue whose effects are sometimes unimaginable. As Andrea Feeser reports, quoting Bob Paoa, 'many of these displaced Hawaiian families moved to "Squattersville" – first the present Ala Moana Park area, which used to be the City's dumping grounds. Later the City relocated the homeless families to a second "Squattersville" location.'[25] Some displaced families, according to Feeser, 'ended up "squatting" not only on a coral reef that had once fed them, but also on garbage'.[26] They would hardly be the first colonised people in the Pacific to do so.

As a visitor – and, I might add, a beneficiary of the life wrought by this consumer economy – I wandered along the beachfront of Kapi`olani Park on a number of occasions. During the day the displaced were not so obvious, but as evening drew in, so did they. In September 2008 about 150 people were living in the park; many dozens more were living along the leeward shore at Wai`anae. Most in the park were men and they were organised, with tents and tarpaulins and designated eating areas under the old trees, where they played music. They did not seem much interested in anyone else; occasionally you might come across one of them asleep on the beach, deep in the buttresses of a Moreton Bay fig.

In a reversal of commonly accepted gender roles in social organisation, it was the women who seemed to be largely alone. Among them were two I still recall. One – a haole American by accent and, ironically, by assertion of American private property laws – 'fenced off' her patch of grass with chunks of rock and challenged any intruder with the assertion: 'You're on my property.' The second – a small, slender Hawaiian, longish grey hair pulled back from a face weathered like a totem – sat on her bench through the heat of day, staring into a hazed horizon.

As it grew dark, the atmosphere changed: some became more proprietorial, some more aggressive. By night the park area is notorious for its drugsters and pimps, and I have seen shabbily dressed people at Waikīkī diving hungrily into refuse bins. Fijian friends who lived near Waikīkī told me that most rubbish bins get rummaged. 'If we have food left over, we wrap it carefully in foil and leave it out obviously in the bin,' they said. These people – islanders themselves – were shocked by what they saw as the indifference and suffering amidst such wealth. Some say that many of the homeless are people of Hawaiian ancestry. According to Trask, it has been this way for some years.[27]

What is clear to the modern inquirer is that when life is out of joint on an island, when systematic suffering and deprivation flourish under the very nose of invasive wealth, all is out of kilter. Even in the heavens the atmosphere is disturbed and before long, insurance premiums – the only thing between an individual household facing a hurricane and financial disaster – become unaffordable.

Pushed to the margins of both wealth and geography, the old ways have never spoken to us more eloquently. The leadership is out there, having gained recent ground in Aotearoa much more quickly than in Hawai`i. The potential in Hawai`i for education and action – through the Kamehemeha Schools alone – is considerable. And the movement back to regenerative agriculture, coupled with changes in water allocation law and fishing practices, is gaining momentum.

The ability of native Hawaiians to organise, protect and advocate around kīpuka, regenerative sites, is being rediscovered. This may prove influential in their efforts to bring pono, the life of balance, into some of the less developed corners of the archipelago. The problem, as the multiple manifestations of climate change keep reminding us, is that many concomitant actions will be required: science, good law and responsive, informed governance all have a part to play.

Defiant biodiversity

The Olympic Pool at the University of Hawai`i Mānoa campus offered another ghostly insight into Honolulu's lost worlds. I tried to exercise here most lunchtimes in the open-air, ozone-treated water, with practically everyone in their own 25-metre lane.

One day as I freestyled back and forth, I sensed something overhead. I looked up through my goggles and there it was, right aloft – a solitary dragonfly, a big, black, hovering pinao. Not as large as New Zealand's giant mountain or bush dragonfly, kapowai (*Uropetala carovei*), but big enough. The dragonfly evolved before dinosaurs: it is one of those delicate-hardy reminders of what once was – another ephemeral trace – but still breeding in tolerable water somewhere.[28]

There are other traces. Down near the Waikīkī Shell complex one tiny spot remains, perhaps half a hectare, that has somehow survived the transformation. Here is some low-lying, ephemeral water, fed by a brackish water table and runoff from the surrounding parkland and residential area. At times of high rainfall it receives some water off Diamond Head. It may be the only such natural place remaining across the entirety of downtown Honolulu–Waikīkī.

The `o`opu, endemic to Hawai`i and so ancient, were laying down their claims to the archipelago millions of years ago before the emergent volcanoes finally flashed their fluked and fiery tails above the surface of the ocean. Hawai`i has five native fish, predominantly Gobiidae (goby) and one of the Eleotridae family, as well as two crustaceans and two molluscs. A traditional saying of Hawai`i speaks to other cultural flows in the Pacific:

> Ka i`a ho`opa `ili kānaka o Waimea.
>
> *The fish of Waimea that touches the skin of the people.*[29]
>
> When it was the season for hinana (`o`opu spawn) at Waimea, they were so numerous that one could not go into the water without rubbing against them.[30]

Miike adds another quote from Pukui about another endemic `o`opu, the nōpili:

> He nopili ka i`a, pili pa`a ke aloha.
>
> *The nopili is the fish; love clings fast.*
>
> This was said of the freshwater goby (`o`opu) of the nopili variety, known to climb waterfalls by clinging fast to the wet stones. This fish was used by kahuna in hana aloha sorcery.[31]

In some protected places, waterfall-climbing kōkopu and young elvers still clamp and sucker their way up remote waterfalls of upland Aotearoa – as once they did everywhere, in hordes, until their river habitat, notably riparian forest, was diminished.[32] Hikaia Amohia, who told me so much about the Whanganui, spoke of the majesty in both the natural beauty of this cycle and its food-gathering heritage at the Ōhura Falls on the Upper Whanganui River.

Today the `o`opu from the Waikīkī beach are still impelled by a tribal memory trace, possibly from as far back as when Kaua`i was still subterranean, still emergent, from that time before Pele, goddess of fire, showed her red tongue to a startled ocean.

In 1998 the Environmental Protection Agency made federal funding available to the Ala Wai Canal Watershed Improvement Fund for a group of volunteers to start clearing and planting the site: 'We had boy scouts, high school students and teachers, even the navy,' said Michelle Matson. Michelle was at the time an environmental resident, and a scion of merchants and shipping magnates of nineteenth-century Hawai`i. One of these, banker and sugar entrepreneur William G. Irwin, was also a philanthropist involved in the extension and protection of Princess Kapi`olani Park.

When Michelle was shown the spot by a friend, she became committed to its rehabilitation. In this tiny pocket of wild land in the park she was excited to discover three species of native Hawaiian wetland plants still growing in abundance, and native birds, too, were attracted to the water. She made a telling observation: 'When ponding occurs `o`opu can be observed swimming in schools as they navigate upstream from subterranean culverts and saltwater channels. The water here is brackish and there is artesian water also in the park – it mixes together,' she explained. Michelle understood that some freshwater also reached the wetland by an underground passage connecting with the notorious canal. 'Things have improved – now there are more reeds where there was just water, now there is more water and basically, left to itself, the wetland has thrived.'

The native fish push through Kapi`olani Park into the small 'waste' area with an ephemeral stream that, while it is rank with invasive plants, refuses to be subdued. Driven by a primal pulse, the fish make their way into the park, pressing stubbornly on as they pick up ancient tracks. In 2019 Wellington's waterfront was reclaimed in a not dissimilar phenomenon, when staff at a foodbar at the railway station discovered that little blue penguins from the harbour were beating the path of least resistance to the compelling scent of sushi.

This is life on earth – fragile tenacity, sensitive chaos,[33] often moving through the unpromising towards the improbable but still with the relentlessness of orbiting stars.[34] Indeed, the nineteenth-century naturalist George Cuvier saw the beauty in the markings on New Zealand's giant kōkopu and named

Awaous stamineus (`o`opu), a species of goby native to the Hawaiian islands, persist in asserting their mauri, albeit in critically reduced numbers, into the tiny remnants of wildness that survive along the shoreline of Waikīkī. *Photo Keoki Stender*

Eleotris sandwicensis, a goby (`o`opu) that frequents the brackish coastal waters, streams and wetlands of the archipelago. It is one of five freshwater fish species endemic to Hawai`i. *Photo Keoki Stender*

it *Galaxias* for its resemblance to a galaxy of stars.[35] Māori called this energy mauri, sometimes summarised in translation as 'spark of life'. In the shadow of Diamond Head, mauri is a remnant pattern in the submerged history of Waikīkī.[36]

When funding for the restoration of the park wetlands fell through, however, efforts gradually petered out. While some real gains have been made since, the restoration has stalled. When Michelle took me there, she was uncertain about what we might find, knowing that invasive species rarely pause. First impressions confirmed these fears – a papaya tree had sprung up; other new arrivals included opiuma trees (*Pithecellobium dulce*, introduced about 1870 and spread by birds), and an ancient date palm that was 'so old it's now historic'. Someone had tossed out a double mattress; and wild chickens, that vocal coloniser of many Hawaiian scenes, had also appeared and were scraping the soil.

Through wire mesh on the park's perimeter a movement caught my eye. A middle-aged man, one of the homeless, sat sorting through his stash. He acknowledged us briefly with his eyes. To me, his presence brought forth a bizarre conjunction – the displaced human being drawn to the last bit of 'wildness' (to use Thoreau's word) on the vast Waikīkī reclamation.

Despite her misgivings, Michelle became excited. Her group's efforts to establish the rushes and the `ākulikuli kai (*Lycium sandwicense*), a wonderful, cleansing plant, were holding out: 'They're using it as a filter on parts of the Ala Wai canal.' The much-loved coastal native, the white-flowered hinahina kū kahakai (*Heliotropium anomalum var. argenteum*) that they had planted was proving elusive, however.[37] Michelle kept searching: 'I couldn't see the hinahina. It was here, tiny, with flowers. It's quite rare and one day it just appeared here.' That day we visited, the `o`opu were not in evidence but all the conditions for life seemed present for them in this one spot. The fight for restoration is not yet over, however.

When I went back in late 2017, I searched for the wild place in the park, but never found it. Even the homeless seemed to have moved on, probably only temporarily. A solitary Hawaiian youth was fast asleep in grass under the midday sun near the wild part. On a side street was a caravanserai of white film-service trucks where the huddle of waiting security guards told me, 'They're making a new series of *Hawaii-5-0*.'

I walked past the university's revivified lo`i kalo, fed by the Mānoa Stream. Two students bearing wet nets had just emerged from the stream.

'What did you find?' I asked.

'A lot of invasive species,' grimaced one.

'No gobies?'

'No, not up here.'

'Any chance for them in the future?'

'Well, it's too channelised for them down below, but maybe we could have them living up here,' he said.

'So these would be fish that, once they were put here, stayed put and never got out to the sea?'

He nodded.

Pulses … I think of eels, tuna as Māori call them, back in Aotearoa, that cannot complete their cycle, growing like fat old eunuchs in the big lakes formed behind hydroelectric dams – unable to venture on their Polynesian-style journey back to their place of death and regeneration somewhere in the seas of the old fighting Kingdom of Tonga where they die like salmon, except in saltwater, in the extraordinary death-dance of procreation. What does their absence from the cycle mean for the ecology of marine life? What species wait out there, unrequited, for the cycle – until the future of their own kind becomes jeopardised beyond relief? A cycle that may have been developed at times in the evolution of Aotearoa when, through ocean inundation, freshwater food sources were restricted, when the need to migrate much further afield may have become an imperative.

Out in the hills and ahupua`a of Hawai`i, thoughtful young men offer prayers to their ancient gods before meals, before planting, before surfing … It is a life that is all of a piece. They follow a path of righteousness, the narrow one that Christ spoke of, that offers the paradox of abundant life. But only through hard toil. The feel of lo`i ooze between your toes, working, working, working. This kind of mahi also goes on in Aotearoa, activating both determined protection and regeneration.

Since 2007, Te Wai Māori, the Māori Freshwater Fisheries Trust leaders group, has pursued Māori interests in freshwater under the Treaty of Waitangi, so far with limited success. Te Wai Māori was set up to fund the education, training, research and promotion of Māori interests in freshwater fisheries. The idea was to enhance and establish these fisheries under the umbrella of Te Ohu Kaimoana.[38] In New Zealand there are crossovers between Māori initiatives and government support to protect freshwater fisheries and habitat. The government has partnered with Māori on initiatives to protect native ecosystems on land and in freshwater.

Since legislation was passed in 1990, Ngā Whenua Rāhui has been the primary agency that supports Māori efforts to save often rare landscapes, plants and animals on their remaining rural lands. More than 170,000 hectares are now under protection. Using a proven trust model that protects indigenous nature on private land in 25-year covenants, the government supports hapū and iwi financially in restoration work. Tangata whenua contribute their labour and knowledge, with a strong focus on training rangatahi (youth). This revival of Māori traditional knowledge is worked in combination with modern pest control measures, with a view to trainees continuing such practices on their own land.[39] Two examples of such work follow – the first in the Wairua River catchment, Northland; the second, Te Hākari Dune Wetland restoration project, about 100km north of Wellington.

Holding Northland's freshwater to a standard

For mariner Millan Ruka, coming home to the ancestral stream of his formative childhood memories was 'a call to arms' for a man of peace. Over a decade Millan, of Porotī and Mangakahia hapū in the Far North of New Zealand, has become a recognised voice of quiet persuasion and hard evidence. He has held the Northland Regional Council – the agency responsible for freshwater monitoring and enforcement – to account over the state of his region's beloved springs, streams and rivers, defiled by poor policy and inadequate oversight.

'I've paddled so many of our rivers,' said Millan, a noted kaitiaki (guardian) of his ancestral northern New Zealand river systems. 'Whenever I worked overseas, I'd always take a canoe, or a horse, to explore rivers – in the States, in Papua New Guinea where I worked for 10 years. It always amazes me how much we've destroyed of our natural life.'

It was the brutal war on Bougainville Island that alerted Millan to the way a multinational corporation and a lax government could trash a great river and the culture it sustained. In Bougainville as a project manager twice during the 1980s, he became known as a hunter who could be called on by the Melanesian locals. 'I'd go down to help locate rogue crocodiles on the Jaba River. I used to love it – I rafted, paddled and got to know the locals, up and down the Jaba and other rivers there.'

Australian giant Rio Tinto's subsidiary, Bougainville Copper Ltd, was plundering the world's largest copper mine with so little regard for the interests of locals, the Nasoi, that their uprising led to 10 years of horrific civil war.[40] Millan saw how unmitigated, profit-focused mining was destroying what he knew as their mahinga kai (designated wild food place). 'The wealth from copper was creating even less return from the river to those people who were losing their fish and plant food – foods like wild sago that grew in the swamps – from the huge pollution by sedimentation and pollutants.'

Millan Ruka has been relentless in his mission to clean up the waterways of Te Tai Tokerau.

When he returned home around 1996 to Porotī, his tūrangawaewae – his place of ancestral standing – on the Wairua River, he learnt that the interests of his people were facing a similar threat. The sacred spring waters of Porotī had been licensed for commercial extraction and were being treated as a commodity for bottling. (The Crown has since purchased the land surrounding the springs and 'land-banked' it for possible redress under a Treaty claim.) At the same time, the river was being subjected to pollution from the dairy industry.

Mining reared its head again, too. An Australian company sought to reopen a century-old gold mine, and this posed a high risk of mercury pollution of the water. 'Even core sampling, then capping of the site – even that damage could interfere with the layers in the surrounding core that nourish the Wairua,' Millan explained. Since then his people have worked with some success to defend the springs from the expansion of the bottling plant.[41]

Millan was so moved by the state of the Wairua River that he spent $100,000 of his savings on a suitable river craft and monitoring equipment to record its 'devastating' pollution levels. He set up Environmental River Patrol Aotearoa with the help of Ngā Whenua Rāhui and Te Wai Māori.[42] Drawing on his considerable reserves of patience, he applied his project management skills with tikanga (correct way) to guide his controversial mission. First, 'I would always be compassionate,' he explained. 'I know that farmers have a lot of pressure on them to make a living. Second, I would keep it very impersonal, so I would collect data, be evidence-based and professional.' In his reports to the Northland Regional Council, he continues to supply evidence of poor riparian herd management: photographs and GPS coordinates, supported by land title references. But in his hundreds of reports to council he has never named names. This style has sometimes won over delinquent farmers to better ways of farming. By doing the monitoring that the regional council is charged by law with doing,

The mouth of the Ōhau River. The Te Hākari Wetland is on the right, adjacent to a small pine plantation. *Photo Lawrie Cairns, Aerial Photography*

his voluntary work has achieved 350km of riparian fencing to keep stock out of waterways. This is the first line of defence against nitrate and phosphate pollution, although he still finds some farmers 'running their cows on the other side of the fence'.

Millan's focus on climate change has also been innovative. He has explored the potential power of river flows for exploitation in ways that would not entail damming. 'My dream for my best project is to harness that power and help shift us from a carbon economy to one in which we can use this energy in the form of hydrogen cells. We need a new form of energy.'

This peaceful warrior brings the rigour and focus that are needed for the restoration of the mauri, wairua and mana of his land and people. Such leadership is also being expressed in a similar set of socio-cultural-ecological landscapes more than 800km away.

Repairing diminished dune country

On the heavily denuded coastal flats of the Horowhenua district, on the southwest coast of Te Ika a Māui, hapū of local iwi Ngāti Tukorehe have farmed a profitable 380ha dairy operation on land enabled by a 1972 diversion of a tidal section of the Ōhau River, known locally as 'the Loop'. The Ōhau then flows into Kuku Estuary, a name that can mean the closing of a shell or the committed coming together of groups. About 25 years later concerns were raised within the iwi at the damage dairying was causing to local coastal dunes and adjacent wetlands. This is a cultural and ancestral landscape, representing the combined works of nature and culture and embodying 'diverse interactions between humans and their environment'.[43] Today the 14ha Te Hākari Wetland has been reclaimed from the farm, bringing back a place of beauty that attracts wildlife and provides a nursery for a variety of native fish. The larger wetland is emblematic of the original cultural landscape, its pioneer plants now rising from the dunes, backed by the hope for more sentinel kahikatea in a once vast wetland and lowland forest that reached across the lowland plain to the foothills of the axial ranges.[44]

Huhana Smith, who has led the project with her partner, ecologist and conservationist Richard Anderson, explains: 'We have got a revegetated forest around a wetland, planted in tī kōuka, karamū, kahikatea, harakeke, all the appropriate natives. We are looking at a flourishing wetland and we have turned Te Hākari Stream's health around to the estuary.' Allowing the dunes to recover their traditional role is also a way of forearming the lowlands against inevitable rising sea levels.[45]

Warming, acidifying oceans and intensifying rainstorms are already changing the coastal zones of Aotearoa. Increased erosion, with its loss of precious soils, generates downstream siltation effects for already dwindling shellfish populations such as the once-plentiful toheroa, Huhana said.[46] Nitrification and e-coli in waterways that flow to the sea have impacted on these taonga species, too. While Ngāti Tukorehe acknowledge the importance of dairying to tribal shareholders, six more areas for recovery, like Te Hākari, have been activated in the iwi's domain, from Ōhau to Waikawa Beach and on to Lake Waiorongomai – all since 2010, when these additions to the plan were simply part of a strategy of aspiration.[47] This has been achieved, Huhana suggests, by calling on what remains of fragmented local knowledge, to build resilience back into a fragmented landscape. In so doing, 'the ongoing powerful cultural relations' can achieve a deeper, mutual restoration:

> The activities of ancestors and their ensuing generations in Kuku have influenced and shaped the lands, waterways and resources over time. [The programme] aims to restore biodiversity and improve relationships between people and their remaining natural resources. The action taken for dune wetlands is a forerunner to a range of projects and processes that aim to reinstate health to the whole tribal environment, from the mountains into the sea – an area significantly modified by intensive agricultural operations.[48]

These two case studies are by no means isolated examples. A four-year government-funded research programme completed in September 2019 has studied a range of estuaries and lakes around the country. Partnering with iwi, its aim has been to generate knowledge for the improvement and greater resilience of these ecosystems.[49]

One of the most telling biodiversity statistics for Aotearoa's freshwater species, however, reflects the extent of the continuing damage to waterways. The number of native freshwater fish species threatened or at risk of extinction has grown rapidly over the last four decades from around 20 percent to 72 percent today. This includes the endemic mussel, kākahi, and Te Waipounamu's freshwater crayfish, kōura.[50] Equally concerning, according to the Ministry for the Environment, 'about one third of our known freshwater invertebrates are threatened with, or at risk of, extinction'. This amounts to 28 species, most of which have 'a high on-going decline'. Just two – the upland bully and shortfin eel – 'have improving trends'.[51] Thirty-four percent of native freshwater invertebrates – the worms, insects and crustaceans that are critical in the diet of freshwater fishes – are threatened. The status of 27 percent of these is unknown.[52] In terms of ecological systems and in terms of Māori species identification, many of these animals are connected through genealogy and story within the context of tribal life.

Old ways and scarcity?

Almost invariably, the journals of Western explorers of the islands of Hawai`i tell of healthy, well-fed people. But not everyone necessarily shared in the abundance. Although James Cook was taken by the general good health of the people, Native Hawaiian Carlos Andrade and Lawrence Miike both strongly suggest that this was an error, arising from the navigator's limited 'sampling' of essentially only the well-to-do.[53] Miike, retired Hawai`i State Director of Health who served on the powerful Commission on Water Resource Management, clarified this: 'The common people lived hard. Some of the original drawings in Hawai`i show the Hawaiians were skinny; it was only the ali`i class who were big and well-fed – culturally that was a sign of their status.'[54]

It is the knowledge of scarcity, written in the mindmaps and stories of the islanders and passed down with strictures that will not be denied, that created the cultural architecture of respect for nature. It was far from perfect. Oceania's new inhabitants wiped out numerous bird species in island after island. But the West, with its illusion of unfettered abundance driving its expansionist myth and its annihilation of other life forms, is no prudent housekeeper, even now struggling to respect the limits and limitations of life-giving water.

In spite of the highly adventurous and adaptive qualities of Polynesians, including a considerable capacity for organisation and hard work, scarcity was often an issue across the Pacific. People of the Pacific were familiar with the struggle to survive. As with most cultures born of the soil, however, among its enduring narratives the Hawaiian world retained stories of generosity – but also admonitions against stinginess, says Andrade.[55] Generosity is a value memorialised in story and song. Having located and settled on a succession of islands from a thousand, sometimes several thousand years before, Hawaiians knew of nature's shortages. Many of the islands they encountered had little if anything to offer in the way of food production or the means for sustained living. The stories of those who are generous of spirit invariably result in virtue prevailing. The story of Aukele is one such, and it relates to water.

Aukele is a high-born man of open-handed generosity, whom Mary Kawena Pukui – that prolific twentieth-century recorder of Hawaiian culture – portrays as forgiving, patient and understanding. Time and again Aukele's jealous brothers try to kill him; time and again Aukele attempts to befriend them, not naively, but with the insight of compassion. Finally, through listening to the wisdom of his wife Namaka and following her advice, he prevails. He must, she instructs him, journey eastward, to where Kāne is believed to have originated from.[56] 'You must fly straight toward the rising sun. Do not turn to right or left or you will be lost. You will wander till your strength is gone and you will meet death.'

Taking his calabash, he goes to the sacred waters of Kāne. The journey is epic, requiring strength, wisdom and courage as he encounters many challenges. Like a number of Polynesian demi-gods, Aukele is able to fly with the aid of his treasures – a kapu (tapu) bamboo stick with an image of Lono, who ensures that his flight stays on course. Eventually, tired from his long journey, he meets the first of three guards, each of whom, recognising his mana and sincerity of purpose, helps him with advice to overcome the hazards that might beset his progress. He comes to a fierce, blind old woman who, after attacking him, instructs him in the ways of outwitting Kamohoali`i, the shark god, the jealous keeper of the water.

For Aukele to succeed, a good measure of Polynesian heroic trickery is required. He must pass through a cavern where the water may be reached through an opening too small for a man – a water pathway. The blind woman helps him to blacken his hands so that the last group of guards will believe that it is Kamohoali`i whose hands are reaching for the water that must be passed up to him by the guards. The subterfuge works, but Kamoho is quick to find him out, pursuing Aukele as he flashes across the sky homeward. By now his brothers have paid the price for their transgressions with death. Aukele finally returns with the sacred water, by which means he is able to restore his brothers to life.[57] Aukele's story is a reminder to the ali`i that their entitlements and privileges come with responsibilities. To achieve greatness, one needs courage, compassion and intelligence – and cooperation with the powers that be.

Californian poet, elder and essayist Gary Snyder, who has a formal Buddhist training and a lifelong engagement with indigenous perspectives of conservation and generosity, makes this point: 'An ethical life is one that is mindful, mannerly, and has style … One must not boast, or show much pride in accomplishment, and one must not take one's skill for granted. Wastefulness and carelessness are caused by stinginess of spirit, an ungracious unwillingness to complete the gift-exchange transaction.'[58]

With colonisation, for 150 years accessible water became a commodity largely for private benefit, and Hawai`i traditions and rights through American law were transgressed on all main islands. The individual took preference over the community; heavy water-using businesses over indigenous rights; and life-giving waters were subject to the strenuous endeavours of heroic engineering. Given the relative limitations of space, water availability and population, the effects were overwhelming.

Now, the public trust doctrine is gradually winning favour in ways that, without excluding the metropolitan, will give more priority to the indigenous – as nature, as people. In doing so, the best interests of all will be better served. However, in places where development, including tourism, is so out kilter with nature, this will be a tough call. Further, unless we can all act with greater urgency, nothing remotely approximating normality can ever be expected in the future now looming over us in the form of runaway climate change.

The three existential threats for life on the planet – and they are connected – are climate change, population increase and unchecked power, be it in democracies or dictatorships. In Aotearoa, the old gift culture of Maoridom regularly steps up to offset the imbalances, as we have seen from the foregoing examples, as well as with inadequate modern social policy. When homelessness in Auckland began to spin out of control in the teens of this century, it was notably a marae that offered food and shelter as government looked the other way. Along with NGOs,Te Puea marae continued this work of aroha for many months – eventually with government support. Other Māori groups also stepped up. Similarly, when rural communities in places where marae influence remains strong are hit by disaster – be it the flash floods that often strike the East Coast, for example, or earthquakes – the cooperatively run marae is likely to be the immediate go-to place for shelter and care.

'Ka mua, ka muri,' as Māori say, 'Look back in order to look forward.' Will history record that these current trends, spiked by modern exigencies, were to become part of the normalisation of societies in which life's needs, in human and natural form, occupy their due and balanced place in our thinking and daily life?

Late sunset, Hawai`i: Storm clouds gather around the lingering light.
Photo Richard Sidey

CHAPTER 6

Hawai`i Island

GODS, GARDENS, FIRE AND WATER

Spinner dolphins, masters of the deep and of aerial display. *Photo Richard Sidey*

Across the bay on the far side of Kealakekua Bay you can make out a white needle obelisk, erected in the nineteenth century by Cook's fellow countrymen. My son Reuben and I get into the sea at the old wharf. The midwinter sun is bright overhead and the water temperature requires no wetsuits. We swim out into the bay, keeping pace at times with some canoeists who've rented a craft. Bright parrotfish and butterfish dart and dodge below us; in places the variety is more spectacular than on the Great Barrier Reef. Pele vents from many spots below.

Out into the bay the water deepens: far below it turns from azure to an inky black, and we can no longer see the bottom. 'There are sharks around the corner, but they never come into the bay,' the locals tell us, nodding towards Ka`awaloa, the old site on the other side of the bay. Out in midstream a larrikan Hawaiian, upright on his board, paddling it across the bay, cannot resist a crack: 'Are you dragging a lure?' he grins.

Beneath its gentle swell the sea is spectral. When I look down into its depths the now-overhead sun seems fixed as a tiny point of light out of which radiate spokes of candid white.

At the memorial, bunches of canoes are drawn up beside a sightseeing catamaran. Close inshore, bright fish flash like decals in the tide. At water's edge a natural wall of boulders rises vertically from the sea floor. Volcano country. There are paths through the trees: people are picnicking, a few have brought children by canoe, some have walked from the road above down a trail for an hour and a half. I check the time with one of them: 11.40am. The swim one way has taken more than an hour and a half of reasonable application, some of it against a current. The return distance is close to 3.4km.

On the short path through the trees from the shoreside memorial a mongoose scuttles away, sending up an electric shriek to its mate, closer to the water. Strange, savage intruders of empire – is this what it has come to? Are these the guardians of James Cook's memory?

Nature offers her sense of continuity next day. Spinner dolphins enter the bay regularly, usually after a night of fishing; these ones are in a pod of more than 40. Reuben, his sister Aliscia and I plunge in, swimming about 300 metres offshore. With a clear sky overhead, it is possible to see through some 12 metres to the bottom. Drawing towards that spot, we can hear the dolphins' high-pitched whistles piercing the water. A pattern develops. As we swim on the surface, they offer a sense of themselves that calls for a downward glance. There they are, directly beneath us, a hand of seven so far down they seemed finger-sized, their contoured bodies and tapered heads aligned in exactly the same direction, a collective, an ōhana of wairua perfectly expressed in design and movement, drawing through their element, aware of and seemingly acknowledging our presence.

Then the epiphany. Staying ahead of our swimming they begin, moving as one, to tilt upwards until they are large and just below us. Then they break the surface and circle, a couple rising completely out of the water very close, huge and spinning.

This becomes a ritual, our sensing the dark shapes below, then the rise and the surface play – dolphins cavorting on the side of the living volcano. Several times, as they come close to us, one of the pod twists sideways, silver belly gleaming, making eye contact.

> Matariki ahunga nui
> Matariki tāpuapua
> Matariki hunga nui
> Ngā kai a Matariki
> Nāna i ao ake ki runga
>
> *When Pleiades,*
> *The gatherer,*
> *Is bright in the sky*
> *The year begins.*[1]

James Cook sailed into Kaua`i in January 1778 on his third and fateful Pacific voyage, breaking yet another seal on the Polynesian quarantine. Almost a year later he returned, this time putting in to Kealakekua Bay, Big Island, the eponymous Hawai`i, in the midst of one of the most auspicious of events in the Hawaiian calendar – Makahiki, the ascension of the Pleiades group of stars. Known as Matariki to Māori, the event is equally auspicious under southern skies. Cook's encounter tells us much in terms of seasonality, signs and gods in the old Hawaiian world.

Makahiki is a time for celebration and thanksgiving, for rest and recreational games in the name of Lono. As god of agriculture he is charged with food plants. His signs are thunder, lightning, clouds, rainbow, rainfall, waterspouts and whirlwinds. Lono is a god of fertility as well as of peace and healing: these latter arts were taught to him by his 'god-brother' Kāne (Tāne as he is known in Aotearoa). For Māori, this is the art of Rongo – rongoā – to which we will return.

Signs of Lono at play: a lightning storm at sea.
Photo Richard Sidey

Cook's ships were greeted by the greatest gathering of people he had ever seen in the Pacific, many of them swimming and paddling towards them. In the oft-quoted words of the great mariner, they numbered 'like shoals of fish'.[2] The white sails of the ships, resembling the ceremonial squares of white tapa cloth borne aloft by parading chiefs for the Makahiki festival, clinched it for the Hawaiians. So auspicious was the appearance of these white strangers at this season, Cook himself was immediately taken as the embodiment of Lono, whose return to the islands had long been anticipated.[3] Initially his fortunes and those of his crew were truly star-aligned. Given his fortuitous timing, and other propitious circumstances, Cook was welcomed and embraced by the celebrants. He and his crew became a focus for days of feasting, parades, games and celebrations commemorating the Hawaiians' god of agriculture and peace. Then came a time for ritual purification with water after which they sailed away.[4]

But only a week later the need to repair a damaged topmast brought the ships back into Kealakekua Bay.[5] Now he was out of season; all parties wrong-footed. With the festivities over, Cook and his complement were no longer enveloped in the heady miasma of Makahiki. The unfolding of minor misunderstandings erupted into a tragedy of cultures at cross-purposes. Travel-weary and easily exasperated, Cook was certainly off his game, already offside with his overextended crew. His orders to flog Hawaiians for petty stealing and his crew's insults to the person of high-ranked chiefs further degraded the reciprocity of the visitor–host relationship. In the climactic mêlée Cook, a non-swimmer, was surrounded in the narrow shallows of the steep shore, then stabbed and clubbed to death[6] – slain like a porpoise. The event, as Thomas Farber writes, 'reverberates still, a tuning fork that will not or does not or cannot stop vibrating'.[7]

Whether it is the characterisation of nature's energies, guardians or enforcers, the Polynesian world view knows no separation; draws no boundaries. This chapter discusses how two Polynesian societies well to the north and south of the Pacific, 7000km apart, evolved from their shared heritage. It explores the roles of critical gods and how these dominated in respect to available resources. These were human societies, not utopias. Offences against nature and against fellow humans – including, on occasion, sacrifice and cannibalism – were culturally ordained and prescribed. Both engaged in such practices. Māori would, for example, eat foes of status felled in war; and Hawaiians used human sacrifice widely, with temples dedicated to this purpose.[8]

I wondered whether Lono had a higher value in the heavily cultivated lands of Hawai`i than was the case for his cognate namesake Rongomaraeroa in Aotearoa. In Ngārangi Rangiuia's estimation there was an overlap in Māori tradition between Rongo of the garden and Tāne of the forests, 'between food and foliage … and for the garden, everything had its place'. While there was a tendency in Hawaiian society to elevate one god above another, this was not the case in Aotearoa. 'If we choose to make one more important than the other, there is no balance,' Ngārangi told me. 'Each atua is pivotal. In Aotearoa they were all elevated to the same platform.' Emphasis on certain gods depending on a tribe's orientation, say, to sea or forest was somewhat inevitable. But that was a different matter.

'Tangaroa in Aotearoa is probably not as important as in some of the other islands,' Ngārangi continued – apart, he said, from the coast-hugging people of the East Coast. 'But if you live in Tūhoe then Tāne's more important' – a reference to the Tūhoe belief that they are of and with the forest, pre-waka. Furthermore, as the new arrivals 'came to the realisation of just how lush and beautiful Aotearoa was, they brought back some of the very old learning and put this knowledge on the land,' he explained. 'The great thing about the ancestors, whatever the environment they came to, they brought that knowledge to the land for their use.'[9]

Water as a god: Rainbows

So much leads back to water – in this case at its most abstracted, as refracted light in mist. In Hawai`i rainbows are generally associated with the presence of gods, and a double rainbow is seen as a union of male and female: 'Of the double rainbow, male and female, the upper, which is male, is called Turehu.'[10] To Hawaiians, the rainbow's arcing display is the perfect expression of pono, balance. It is a balance in Māori between Uenuku, the female energy and Kahukura, the male: 'Kahukura is the messenger of death – he takes away,' explains Ngārangi Rangiuia, 'Uenuku is the one who nurtures and gives back.'

The curve of the rainbow, Ka Pi`o o ke Ānuenue, is a manifestation of the rainbow god who, as sister of Kāne and Kanaloa, is of highest rank herself. She may hover over those who are of god-like rank in Hawai`i, including children, and she acts, too, as her brothers' messenger.[11] Throughout Hawai`i rainbows may betoken the presence of high leadership. The old story of 'the rainbow maiden' of Waimea, from the southern coast of Kaua`i, tells how she reveals at the end of a suitor's quest that her birth name is indeed 'Ua, god of rain'. As an infant, Ua was delivered from certain death by, appropriately, the god of waterfalls.[12] Ua is the word for rain in Māori and Hawaiian.

The rainbow gods bestride Polynesia. But there were humans so-named. The original Uenuku was an ancient warrior of early Polynesia who lived long before the time of the waka explorers out of those islands of origin, Hawaiki. 'All can whakapapa to him,' says Ngārangi. 'He was a great chief in Polynesia.' Māori unite under this warrior chief.[13] From his perspectives of both Māori and wider Polynesian culture, Peter Buck (Te Rangi Hīroa) quoted, in English, a rhyme of antiquity, bringing together the whakapapa of some, though not all, of the known waka under this Uenuku:

> Behold Tainui, Te Arawa, Mataatua,
> Kurahaupo, and Tokomaru,
> All float on the ocean vast,
> The tree trunk was hollowed in Hawaiki
> And so Takitimu took form,
> A night was spent at Rangipo
> And Aotea took the sea at dawn.
> These are the canoes of Uenuku
> Whose names resound unto the heavens …[14]

This mortal Uenuku took his name from an auspicious rainbow, for it was he who journeyed far to find his beautiful wife, Hinepūkohurangi, goddess of the mist.[15] 'While searching he passed [away], and the people knew that he had passed because of the rainbow that was shown when he died,' explained Ngārangi. Hinepūkohurangi also refers to mist as humans may experience it, he added, 'gathering and collecting it, as one does when walking through mist'.

The name Uenuku has become widely recognised by New Zealanders after *Te Uenuku*, an early Tainui carving, was rescued after two centuries lying buried – 'embalmed' and safe from ransack – on the shores of Lake Ngāroto. This became the symbol for the celebrated 1984 *Te Maori* exhibition, mounted initially in New York and subsequently in Aotearoa. The exhibition contributed to the paradox of 'the overseas effect' – it was New York's enthusiasm that stimulated many Pākehā New Zealanders into a fuller appreciation of Māori culture.[16]

For many iwi, ko Uenuku tāwhana i te rangi (curved or bow-like Uenuku in the heavens) along with Kahukura, who is of the same family as Uenuku, hold deep significance.[17] Kahukura, the rainbow god, is referred to in the name of Ngāti Korokī Kahukura iwi, based at Maungatautiri, who have profound associations with Tainui's Waikato river.[18] This Kahukura is 'a water being', Ngārangi explains, 'but he may also be associated with mākutu and malice.' He is 'the god of war who stands with one foot on the ocean and one foot on the land'. By contrast, 'Uenuku straddles land and the river, the land and lake or glacier. If you see a rainbow like that, that's Uenuku.' Rainbows therefore signified all manner of events, from the prospect of tribes advancing on the warpath to the birth of a great chief.[19]

Kahukura made the voyage to Aotearoa in the waka *Tākitimu*, so it is not surprising that wooden depictions of him are also found in Ngāi Tahu culture, in Te Waipounamu.[20] Uenuku tribal groupings are associated with the Upper Whanganui River, too, in the Waimarino district, where Tainui influence is present. However, rainbow associations and allusions are also common in the misted forests of the eastern side of the island, where Āniwaniwa, on Lake Waikaremoana, is a hallowed rainbow for Ngāi Tūhoe. Their neighbours, Ngāti Awa, uphold the tradition of the double rainbow – male upper, female lower – found in its place of origin, Tahiti, as well as Hawai`i.[21]

Gods in life and custom

In *Hawaiian Mythology*, Martha Beckwith has recorded five prayers to Kāne, one of the supreme gods along with his brother Kanaloa (Tangaroa). Each prayer might be summed up – inadequately – as being about awe and gratitude for everything from clouds and rainbows above, to coral life below in the ocean. Kanaloa's domain is the ocean, but today science tells of us of linkages such as freshwater discharge affecting the health of coral, the lives of fish and marine mammals. The familial partnership with Kāne is, of course, founded in observation.[22] Most specifically, there are prayers of thanksgiving to Kāne as the interceder for plants – similar to Tāne Mahuta's role as god of the birds and the forest in Aotearoa. His special Hawaiian plants are named: `ie`ie (*Freycinetia arborea*), lehua (the flower of the `ōhi`a tree), `awapuhi (ginger), kapalai (fern) and maile (a fragrant-leafed vine favoured for lei).[23]

Western culture has no equivalent for ahupua`a cultivations, with their combination of permaculture[24] and aquaculture essential to life in pre-contact Hawai`i. Kāne is the god or akua of daylight and water and, above all, he is god 'of the streams and waters [lo`i kalo] that gave life to the irrigation systems'.[25]

As for the god Lono, his powers throughout the Hawaiian archipelago differed according to an island's bountifulness or, more precisely, its suitability for cultivation. The islands of O`ahu and Kaua`i had the benefit of being sufficiently geologically aged and exposed for permanent streams to have formed. On their windward coasts these landscapes lent themselves to ahupua`a for the tiered cultivation of water-fed taro. These pie-cut divisions of occupied islands from mountain to sea are found 'in tropical Polynesia and Micronesia'; however, this distinctly Hawaiian term dates from the fifteenth century, and has evolved as a means for kin groups to offer 'a tribute to centralised government' from within defined territorial limits.[26]

It is estimated that there were in excess of 2000 ahupua`a[27] of greatly varying sizes run by extended kin groups throughout the archipelago. Half of them were in the 'young volcanic flatlands of East Maui', according to Lorenz Gonschor and Kamanamaikalani Beamer. 'Two Polynesian societies, Hawai`i and Tonga, had evolved from tribal, kin-based systems of social organization and land tenure to centralized states with feudal-like systems of land tenure, while Tahiti was somewhere in the middle of an evolution between the two.'[28] The ahupua`a was a 'culturally appropriate,

Taro crops growing in the Hanalei Valley of Kaua`i, where there is ample rain. *iStock.com/Don White*

ecologically aligned and place-specific unit'.[29]

The largest islands in the group, Hawai`i and Maui, were newer landscapes that 'had to depend to a far greater extent on rain-fed cultivation' – hence the strong religious adherence to Lono, with prayers to him for rain, was well entrenched on the west of Big Island Hawai`i, where Cook landed.[30] Makahiki was the time in this feudal system when tributes were paid, together with ceremonial thanks to the gods. Eighty years ago, Handy identified stream systems determining land divisions for ahupua`a on Kaua`i, O`ahu (Hanalei, Wai`oli, Waipā) and West Maui (Waihe`e and Waiehu, and others) and in rugged Moloka`i. Though he gravely understated their numbers and ubiquity, Handy understood that in each of these places Kāne was venerated as supreme god – specifically for the stream water, 'which was life to his staple of life, his taro'.[31]

The sacred realms of these places spread from Lihu`e on the central plain of O`ahu, tucked between two mountain systems east and west from where, under the king Ma`ilikākahi, 'O`ahu's golden age' emerged in the fifteenth century. This was a time of peace and prosperity, when chiefs from the other islands visited to learn how 'to emulate what O`ahu's rulers had accomplished'. Only Kaua`i rivalled O`ahu's dominance, but it was disadvantaged by its distance from the other main islands.[32] Shortly after Cook's death, island kings became engaged in a process of warfare and alliance-building that led to the emergence of the Hawai`i-wide Kamehameha kingship. This lasted nearly a century until the sudden end of Queen Lili`uokalani's reign in 1893: she had risked everything by trying to restore the more equitable 1864 constitution over the militia-driven Bayonet Constitution that had disenfranchised her people.[33] By then the traditional structures built up around freshwater, cultivation and fishing – the communal organisation of the main islands – were rapidly breaking down. The Organic Act, passed in 1890, democratised general fishing rights to all, meaning that rāhui and sustainability were gone and, along with them, any native autonomy over these matters.[34]

Its dénouement was familiar, played out with local variations in other parts of the colonised Pacific. The determination of white opportunists to own good land through legal title contributed to the upheaval of the old Hawaiian social and political order. It was not so different in Aotearoa. Even King Kamehameha gave up rights to all lands except for those designated as 'Crown' (Crown lands in New Zealand are government lands). Princess Bernice Pauahi Bishop's bequest to 'Kamehameha schools' of Kamehameha lands was not simply about lands dedicated to their customary heritage of regenerative agriculture or their viable commercial properties. Some 47,000 children of Hawai`i heritage are educated annually at public schools, and numerous charter schools and preschools.[35]

Hawaiian scholars find that, as the institutional stewards of these lands, the schools are 'tied to a lineage of traditional ali`i' with the wherewithal to generate 'a perpetual vision for its ancestral lands and communities'. Further, wherever traditional lands exist, that restoration of these traditional ways needs to be a model for cultural-based regeneration that offers resilience to climate change.[36]

While no one can be definitive about the size of the pre-contact population of the Hawai`i islands, David Stannard's relentless logic makes a case on historical and other evidence – some archaeological – that it could have been as high as 800,000.[37] With a landmass almost 17 times less than that of the combined main islands of Aotearoa, even the moderate estimate of Hawai`i's population at 600,000 in 1779 is six times Cook's estimate of the Māori population in 1769. Through the same effects – including introduction of disease and land loss – as those suffered by Hawaiians, the Māori population fell to around 40,000.[38] But at the time of contact, by any measure, the collective Hawaiian response to the landscape in which they found themselves was impressive. Recent research suggests that regenerating the old systems of production based on cooperation and regenerative agriculture would provide sustainable ways of meeting modern Hawai`i's annual demand for one million tonnes of food in self-sufficiency.[39]

In Hawai`i's conditions of climate, soil, water and the demands of rising population, the critical ingredient for survival became organisation – of landscape and of people. These organisational structures centred on the taro, with its numerous cultivars – between 368 and 482, with some 676 native names – and the waters it grew in.[40] Depending on the mercies of climate, the vagaries of soil and the nature of the terrain, taro might be produced on 'dryland' farms, too. For those taro cultivars that were dependent on rainfall, or that grew in ahupua`a, where rains were more regular and supply more reliable, the gardens' function followed form. In the lo`i kalo, the taro gardens of the Hawaiian archipelago to both windward and leeward, in wet and dry situations, were the engines for what became a staple. As the officers of HMS *Blonde* recorded in 1825:

> Their chief subsistence depended on agriculture. The taro (*Arum Costatum of Solander*) does not grow in the perfection required for food without considerable pains, especially in situations where irrigation cannot be practiced: where it can, they formed the taro fields, after the manner of the rice grounds of the East, dividing them into small squares which might be easily flooded, and planting the roots in rows, keeping the whole carefully weeded.[41]

Under the command of Lord Byron, the poet's cousin, the *Blonde* had repatriated the bodies of the Hawaiian royal couple King Kamehameha II and Queen Kamāmalu, who had tragically succumbed to complications from measles while on a goodwill visit to England.[42] The Union Jack has remained in the Hawaiian flag as a symbol of Hawaiian–British friendship ever since.

Allocation of irrigable water in Hawai`i seemed to be strictly enforced from the highest level. The *Blonde* entourage were privy to a water ruling from Kamehameha. A woman complained that a 'rich neighbor had kept the water of the public watercourses flowing in his own grounds beyond the regular time allotted to him, and that therefore her fields had been injured by drought'. Treating it as a matter of considerable seriousness, the King immediately dispatched an emissary with the complainant to sort it out.[43]

Taro growing in freshwater ponds. *Photo supplied*

The value of taro – and that of water – in customary Hawaiian life is intertwined into Hawaiian tradition and language. The stem of taro is the hā – the word for breath. As Hawaiian story explains it, the hā lifts the taro leaf, to connect the plant world to the human, 'so that each stem is an umbilicus'.[44] In this sense, it is not unlike the Māori use of the word whenua, or land, serving also as the term for the placenta. In Māori the word hā breathes also with a water association. Missionary Rev. Richard Taylor wrote: 'the beautiful word haruru [halululu in Hawaiian] is taken from the shaking or vibratory sound of water, the noise of the surf; ha is the breath and is the natural sound of an aspiration'.[45]

As in all Polynesia, cosmic origins begin with the birth out of Te Pō, the eternal darkness, created by the original parents – in Hawai`i, Papa and Wākea. They held the world and all its beings within the shuttered confines of their mutual embrace. This epic narrative, whose fabric binds together many of the peoples who dwell across the woven mat of the Pacific, takes on a slightly different design depending on the location.

In Hawai`i, Papahānaumoku 'begetter of islands' and Wākea 'the bright one' – the equivalent of Papatūānuku and Rangi in Aotearoa – give rise to islands. Their firstborn is the kalo (taro), whose name is Hāloanakalaukapalili ,'Hāloa of the trembling leaf'. He is closely followed by the earliest human ancestor of the Hawaiian people, Hāloa, 'the stalk of the ancestor'.[46] Wākea is the ancient god-progenitor of the Polynesian people, father of Pele. Wākea is not the same as Rangi, the Māori progenitor of all, whose name is not found in Hawai`i. However, the name Wākea as Wātea – joined, unsurprisingly, by Papatūānuku – representing the space between earth and sky, does figure in Māori mythology.[47] Ngārangi Rangiuia offers a different dimension: 'The tua in Papatūānuku is "tuawhakariri" – from that ancient realm and to do with space and time and the amount of time it has taken for things to get to where they are now.'[48] Indeed, Wākea and Papatūānuku both represent the space–time element, 'the idea of how long things have taken, and the layering of things. The "tūānuku" of Papatūānuku is about the nature of life, constantly changing and growing.' In Māori thought, the firstborn of Rangi and Papa is the kūmara,[49] whose role is similar, but not as comprehensively or consistently in Māori culture as is the taro in Hawai`i.

To this day, where taro is harvested in Hawai`i it is ceremonially pounded and mixed with water to make a dish called poi, and served for its high food value and taste. In other parts of the Pacific it may be mixed with coconut milk – for example in the Marquesas, where it is called popoi.[50] The officers of the *Blonde* found poi to their liking, 'extremely good and rather like Scottish sourens with milk and sugar'.[51]

Taro cultivation and poi preparation are possible only with a plentiful supply of water: the taro is nothing without it. According to legend, when Hāloanaka was born, he was so misshapen his parents believed him to be dead. Wākea buried him. Next morning the child had sprouted and continued to grow. Thus began the close relationship for Hawaiians with a cargo plant that has sustained them in these islands for 700 years.[52] And as if that were not wondrous enough, the taro story celebrates the Polynesian role of the elder brother, kua`ana (tuakana in Māori) in human families, who provides for his younger brothers and sisters. This is recognised in the idea that the corm or root of the plant is the `ohā, and from this grows the `ohana, the family, as happens when the corm grows and multiplies.

A Māori account of the kūmara ties that plant's genealogy in with humanity's – and that of the kiore or rat – but it does so through the medium of the man-god Rongomāui and his wife, Pani.[53] 'He went to the stars to get it,' Ngārangi said, taking up this account with a slightly differing inflection from the kiore narrative. 'While there, he stole the children of Tangotango – the kūmara and the taro – and stowed them in his scrotum and brought them back to place them in a spring. It was Pani who grew them.' She ascertained the best types of soil for them, which included a form of biochar, widely practised by Māori. 'The old people dug as little as possible into the earth – this is our earth mother,' he said. 'They would make just a small hole for the tuber, heap up the surrounding earth and build a mound. The hole was there so that the roots of the plant could spread out.' Moana-ariki, a dark, clinging, crumbly sand, was one of the preferred soil types. Ngārangi suggests the well-known photo of a row of Māori gardeners with their kō (digging sticks) tilted at the ready to open up the earth is an image designed for English cultivation practices and is therefore misleading.[54]

The arduous work of soil conditioning for cultivation was undertaken throughout much of Aotearoa, some of it informing new science. In the 1920s when scientists Rigg and Bruce surveyed about 400ha of modified garden soils on the Waimea Plains near Nelson, they described what they called 'Maori gravel soils' as 'conspicuous by its dark colour and the presence of small stones and coarse sandy particles, in fairly constant proportions'. This modified soil – one paraumu – was richer in plant-available minerals, higher in phosphoric acid and lower in lime demand than any adjacent Nelson soils.[55]

What the scientists suspected was that the traditional application of charcoal soil amendments, or biochar, was what made the difference. To Māori, their crops were tapu and human waste as fertiliser was forbidden. Instead, ingredients of nitrogen, potash, phosphate nutrients, charcoal and fish were mobilised. Local peat and biochar, probably made from local mānuka, were added to improve soil moisture in typically dry summer spells.[56] Best records that for Tūhoe, the char probably came from the umu or earth ovens.[57] The fact that even today the Waimea char is still oily to the touch indicates the locals used the same technique.[58]

Taro cultivation and poi preparation are possible only with a plentiful supply of water

Scientist James Lovelock has asserted that, by locking up carbon in biochar and bio-oil, humans can save themselves from runaway climate change.[59] The application of biochar to soil can lead to a marked reduction in methane emissions from the decomposition of animal waste in waterways. Methane currently represents the more intractable half of Aotearoa's total greenhouse emissions.[60] Big storms and ex-tropical cyclones now regularly wash the spills of 'slash' out of New Zealand's cutover pine plantations, and can cause massive soil loss, leading to the destruction of watercourses, bridges, farms and whole villages. Instead, pine forest and orchard green 'waste' could be carbonised and, with nutrients added, made into a fertile soil conditioner. This doubles as sequestered – but usefully active – carbon. Compared with open firing, the entire biochar process significantly recovers discharge of CO_2 into the atmosphere by as much as 30 percent. The char remains active in soils for thousands of years.[61]

Biochar is but one practical, circular approach to a number of modern land issues. In Aotearoa, it may provide a partial alternative to the financial and environmental burden of industrial fertilisers and projected land uses for animal farming and horticulture that exceed the capacity of soil health and the water cycle.[62] Some Māori groups in the regenerative agriculture movement are re-examining this approach as part of their tikanga.[63] Interestingly, colonial biochar, made by baking cut willow piled under a blanket of earth, was used by early Nelson settlers in their hop kilns.[64] But how quickly memory is lost.

Just along the road from the old fields of Māori char and the old kilns of the hop-growers is an example of late modernist thinking. This is a planned 53-metre-high dam in the Lee Valley, the Waimea Community Dam, designed mainly to deliver a more reliable water supply to the horticulture/viticulture industry on the Waimea Plains. The first of its size anywhere in Aotearoa in more than 40 years, with a cost of well over $100 million, this structure has been presented to ratepayers as also being a response to water demands in a changing climate for a growing population. Planning has been deliberate, but no serious consideration was given to such holistic, but scientifically credible alternatives as those practised by their predecessors.

Always labour-intensive, the production of biochar alone cannot meet all the dam's goals. But for a country that prides itself on innovation, the inability for the land and governance sectors to look outside convention to a range of already-proven solutions is costly – and not just financially. This was one decision that might have been guided by an integrated range of approaches, including recourse to local history, upgraded pipelines, more localised water storage, and education in better water use and conservation. Taken together these factors might have proved more viable for the long-term.

CHAR – THE RETURN OF AN OLD TECHNIQUE FOR NEW CHALLENGES

The ancient method of pyrolysis, choking the flame to create charcoal for soil enrichment.

Above: iStock.com/Sofiia Kazarina; Below: iStock.com/ronemmons

The old adaptive wisdom in using 'anthrosols' proves to be a direct forebear of the biochar process, pyrolysis – wood burned in an oxygen-starved retort that produces a light, microporous carbon.[65] 'The porous nature of charred plant matter offers remarkable regenerative properties for soil,' says biochar and soil specialist Don Graves. Applied to soil, it helps to engender a protective habitat for essential soil microorganisms – symbiotic fungi and beneficial soil bacteria. 'Porous structures of biochar absorb and retain soil water in a manner like a sponge,' he adds.[66] Like mulching, such moisture retention can be useful in drought-prone districts such as Nelson.

As a soil additive, with balances of good compost, biochar holds remarkable regenerative properties. Its ability to provide physical protection to soil properties and high moisture retention works by neutralising or mitigating effects of excess phosphorous and nitrates in soil and water – at comparatively low cost. With soluble artificial fertilisers or quality compost, not only is biochar good for damaged and depleted soil, it also removes residual levels of carbon dioxide in air and water. Biochar offers considerable potential for agricultural systems struggling to become more circular in their management.[67]

Water, gods and governance

To a modern mind, at first sight there is nothing strikingly logical about old Polynesian myth. Unfortunately, however, it is the linear logic that separates modern culture from a recognition of the often submerged energies in which nature is provider and healer in what is known as the circular economy.[68]

The old Hawaiians have tens of thousands of gods representing the make-up of these energies. Of the big four, two – the supreme god Kāne and his travelling companion Kanaloa – have been introduced. They sit together on the same level as Lono and Kū. In other parts of Polynesia – in Aotearoa, for example – Kū (Tū) is always the god of war, but also the god of humanity;[69] his authority extends to hunting, cultivation and fishing.[70] In Hawai`i he is god in a symmetry of war and prosperity, and is often described in conjunction with the balancing qualities of his wife Hina.[71]

Prayers to Kāne are as heartfelt as any; among the reasons are those set out in the prayer below, recorded by David Malo from his home in Lahaina, Maui:

> Make Kane ia hii
> Hii luna ka lani o Kane …
>
> *Kane wearies himself in death with care,*
> *Care for the government of his own heavenly kingdom,*
> *The earth is governed by Kane,*
> *Kane cares for the mottled cirrus clouds,*
> *Pray to Kane for life.*
> *Kane is the god of life.*[72]
>
> Ea ia Kane ka waiola
> E ola i ke kini o ke akua! …
>
> *Kane comes with the water of life*
> *Life through the multitude of the gods!*
> *Sacred! Sacred! Life! Life!*
> *Life through the chiefs! Life through the gods!*[73]

Through Kāne the interaction of light and water is defined and taken care of. His being is a way of explaining the base ingredients for photosynthesis, whereby trees take in sunlight and water, embed carbon in their timber and produce life-giving oxygen.

Deep in the southern Pacific, Tāne is god of the extensive forests of Aotearoa. He is photosynthesis incarnate. It was Tāne who, with his heavenward thrust, forced the original parents Rangi and Papa apart – thereby bringing in te ao mārama (daylight) and delivering all beings from what hitherto had been eternal darkness. He retains his power in legend as the bringer of light – which may also be seen as bringing knowledge or consciousness to the world.[74]

But Tāne is also Tānetewaiora, the one who, in his thrusting, 'opened the fountain of living water' to all beings, as 'giver of life'.[75] Tūhoe tohunga and Presbyterian minister Ron Matthews understood how Tāne retains fundamental links to water. As the offspring of the two celestial beings, Rangi and Papa, Tānetekapurangi, god of fertility, is called on to assist with a birth. (In Hawai`i his focus was on better health.)[76] 'Since creation was brought into being through a cradle of water,' Ron explained, 'everything brought into being is created through the sustenance of water.' As Tānetewānanga, he is 'the source of all knowledge'.[77]

Elsdon Best takes two pages to list Tāne's many manifestations. Perhaps most fundamentally, Tānetokorangi features in reference to the mist, says Ngārangi.[78] Among his many other roles he is Tānetuhunui, god of the fine weather of summer. (Where I live, under certain conditions, fine mist and fine weather are not incompatible.) Tānemuriwai controls the freshwaters of the earth. Tānetemuriwaihoe represents floodwaters and deluges that cover the earth. Tānetāwhirimātea

Pā tuna (eel weirs): a common sight up and down the Whanganui River before the advent of the river steamer deemed them to be a hazard to navigation. *Whanganui Regional Museum*

is god of winds, who in this role figures strongly in the creation story, also finding expression in clouds and lightning.[79] Tānetāwhirimātea is most familiar along Aotearoa's windy coasts and mountains, while for Ngāi Tūhoe, Hinepūkohurangi, goddess of the mist, who represents the earth mother's greeting to the sky, is also an ancestor.[80]

These are key players in an understanding of the elemental nature of water, each and all intricately interwoven. In many respects, the Tāne of Aotearoa is still the Kāne of Hawaiian fame. In an age of skies becoming more unstable and unpredictable under ballooning anthropogenic CO_2 levels, a deeper comprehension of the abiding significance of what Tāne/Kāne represents to these respective indigenous cultures can enrich our insights in ways as helpful as the science.

So what are the differences between the 'cousin cultures' of Hawai`i and Aotearoa? Are they the differences between one culture that grew to harvest water and one that almost never needed to? There is never a single answer to such a question, but its asking does throw light on their belief systems and organisation, then and now.

For Hawaiians, their streams became source and vessel for some of the most seamless, ecologically efficient and extensive horticulture-aquaculture practices the world has known. Some might call it permaculture. In Aotearoa, with more land, more resources and fewer people, water management was rare: the rivers were used largely for fishing, and a few, such as the Whanganui, were also important transport routes. Their pā tuna (eel weirs) and utu piharau (lamprey weirs) were intricate but flood-proof, extensive yet canoe-navigable, yielding great bounties of fish.[81] In the central volcanic region these fish were largely absent; but even there, Rotoaira was a lake abounding in kōaro or whitebait, *Galaxias brevipinnis*,[82] while eels were taken at certain times using stone-walled traps with flax and supplejack-framed poha or nets – even out of the sulphur-prone Whangaehu River – by Ngāti Rangi. Their kaitiaki, however – the giant eels who lived and roamed in these waters – were not to be touched.[83] Using a variety of methods, the galaxiid and other species, including panoko (torrent fish), toitoi (bully), pātiki (flounder), kōura and kākahi (freshwater crayfish and mussels) were taken throughout Aotearoa.[84]

The Polynesians on both island groups organised into extended family living and working units. With their growing numbers and more restricted supplies of water, the Hawaiians, as we have noted, created much more formalised structures. By contrast, with an overall smaller population and greater landmass, divisions were still real for Māori but nothing like as rigid or formalised. And they remained tangata whenua, the people of the land. For them it was the iwi and hapū, identity triangulated on the eternals of rangatira (blood), mountain (place) and river (life). Whakapapa, the cultural and genealogical lifeblood, with its expression vested down through time and space to all but the lowly, remained, quite literally, both code and messenger for every transaction, be it with society or nature.[85]

Celebrating water

Of all the Hawaiian chants, one in particular is replete with Hawaiian gratitude for water. It was recorded mid-nineteenth century from a hula given in Kaua`i, an isle lush to windward and where, since Kāne and his brother left behind few springs, they are not worshipped.[86] If the most fundamental, romantico-mystical aspiration forged largely out of English landscape is, arguably, the story of the Holy Grail, the equivalent in this archipelago, e Mele no Kāne, is still with us, thanks to one of Hawai`i's foremost collectors:[87]

> He ú-i, he ninau:
> E ú-i aku ana au ia oe,
> Aia i-he`a ka wai a Kane?
> Aia i ka hikina a ka La,
> Puka i Hae-hae;
> Aia-i-laila ka Wai a Kane …
>
> *A query, a question,*
> *I put to you:*
> *Where is the water of Kane?*
> *At the Eastern Gate*
> *Where the Sun comes in at Haehae;*
> *There is the water of Kane …*

As each stanza opens with that same question, the chant moves from east to west invoking the sunset out to a jutting basalt pillar 40km west of the Kaua`i horizon. This is tiny Nihoa. The chant turns then to the mountains and valleys of larger islands. In the penultimate verse, with typical Polynesian comprehensiveness it answers first to sky then earth:[88]

> *Up on high is the water of Kane,*
> *In the heavenly blue,*
> *In the black piled cloud,*
> *In the black-black cloud,*
> *In the black-mottled sacred cloud of the gods …*

Then to the final stanza, which answers:

> *Deep in the ground, in the gushing spring.*
> *In the ducts of Kane and Loa,*
> *A well-spring of water, to quaff,*
> *A water of magic power*
> *The water of life!*
> *Life! O give us this life!*[89]

To that great scholar of myth, Martha Beckwith, such an account of the gods' quest for springs was never intended as a practical guide for the skills required for irrigation. Rather, its purpose was to 'simply express the mystery which even to an old Hawaiian today [1930s] belongs to such a phenomenon'. Part of the mystery is held in its fundamental symmetry. Kāne is the thrusting male, the spring of water, while the source is held by Papa, the earth mother and source of all life.[90]

For Hawaiians, the practical mystery remained – how to locate and obtain water? After contact, technology changed that. Once iron tools were introduced to the islands, Kamehameha III, who grew up on Big Island, was able to cut a road there through 30 metres of rock, thereby opening access for canoes to a small bay for fishing. A son of the original king of all the islands, he was also a royal enabler of gardening. In an echo of the famous 'where are the waters of Kāne?' this adaptable leader also attempted to dig through lava rock for water, 'but the want of tools forced him to desist. The fields which he had reclaimed from the waste, and brought into high cultivation, still bear his name.'[91]

In the mountainous areas of Maui, Kāne and Kanaloa, lacking water, discuss whether water exists at such heights. Kanaloa commands; Kāne thrusts with his fine-grained kauila (*Alphitonia excelsa*) staff 'and water gushes forth'.[92] Kāne digs his spear first into one hole and then into another, saying, 'This is for you, that for me.' The gushing water from these strikes is still known as 'the water of Kane and Kanaloa'.[93] The accounts from early travellers and settlers, given below, describe how belief in these myths was still guiding daily life on Big Island in the first few decades following European contact.

A missionary view

Travel east for 20km from Kealakekua Bay, Big Island, and you come to the thriving village of Kailua, home to King Kamehameha I and his capital, with a population of 3000–4000. Kailua was near the original home of Sereno Bishop, an American-educated scientist and Presbyterian minister, born to missionaries in 1827. Later, in the campaign to annex Hawai`i to the United States, Bishop recorded that water was scarce in this district, at the leeward base of Hawai`i's westernmost volcano, Mt Hualālai (2522m): trees were limited to a few coconut, kou and pandanus:

> There were no gardens, for lack of water. Heat and general aridity characterized the place. But it pleased the natives, on account of the broad calm ocean, the excellent fishing, and the splendid rollers of surf on which they played and slid all day.[94]

There was, however, 'ample cultivable land in the moist upland' several miles inland, at an altitude of 300–700m. Bishop wrote, 'While one shore may be absolutely rainless for months, gentle showers fall daily upon the mountain slopes.' With the fertile volcanic soil here, 'The natives found their chief means of substance, and in all seasons, were sufficiently fed.'[95]

William Ellis, who had visited northern New Zealand briefly and served as a missionary in Tahiti for six years, arrived at Kailua in 1822 and settled his family there the following year.[96] Ellis began by touring several Hawai`i islands, and set down some of the earliest detailed accounts of taro growing, including ahupua`a. Aided by a lively curiosity and, successively, two intelligent wives, Ellis exhibited a deep interest in Hawaiian life – together with a shrewd appreciation of possible sites for missions, with a good supply of water.[97] His writing describes life on an island where water strongly influences the impress and expression of human life on the land.

Reflection of a pond near the Black Sand beach, Kaimū, Puna district, Big Island, Hawai`i. Kaimū was temporarily destroyed by an eruptive flow of lava from Mt Kīlauea in 1990. *Photo supplied*

In Kealakekua Bay, Ellis recorded that the chief 'offered us a calabash of fresh water, a great luxury on this [western] side of the island'.[98] As the party headed towards Honomalino Beach, water was still scarce; along the coast to Kona they travelled with a single calabash of brackish water.[99] After about an hour they came to the Waiohinu district, where rainfall was 'generous', despite it being in the lee of the mountains and so protected from trade winds. Here they saw the first freshwater stream since they had landed in Kona many days before, as well as smaller spring-fed streams.[100]

> Our road, for a considerable distance lay through the cultivated parts of this beautiful valley. The mountain taro, bordered by sugar cane and bananas, was planted in large fields on the sides of the hills, and seemed to thrive luxuriantly. On leaving the valley, we proceeded along by the foot of the mountains, in a line parallel with the sea, and about a mile and a half from it … The country appeared more thickly inhabited than that over which we had travelled in the morning. The villages along the seashore were near together, and some of them extensive.[101]

In the southeast corner of this manta ray-shaped island is the Puna (spring) district. When Ellis visited Kaimū, the sharp `a`ā lava had broken down into light soils 'adorned with plantations, groves of cocoa-nuts, and clumps of kou trees'. The village, with a 'fine sandy beach' where waka could launch, was home to about a third of the 2000 or so inhabitants along the coast. He thought it an ideal mission site, apart from limited access to freshwater:

> There are several wells in the village, containing brackish water, which has percolated from the sea, through the cells of the lava, undergoing a kind of filtration, and is collected in hollows scraped out to receive it. The natives told us that, at a distance of about a mile, there was plenty of fresh water.[102]

Further south, at Kapoho, Ellis was delighted to come upon a landscape of 'luxuriant gardens' that were well irrigated from top to bottom. The party walked uphill through the trees above the lava plain, into gardens at the base of a steep rise; there were more gardens, a climb, and then 'another charming valley' whose sides were laid out in gradually sloping plantations, surely an ahupua`a, 'in a high state of cultivation, planted with taro, bananas, and sugar cane'. In the verdant valley floor, along a narrow border by a lake, grew breadfruit, kuku and `ōhi`a trees, sometimes 'with a lowly native hut standing beneath their shade'.[103]

It was a highly productive, deeply integrated horticultural system that Ellis called the 'smiling gaiety of the whole'. Since cosmology, lineage, nature and story are woven together in myriad bindings throughout Polynesia, such a manifestation of their thinking in nature management should not surprise us.[104] Nor should their resilience – and their adoration of Pele, goddess of fire, even in the face of her handiwork. Pele traditionally drew the forces of death and devastation into tension with her other capacity for new land, life and creativity.

Despite the destruction as a result of the 1960 eruption in the district, by the turn of the century many palatial homes were being built on the smooth pāhoehoe lava. This is the perfect example of 'unreal' estate, set among the regenerating kīpuka – tiny, sometimes extensive mounds standing untouched above the hot, labile lava from which the remnants of plants begin the task of ecological regeneration.[105] In a series of eruptions in June 2018 some 400 of these houses were destroyed, wiping out 'Vacationland' entirely; Kapoho Bay filled with lava 2000m out to sea.[106]

At dusk … we could see Pele's crucible showering flame and sparks with flumes of active steam and smoke rising from the coast, not far away. To see her up close we took a narrow, winding road south along the craggy coast. We reached barriers where until recently the main road, now devastated, had headed. Continuing through warning signs on the make-shift road we drove on through fresh black `a`a lava towards what we could now see was a red glow underscoring its burning light onto the billowing cloud plume above.

As we passed near low coastal trees a large white owl, a short-eared Hawaiian owl (pueo) kept pace, flying for more than a kilometre at window height, just a metre out from the car. Here the owl is seen as an amakua (spirit ancestor) as it is for Māori. At the road's end hundreds of cars were parked and as if for a rock concert, marshalls [were] directing traffic to more parking areas. People then walked along a marked route. Close to the ocean's edge, showers of flame and sparks flew up towards the new moon rising above a towering plume.

Finally we stopped walking at a point close enough to witness the island-making industry of Pele, her pulsing fire erupting at the lapping sea edge, lava flaring as it vented into the water.

Between Puna and Hilo, this is the contemporary place of active volcanism – domain of Pele and her eight fiery sisters and four brothers, whose names reflect different qualities of volcanism.[107]

In the impressive beauty of the high rainfall district of Hilo flow a number of rivers – the Wailuku, Wailoa, Wairama and Waiākea. Proverbs here speak of the rains.[108] Dennis Kawaharada gave the background to one told by Mary Pukui for October:

> Cold fronts generated by low pressure systems in the North and Western Pacific move eastward along storm tracks, sweeping over the Hawaiian Islands from the direction of Nihoa, a small island WNW of Kaua`i, and Lehua, a small island west of Kaua`i, just off the north tip of Ni`ihau. At the leading edge of these cold fronts are storm clouds that bring heavy winter rainfall and strong winds: 'Pa`apa`a`ina ka malo i loko o `Ikuwa'. 'The loincloth flaps in the month of `Ikuwa.'[109]

In the northeast, on the edge of modern-day Hilo is a well-watered district with three rivers; one of them, Wailuku, rises on Maunga Kea. Lake Waiākea covers just 10ha, but is the largest in Hawai`i. Ellis found it 'tolerably deep', and 'navigated by canoes and boats a considerable distance inland'. Today the lake sits within the small but enchanting Wailoa River State Park. Ellis waxed lyrical:

> The face of the country in the vicinity of Waiakea is the most beautiful we have yet seen, which is probably occasioned by the humidity of the atmosphere, the frequent rains that fall here, and the long repose the district has experienced from volcanic eruptions.
>
> The light and fertile soil is formed by decomposed lava, with a considerable portion of vegetable mould. The whole is covered with luxuriant vegetation, and the greater part of it formed into plantations, where plantains, bananas, sugar-cane, taro, potatoes and melons, grow to the greatest perfection.[110]

Coconut and breadfruit groves abounded 'in every direction', and on heaps of lava rubble, Ellis and his party saw potato vines 'growing very well in the midst of them, though we could scarcely perceive a particle of soil'. It was a prosperous place: the houses were well built and 'the people generally industrious'. Ducks were plentiful in the streams and a couple of large fishponds swarmed with fish near the shore. Ellis estimated the place, no larger than half a bay, as he put it, supported a population of at least 2000 people.[111]

A day or so later, after crossing Te Maunga Kea Ridge, his party were deep in the Waipi`o River valley. The valley floor was 'one continued garden, cultivated with taro, bananas, sugar-cane and other productions of the islands, all growing luxuriantly'. These cultivations continued for 8km, with well-maintained and well-stocked fishponds.[112]

The hills surrounding the broad valley were 'nearly perpendicular, yet they were mostly clothed with grass, and low straggling shrubs … A number of winding paths led up their steep sides, and, in several parts, limpid streams flowed in beautiful cascades, from the top to the bottom, forming a considerable stream,

A waterfall near Hilo, Big Island, Hawai`i.
Photo Aliscia Young

Lava flow on the coast of Kalapana, Big Island of Hawaii, where the rivers of lava meet the ocean and throw up sulfurous steam clouds. *Unsplash: Buzz Andersen*

which, meandering along the valley, found a passage through the sand-hills …'[113]

Not all extensively cultivated regions were the same, but there are similarities. Handy and Pukui described the area of Ka Lae, Big Island's southernmost point:

> Beyond the shore the immediate landscape is one of rough irregular exposed black lava interspersed with small bushes and sparse tufty grasses, green from December till May, dry for the rest of the year. In the old days many footpaths meandered seaward through this wide wasteland from the higher slopes miles inland, where there was soil enough for gardens. The scattered homes and gardens of this lower zone of habitation were those 'of the seaward slopes' (*ko kula kai*), where sweet potato and gourds were cultivated, but little else. These households had relatively easy access to the sea and consequently depended on shellfish, seaweed and inshore fishing, and had these for exchange with relatives living farther up the slopes.[114]

With rising altitude, moisture increases and evaporation decreases, affecting cultivation. Handy and Pukui explain that each kula or level had its name:

> Beyond the *kula kai* (the lowest habitable zone) were the dwellings 'of the upland slopes' (*ko kula uka*), less accessible to the sea, but increasingly favourable for gardening. In addition to sweet potato, dry land taro of a variety called Paua was planted, and sugar cane flourished … Beyond this the open slopes (*kula*) become fern lands, then gradually merge with the lower forest (*wao*). In this zone where fern, bushes and small trees prosper other varieties of upland taro requiring more water were cultivated, under mulch to keep in the moisture. This continued right back into the lower forest. Here were the wild bananas, wild yam (*Dioscorea*), pia (arrowroot), and tree fern (*Cibotium*), whose starchy core was eaten, extending down into this zone from the rain forest.[115]

Altitude alone did not determine how land use was allocated. Distribution of rain, too, was significant, they noted. To the east, 'the wet uplands were wetter and extended lower than on the west, which was both beyond the range of heavy precipitation from trade winds and cut off somewhat by the shoulder of Mauna Loa running to Kalae'.[116]

Seventy years prior to the 2018 eruptions Handy and Pukui set down their understanding of Big Island's Mauna Loa:

Centre: Volcanic gases escape from the earth through sulphur-laden vents on Whakaari (White Island), shown above, an active volcano in Aotearoa's Bay of Plenty, now of tragic notoriety. *Photos Richard Sidey*

> They took unto them this huge Mother mountain, measured their personal dignity and powers in terms of its majesty and drama. They named their land 'The Breast' (*Ka-`u*). They loved Pele, whose home was their land: they endured her furies, and celebrated the drama of creation with which they lived so intimately in the songs and dances of the sacred hula, which dramatizes the myth of the 'Woman of the Pit' (the crater, Kilauea) and her 'family', embodied in cloud, thunder and lightning (Lono), in the forest and verdure (Wahine O`ma`o, 'Green Lady') in Hi`iaka 'of living waters', the healer, and other cosmic terrestrial forces that encompassed them.[117]

It was entirely consistent, then, that it was again two women who, from across the Pacific, responded to the plight of Ngāti Tūwharetoa's ancestor Ngātoroirangi and his need for warmth. During his expedition of exploration and claiming, Ngātoroirangi nearly perished on the Central Plateau of Te Ika a Māui in a lacerating snowstorm driving out of the Southern Ocean. Sensing his distress, Kuiwai and Haungaroa in Hawaiki sent him a sign, raising smoke over a fumerole so that he could locate warmth in the blizzard. This was done through the agency of two underground-dwelling taniwha, Te Pupu and Te Haeata. The underground transmission of fire is described in a manner broadly consistent with the volcanic connections across the thinly crusted central North Island.[118] The path progresses through Te Puia o Whakaari (White Island) on the East Coast, across the Rotorua district to Tongariro, including Okakaru near Kawerau, Rotoehu, Rotoiti, Rotorua, Rotomahana, Paeroa, Ōrakei-kōrako and Taupō.[119] A now warmed and repotentised Ngātoroirangi struck the earth, after the fashion of Kāne with his staff: he created Ketetahi Springs, the volcano Ngāuruhoe and, by his stamping, Ruapehu. His naming continued with the other springs and waterways of the region.[120] So, too, did other tribes of the mountain – not least to the southeast, Ngāti Rangi, devoted guardians of Ruapehu's sometimes explosive crater lake and their cherished streams and rivers, some now dam-compromised.[121]

For the Polynesians, especially those guardians living at the very lip of the Pacific ring of fire, the relationships between Pele and her elder brother Kamohoali`i, and other such

representations, are seldom far apart, even in genealogy. The locals in Pele's district gave Ellis names of gods connected to the volcanism: they explained that Kīlauea had emerged from a volcano family, dwelling at the old placename Kirauea, Tahiti. They, too, had figured the earth's fiery story of trans-Pacific connection into their genealogical matrix.[122]

For all the order and industry Ellis observed and admired, as a devout missionary his evangelism gave him no pause for the changes he was about to render. The European model of colonisation began with colonisation of the soul. The allure of new ways – including capitalism, child of the new religion – and the unseaming of the old was, usually, extensive. But it was never comprehensive.

Just before we left Kealakekua Bay, the Lono of Beckwith's chant spoke: this was 'Lono, the rolling thunder/The heaven that rumbles/ The disturbed sea'.[123] After the dolphins, as evening fell and humidity rose, Lono took full charge. The fan palms rattled noisily against the sides of the house, the rain came in and the thunder cannonaded. It was Boxing Day. This was within the four-month observance period for the rising of Pleiades, Makahiki. This is when Lono, paramount to the dryland farmers especially, as god of thunder, lightning and the delivering rains, as well as of `uala (kūmara), must be acknowledged.[124] Lightning flicked on and off over Kealakekua Bay – fairylights on the blink – this first night after Christmas. It's a leeward shore and thus one of the districts where Lono really came into his own.

The pulses of lightning came in intervals of 5, 13 then 23 seconds and then even longer. Towards the end, one brief flash clung to the sky in a way that stole the day's last secret – the lost sun's now hidden handiwork lit momentarily by the wildfire on a still orange cloud, high enough to catch the sliding fingertips of what was an already departed sunset.

The storm lasted at least an hour. When morning came, the air was still crackling in the bay. Locals said that it was an uncommon occurrence, in that part of Big Island where potable water was always an issue.[125]

Links in the trans-Pacific Ring of Fire: the glowing light of erupting Mt Yasur (left) and Vanuatu's twin lava lakes on Ambrym Island (right).
Photos Richard Sidey

This chapter has considered the cultural context of several leading gods in these two archipelagos, especially pertaining to the customary uses of water. Despite their being increasingly compromised during the past 200 years, these patterns of the past are still in the energy of these landscapes and their Polynesian peoples.

We cannot all become Polynesian, but we can pay respectful attention to some Polynesian ways of being with and of nature. Just as a treasured place of nature can be granted protection, indigenous cultural knowledge in both Hawai`i and Aotearoa can be recognised as adding to the benefits of Western science. A modern culture that better knows its original people's stories and perspectives will be a more resilient society both physically and socially. These gifts of relational, integrative thinking have been held out for a long time. Now, as pressures on water, ecosystems and life itself become weightier, the need for better understanding of old ways has become imperative.

The southern sky above Mou Waho Island, Lake Wānaka.
Photos Richard Sidey & Aliscia Young

The sinuous Whanganui River, layered in history and mystery. *Photo supplied*

CHAPTER 7

Keepers of the Long View

WHANGANUI RIVER

Māori living in Aotearoa, with the longest rivers in Polynesia, gradually ascribed particular cultural values to them over 700 years. This may explain in part the rivers' mystery and allure, adding depth to the experience for today's river travellers. Part of that association was the presence of traditional guardians, many and varied in Aotearoa waters. Indeed, resident taniwha in rivers and lakes were ubiquitous in Aotearoa. Similarly, Hawaiians connect to water lizards, or mo`o – but Hawaiian references to such entities are rare.

Throughout Aotearoa taniwha on land might take on the form of a lizard. In water, certainly in the Whanganui, they could assume, or occupy, the body of a particularly large eel or a floating log. T.W. Downes wrote: 'it is quite possible that the idea evolved from the lizard, as these taniwha with a few exceptions seem to have borne all the characteristics of the lizard'. Karanga Ngātaierua, daughter of the poropiti (prophet) Te Kere Ngātaierua from Tawatā on the Upper Whanganui, told Downes, 'Lizards formerly were water-creatures and lived entirely in the sea. The fish also lived in the same waters, with the shark as chief or king.'[1] As is the case with the eel and the salmon of the northern hemisphere, some taniwha are, like eels, capable of living in salt- or freshwater, depending on what stage of their lifecycle they are at. Their associations with old Hawaiki brought them considerable mana. One such guardian was Tūtangatakino.

More than 35 years ago, when I interviewed the oldest tohunga on the river, Titi Tihu, grandson of Te Kere Ngātaierua, he gave an account of an entity from his boyhood in the 1890s, living at or near Tawatā. One evening when he was about eight his mother asked him to paddle up the river to Ōhura for medicine that was urgently required for an ailing uncle. The boy protested that he was frightened, but his mother refused to take no for an answer.

A pātaka kai (food storehouse) at Tawatā on the Upper Whanganui. Tawatā was once home to followers of the nineteenth-century prophet Te Kere. *Whanganui Regional Museum*

The lair of the taniwha Tūtangatakino was said to lie beneath this prominent rock, Petipeti a Aurangi, on the Whanganui River. *Watercolour by Thomas William Downes, 1937–39. National Library, Wellington*

So he hauled his dugout waka from the bank and launched himself apprehensively into the current. This was not a reach that anyone would volunteer to paddle a shallow eggshell of a craft on in the darkness, the river oily in its riffles and quick-silvered, its steep forested banks giving away nothing. He'd not gone far before he saw two huge eyes, like the signal lamps on a riverboat, moving towards him, the creature's head raised above the glassy surface of the river. The frightened lad knew immediately that it was Tūtangatakino, whose lair lay beneath the prominent rock called Petipeti a Aurangi.[2] Today, canoeists still leave a placatory sprig of leaves in this spot as they pass the tapu rock.[3]

Instead of being a menace, however, the taniwha, duly exercising his guardianship role, not only allowed Titi to pass but acted as his guide and protector – a pathfinder on a dark night. This was an acknowledgement of Titi's mana: his ancestry, his intelligence and self-possession[4] – and of things to come.

His grandfather Te Kere was a deeply knowledgeable tohunga and healer. In the 1890s he attracted to his village, Tawatā, 1000 followers of his Pae Tiuihou movement, an expression of sovereignty, mana Māori and opposition to the alienation of Māori land.[5] According to Downes, Te Kere had first encountered this same guardian off the Northland coast when it delivered him as a young man from certain death while he was crewing on sailing boat near an islet named as Motu Kākaho.[6]

Mana is also about potential. The water taniwha's response was an acknowledgement of the destiny of Titi that would see him go on to undertake what became the longest-running court case in Aotearoa's history. In a succession of hearings between 1938 and 1964 the Crown's assumption of control, use and delegation over the river was challenged. The litigation finally foundered on another rock, this one of apparent ignorance in a so-called specialist court of what were then all white senior judges – the Māori Appellate Court.[7] There the case languished until the 1999 report of the Waitangi Tribunal.[8]

Years after Titi's first encounter with Tūtangatakino, the pair met again. Titi was building a meeting house with his relations at Tarata near Waitara, in the Ngāti Maru district of adjoining Taranaki. Again he was assisted by the taniwha. Overnight, as the builders slept – and much to the astonishment of fellow workers – the house's pou tokomanawa (ridgepole) was mysteriously installed. Titi ascribed this to the endeavours of Tūtangatakino. Titi, the old man from another world who could magic the birds and insects to safety from a patch of bush before it was fired, and summon with chants from the river the flicking tail of a taniwha, gave a priority to this story I couldn't begin to comprehend. Why was this so important to him, and why would he inform an ignorant stranger of such details?

The message, I now understand, was that Tūtangatakino, who ordained all such matters, was a veteran of warfare in Hawaiki: he was a guardian who had accompanied the *Aotea* waka on its migratory journey from Ra`iātea's dramatic landscape, in French Polynesia.[9] In this place of ardently expressed volcanism, possibly located in the Fa`aroa (Wha`aroa) River, he was deployed by tohunga to annihilate enemy war parties.[10] On arrival in Pātea, Aotearoa, he had taken up residence on the Upper Whanganui. From there he exercised his wider protective role, by dint of his powers of flight, over the Taranaki river people, united by the kinship of the *Aotea* waka.[11] This story, then, was about a sea- and freshwater-capable guardian that continued to uphold his mission for hundreds of years.[12] But it was a story also coded for the mana and bloodlines of its teller; of the influence and reach of that mana.

The old proverb, 'Ka mua, ka muri', 'look backwards in order to look forward', might be seen as one of those lying at the heart and beauty of Polynesian reciprocity, then and now. It also characterises Te Kere's life plan of service to his people and, especially, the restoration of old ways. Across the motu (land) his life continues to inspire many of his numerous descendants of a younger generation who seek the knowledge of this somewhat hidden ancestor, source of insight and poetic wisdom.

A generation before, Gerrard Albert and a group of leading men from different groupings of the river had become immersed in the whare wānanga teachings led by the kaumātua Rangitihi (John) Tahupārae, who had trained them in the old ways, the old knowledge. This group had then spent many years driving toward a solution to the wicked problem that had left their river – and numerous significant tributaries that were also tribal bloodlines – beheaded for an upper river hydroelectric scheme; denatured of much of its surrounding forest by farming and settlement, its waters clotted with soil runoff. All of this had impacted their mana and the tribe's capacity for physical and spiritual sustenance from their ancestor.[13] Like Titi Tihu, they never gave up.

The Crown and iwi negotiation towards a settlement under the Treaty of Waitangi continued for more than a decade. At one

Titi Tihu, grandson of Te Kere Ngātaierua, at Cherry Grove, Whanganui River, 1984. Tihu was a highly influential tohunga, seen as a pioneer litigant in the process that eventually won the recognition that gave legal standing to the river. *Photo Bruce Connew*

point, spokesman Gerrard Albert indicated that an agreed solution was closer. When it came it would be a 'creative' one. He was not exaggerating.

Almost 20 years after the Waitangi Tribunal's Whanganui River report, and nearly 30 years after Titi Tihu's death, the agreement signed at Rānana finally broke the legal deadlock of some 150 years between Crown and iwi. This was no ordinary settlement, and it happened with a world-first solution.

The interminable Whanganui iwi v Crown ping-pong of claim and legal counterclaim over who owned the river was no more. In a rare spirit of imaginative law-making, differences between Crown and iwi would be set aside. Instead, both would recognise the river, integral from mountains to sea, not quite as a law unto itself but as a legal entity; in corporate speak, 'a legal personhood'. Accordingly, a duty of care to the river is now uppermost: the river is recognised in law as a source of spiritual and physical sustenance, and 'including all metaphysical elements' for the hapū and iwi. Further, the settlement recognises: 'The iwi and hapū of the Whanganui River have an inalienable connection with, and responsibility to, Te Awa Tupua and its health and well-being.' A payment of $90m is being made over 30 years as a token recognition of losses suffered. A guardian representative of both Treaty parties has been appointed to ensure that expenditure for river restoration is effectively deployed. This agreement, Te Awa Tupua (Whanganui River Claims Settlement) Act, became law in 2017.[14] It now attracts ongoing international attention. While sometimes described as an assertion of nature's rights, it is actually more a recognition of the indigenous relationship to nature in a particular location.

Tūtangatakino then, according to Whanganui lore, was a ponaturi (male energy of the marakihau partnership), a hikuwaru (eight-tailed) creature 'configured,' said Petuere Kiwara, another descendant from Te Kere who is also inspired by his forebear's uncompromising commitment to the river, 'to kill those who did not respect his wisdom and presence'. However, his opposite number, the marakihau, remained directly under Tangaroa's powers at sea. The female element of the marakihau dominate the oceans and ponaturi, the male element, dominate the rivers and streams and lakes.'[15] (This loving–punitive entity serves a similar role in Inuit culture.)[16]

Tūtangatakino was a special kind of taniwha called marakihau – depicted in pairs in meeting houses of the East Coast region, including those of Ngāi Tūhoe in the Urewera. Carvings show marakihau full-faced, with their bodies turned to the side. Judith Binney describes three houses featuring such carvings, all executed by followers of the prophet and guerrilla fighter Te Kooti Arikirangi Te Turuki. When Te Kooti visited the Whanganui district towards the end of his life, he debated fiercely with Te Kere Ngātaierua on delicate matters of mana and association.[17] Te Kere's descendants say that Te Kere was warning Te Kooti that it was fine to adopt the teachings of the Bible, but that 'we must keep these things separate from the old teachings'.[18]

The original Pākehā historian of the river, Thomas Downes, described a site that was associated in ritual with this ancient entity: Arimatea, once a kāinga situated on a small isthmus, its tongue lapping the river 32km below Taumarunui. 'Tutangatakino,' Downes wrote, 'lived a few miles downstream.' The site is steeped in significance. It was one of the five places on the river where niu poles were erected by the Paimārire movement for symbolic and ceremonial purposes as they fought to reclaim lands, forests and water from the colonial forces in the 1860s. It was also the site of a three-metre-high tūāhu or raised altar, whose platform, by 1928, had been broken down by cattle.[19] When I visited with elder Hikaia Amohia in the 1980s, the niu stood in a glade of ponga treeferns and beech. There were the signs of excavation that formed the tūāhu site. The cattle, too – Māori-owned – were still there. Germane to this discussion is that the tūāhu was named Tawhi, and, as Downes wrote, 'its guardian spirits or atua were Moko-hikuwaru and Tu-tangata-kino'.[20]

Life-long resident of Taumarunui, the highly respected kaumatua Anihera Henry, at 76 gave evidence before the Waitangi Tribunal hearing on Tūtangatakino's fundamental role:

> Tūtangatakino guarded and cared for our ancestors [of] the *Aotea* waka, from whom we also descend. Today he resides within the Whanganui River, at the upper reaches. He guards the passage for us, the descendants of all hapu of the tribal people of Te Ati-hau-nui-a-Paparangi …[21]

Downes understood that Tūtangatakino had moved to the upper Whanganui after many generations of presence at Pātea, northwest of Whanganui. He took retribution from a Ngāti Maru chief, Kohau, who refused to formally acknowledge him as he paddled the upper Whanganui. The waka was overturned and Kohau's crew suffered significant loss of life. This is essentially a story of two related iwi tied to the same descent lines and founding waka, and of the importance, nevertheless, of not presuming on each other's territory or waters. [22]

But these were not static creatures. Te Kere, tohunga, healer and keeper of old wisdom, became an active pacifist from the time of the creation of the Kīngitanga in 1858. Tūtangatakino's energy is believed by some of the whānau to have become more peaceable. But from the arrival of the waka in Aotearoa he

River of history, river of mystery, Upper Whanganui. Now a river with its own legal standing.

was always a kaitiaki, original, continuous and now contemporary.[23]

The guardian was also responsible for enforcing Te Kere's boundaries. Titi Tihu had explained to me how, again as a lad, he had transgressed a boundary marked in stone and was hit by a beam so powerful he cried out to his grandfather for mercy, 'Koro, please help me!' and was released.[24] In such ways Tūtangatakino also seems to align with the power of a rāhui, or prohibition, that by definition was territorial. (Rāhui in relation to water are addressed elsewhere in this book.)

Tūtangatakino was arguably the most dominant of many local taniwha throughout the waters of Aotearoa, many of whom patrolled the Whanganui River. Best-known in the township, still, is Tūtaeporoporo, raised by a tohunga as a young shark in a pool of the Rangitīkei River near Pārewanui. Enlarged by care and incantation, the shark then made its way – being both salt- and freshwater-capable – to the lower reaches of the Whanganui, where it engaged in plunder and carnage until brought to heel by Ngā Rauru chief Aokehu from the western bank, who worked in cooperation with a local chief, Tamaahua. An act of courage and ingenuity by Tamaahua ended the taniwha's disruptive life. The code of this narrative allows a way of speaking of conflict and its resolution between the tribes of the Whanganui and immediate neighbours on either side that, because of their intermarriage, are often simply too painful to recount as battles, be they won or lost.[25]

An ancient tekoteko (carved figure) framing the door of Te Paku-o-te-rangi beside the lower river reaches at Pūtiki marae, Whanganui, also

Taniwha Tūtaeporoporo was raised by a tohunga as a young shark in a pool of the Rangitīkei River, near Pārewanui. *Photo Bruce Foster*

characterises Tūtaeporoporo and Aokehu. The carvings originate from an old meeting house just up-river that once stood at Matahiwi, and incorporates a marakihau from a very old local tradition. This marakihau, too, suggests a unity among the people of the river.[26]

None of this is to suggest that the many taniwha of the river are just Māori devices of storytelling.[27] Their force, together with mākutu (sorcery), remains embedded in the culture, framing perceptions in amazing ways. A member of Titi Tihu's extended family made this point to me, in memorable terms, 35 years ago at a hapū ceremony near Taumarunui. I was only at the beginning of my understanding of te ao Māori and, as I spoke admiringly of that world, the man seated beside me, who was educated and church-going, gestured with a wrist set in plaster: 'I defied the old man [Titi Tihu],' he said. 'The car was a write-off.' In such cases, ancestral waters, ancestral guardian and the responsibilities of those whose lives are touched by them become inseparable.

Water for healing

Over time, Hawaiians and Māori became more distinct – in certain matters of culture but also, as they went about the business of adapting to the land, in practice. While their practices were still Polynesian, the ingenuity and enterprise Hawaiians brought to water supply and use was something else. However, despite their necessarily more managed approach to water on a daily basis, the Hawaiians engaged in rituals that in many ways were little different – if anything, even more elaborate – than those described to me many years ago by elders on the Whanganui River.[28] Once again, Makahiki was a part of this:

> It was during the *Makahiki* season (near the end of *`Ikuwa* [October] or the beginning of

Welehu) that the water purification ceremonies called *Hi`uwai* were performed by the people. At about midnight, the people would take off their clothes and bathe in salt water or fresh water for hours. They would warm themselves by a fire, and dress in their finest *tapa* (clothing) for the feast and games the following day.

The practitioners (*kahuna*) would have a more spiritual *Hi`uwai* in spring water. The location of the spring would be sacred, and the ceremony was done in private, with their apprentices.

I was told by my grandmother that the kahuna would first bathe their *piko* – the three navels – the genitals, the navel, and the crown of the head to open the pathways to the gods, the ancestors, and the future. They would then wash their eyes, nose, and mouth to be able to see more clearly, hear more distinctly, and speak more wisely in the coming year.[29]

For comparison, Te Kere's great-grandson Arthur Anderson (born in 1917) took his teaching from his grandmother, Karanga: 'Whenever one of us children was sick she was always able to come up with the right remedy, no matter what it was, which she obtained from nature, either from the bush or out of the ground. Karanga was a firm believer that when you are ill you sit down and discuss it,' he said. 'You have to clear your conscience with your mother or grandmother, then the medicine will do its work. But if you have something in the mind, that acts as a bar to the effects you would get from the medicine.' If there was a blockage, 'She would take us to the water and she would have a karakia, and one of the things she always did was pick up a stone and sign the cross on the forehead and throw the stone back in the water. We would come away feeling a different person altogether.'[30] The family were Catholic, but at its base theirs was a Māori tradition into which Catholicism had been strongly grafted.

The marakihau was also recognised in the Motunui panels from Waitara, near today's Te Ātiawa Ōawe meeting house, a place that is within the tribal range of the *Aotea* waka rohe. These exquisite epa are intricately carved panels that usually form the back wall of a meeting house – in this case they served as the support for a long-vanished pātaka. Carved between 1750 and 1820, they were clearly highly valued and during the turbulent intertribal wars of that period were hidden for safety in a nearby swamp, where they became lost.

In the 1970s they were discovered by chance in a private collection in Europe and 40 years later, after extensive negotiations and a $4.5m payment by the New Zealand government, were brought back to Aotearoa. They are now honoured within Puke Ariki Museum in Taranaki. There is a sense within the family that the 140-year span between their loss and return represents the beginning and end of colonisation.[31]

Now, as recovery takes place in lives and locations of the poropiti (healer and sage) Te Kere, his whakatauākī still speak to his mokopuna. They resonate as a new mārae goes up at Tawatā, where he conducted much of his teaching and healing until his death in 1901. One whakatauākī given to me by a contemporary leader within his family appertains to his old ahi kā and seems especially apposite:

Nō hea te kakara e hau pā roa nei? Ko te auahi o te ahi kā o Te Kere. (From whence does this sweet fragrance upon the winds come? 'Tis the smoke from the home fires of Te Kere.)

Limahuli valley behind rugged mountain ridges near Ke`e beach on the Nāpali Coast, Kauai.
Photo Braden Jarvis

CHAPTER 8 *Kīpuka of Influence*

LIMAHULI LOOKS BACK TO SEE FORWARD

`Ikuwā is the month which breaks out in storms, the sea roars, the thunder roars, the birds make a din.[1]

The quotation speaks the old climatic verities that, surviving into the late Holocene, set a rhythm for life in Hawai`i. They held true even in 2008 when we first visited Limahuli. Our introduction to the place was at the beginning of the winds of `Ikuwā (October), about a month later than usual. In 2008, the National Tropical Botanical Garden, established in the mid-section of the Limahuli valley, was little known outside of the district, except to specialists.[2]

Eventually I tracked down Kawika Winter, the centre's young director, who welcomed my suggestion that I write about the ahupua`a at Limahuli: 'I've been wondering when someone would write about this.' Like several younger Hawaiians I met, each of them with a 300-year gaze in their eyes, Winter is poised and still, and steadfast in his dedication to restoration of the old ways.

Limahuli, one of the properties owned and managed by the National Tropical Botanical Garden, is on the garden island of Kaua`i, where most of the state's taro is grown. The 400ha site is a special place and one of the finest examples of pre-contact Hawaiian management of their land and water for food.

An ancient ahupua`a under advanced restoration, the Limahuli valley and surrounding mountain range rises in the remote Hanalei coastal region near the far end of the northern road that runs from the airport through Princeville, finishing on the sheer, rugged and barely tracked Nāpali east coast.

No sooner are we on the road from Lihu`e, Kaua`i's airport, than the rainstorm that has been menacing in ways for which the island's leeward side is famed, begins to deliver. This, the start of the winter rains is, we later learn, the first rain in six months. Suddenly the closeness of the atmosphere lifts and there is a slight chill, so pleasant for New Zealanders struggling in the humidity. The intensity of this rain drives everything before it, which is why parts of Kaua`i Island, such as Wai`ale`ale, the mountain, are among the wettest places on earth.

There is a place in southern Fiordland, New Zealand, the Cropp River, which averages over 11,500mm of rainfall a year – one of the five highest annual average rainfalls in the world, and not far behind the leaders, two almost contiguous sites in monsoonal India. But also in the top 10 for rainfall are two sites on Maui and the one on Kaua`i.[3] Much of this water washes back into the sea. However, a great deal is trapped in the capacious complexity of the aquifers of the archipelago, some on land and some in the depths of the ocean. Apart from Hawai`i, many places in the Pacific, including New Zealand, Tonga and Sāmoa, have physical structures that hold vast quantities of freshwater.

When we call at Princeville, a resort of sought-after views, golf courses and other appurtenances of a steel-jawed tourist trap, the shops are filled with holidaymakers seeking refuge as roof guttering struggles under the weight of runoff, and excess stormwater explodes out of overworked pipes. The man in the shoestore wants to close up and get home: 'There's a travel advisory out and they will review the safety of the big bridge just down the road at 3.30pm.'

It is 3.20pm, and we have already come some distance on the 'wrong' side of the road; I elect to head off across the bridge and up the coast to Hā`ena before it gets any worse. The bridge is a modern concrete structure, curved on its horizontal axis, and the Hanalei River runs high against its pillars. The tree-lined coastal road is busy, and often the oncoming vehicles take up more than half of the road; a couple of times there is a sharp intake of breath from my wife Scilla, as our sedan straddles the very edge of the road ditches. Fortunately, few are travelling at speed.

The forest edge continues to draw in until we see, through the rain, a tall man in a limp, soaked raincoat cleaning out a drain at the roadside. I stop the car. He pauses on his spade. 'You've got here,' he reassures us. 'It's right around the corner.' His wet face beams – he loves water, this man. 'I'm Chipper's brother, John.'

The flowers of lei (floral garlands): vibrant Hawaiian frangipani (*Plumeria apocynaceae*).

Photo: David Young

Chipper is a founding mover of the Limahuli project. On the cusp of the Hawaiian renaissance, Chipper and John's grandmother, Juliet Rice Wichman, acquired the original property in 1967, with a dream of protecting and restoring the valley. At about the same time, singer–collector Eddie Kamae was rummaging in Honolulu backstreet book stalls for old Hawaiian sheet music. Juliet, with her son Charles (Chipper) and his wife Hau`oli, took the first steps towards restoration. Almost a decade later Juliet gifted her 6.8ha garden, the section of the valley closest to the ocean, to a nascent organisation chartered by the US Congress as the National Tropical Botanical Garden (NTBG). In 1994 Chipper and Hau`oli gifted the remaining 400ha,[4] seeing this as the option that best fulfilled their grandmother's dream. Today the NTBG abuts the Hā`ena State Park boundary.

As Kawika says, the word 'Hawaiian' was then barely respectable. The Wichmans knew it would take another generation before their bold act of generosity might stand as a model for those Hawaiians seeking to bring back the old ways. It took all of that.

Today Limahuli has become a beacon for a renaissance in indigenous resource management, restoration and regeneration of both the natural and the cultural; of education, outreach and hope. The NTBG is a kīpuka – a word for an often small island of vegetation raised just above the path of a lava flow, with enough elevation for surviving plants to regenerate in Pele's old scars.[5] It is from this growth that groundcover, new forests and wildlife can reclaim their place on the island. Conspicuous amid the new `a`ā lava beds of Hawai`i, the kīpuka bears a literal truth with a metaphorical power that ripples out to many of the islands in the Pacific. It is most certainly a concept readily understood in the regeneration movement of Aotearoa – pre-Covid, indigenous exchanges between Hawai`i and Aotearoa were part of the norm.

Limahuli is run from a house built in 1992 by a millionaire, who sold it to the trust. The house stands near the entry to the complex, above the path of the stream that today canters down the steep hillside to the rear. The stream feeds the heart of the project – the revived archaeological terraces of the lo`i kalo. At the back of the property, topping the skyline ridge, is the startlingly abrupt peak of Makana, a sacred landmark thrusting high above Limahuli's glorious stepped waterfall. Below the falls run the old `auwai or water channels, built to zigzag across the valley face to slow the rapidly descending stream on its path to the sea. In the old days the coastal reef along the 6km-wide bay was penned.

Kawika explains that this is right in the middle of the food basket of the old Hā`ena district. It is one of four such designated gardens with taro and tropical plants – some desirable indigenous and others strictly invasive – on display to the public. Kawika notes that 99.3 percent (63 species) of the forest immediately above the taro planting is non-native species, including seven Hawaiian-imported plants, as well as invasive species such as Java plum and autograph tree (*Clusia rosea*). The higher elevation areas are still mostly native and are one of the last refugia of endemic diversity in the archipelago's most biodiverse ecoregion.

Limahuli Falls. *Photo Aliscia Young*

Long ago Hawaiians had converted the valley floor and associated alluvial plain in the ahupua`a for flooded-field agriculture – including lo`i kalo and `auwai – but most of

Water shaping and defining its host landscape – in the Southern Alps of Aotearoa. *Photos Richard Sidey*

the forest was managed to retain the ecosystem services vital to sustaining a productive cultural landscape.

When the new waves of colonists arrived, valued hardwoods were targeted and ungulates, notably cattle on Big Island, were released to graze the islands. Hawai`i's modern deforestation didn't begin until the abolition of the ancient religion in the nineteenth century took with it the kapu system – the sociocultural mechanism for regulating resource extraction. Trade in sandalwood, which was carried in American vessels to Canton to meet repayment of mounting debts, continued until accessible supplies were run down in the late 1820s: at its peak in 1822, 1600 tonnes was exported. Merchants had flooded the market with various goods on credit for promises of sandalwood, but once the resource was exhausted the chiefs could not repay. By the turn of the century the demand for wood to fuel steam engines – trains and ships – resulted in millions of cubic metres of hardwood timber being removed from the landscape annually for several decades. Eventually the streams started drying up and the territorial American government put stop-gap protections in place. This is a familiar story in colonisation. Chiefly debts were claimed by Americans for decades, sometimes at inflated charges.[6] The old order was under attack from the inevitability of the new, and Hawaiian leadership had become agents of this change.

That was some 200 years ago. Recent surveys indicate that high in the Limahuli valley many varieties of native trees of stature and girth still stand, including the Hawai`i pōhutukawa, the `ōhi`a. Natural regeneration of such plants is now assisted by rat control measures and a predator exclusion fence, to enable endangered seabirds to nest on the giddy heights of its upper reserve.[7]

'In the minds of our ancestors,' Kawika told us, 'water is considered to be the most important resource. One of the reasons is because it connects everything.' The Hawaiians of old passed strict laws for upholding the respect due to the high value of water, 'and death was the penalty for non-compliance'.

A physical focus for the Limahuli project is the restoration of wetland taro fields. Tradition tells of a number of other 'desirable' plants, brought on the voyaging waka, forming what Kawika calls 'the botanical toolkit' – banana, turmeric, sugarcane – and paper mulberry for making tapa.[8] Of the 27 species they imported, the niu or coconut tree also ranked highly: it was 'one of the most important plants on their voyaging canoes,' according to Kawika. To Hawaiians, niuolahiki was a tree with its roots in heaven – a place they called Kahiki (Tahiti), but a near equivalent to the Hawaiki of Māori lore:[9]

E Niuolahiki o
Life-giving coconut
I kupu i Kahiki
That budded in Kahiki
I kumu i Kahiki ilau i Kahiki
That formed a trunk in Kahiki
I mole i Kahiki,
That rooted in Kahiki
I hua i Kahiki
That bore leaves in Kahiki
I o`o i Kahiki e!
That ripened in Kahiki[10]

One of Kawika's teachers was a keeper of ancient priestly knowledge, descended from the remote, long-inhabited Moloka`i. He told Kawika, 'there would come a time when I would walk and speak and no one would know what I said …' He often talked about 1000-year events. Kawika's teacher himself had been taught by a 93-year-old Hawaiian woman who was 189cm tall: 'When she chanted to wind and rain, her voice emulated those sounds. She danced animal dances, she passed on the knowledge entirely to him – the last one.'

'Ahupua`a is my interest and hula was my guide,' Kawika continued. 'My teacher taught me an amazing amount. Leaving the hālau [the longhouse for hula instruction] to move to Kaua`i and take this job was so hard. One year later I went back – we were trying to restore that ahupua`a; there is no way that I could do this without those teachings. Yet there was still something missing … I knew there was one last teaching.'

'All you have to do,' Kawika's teacher told him, 'is listen to the land – you have to know how to listen.' 'That's what I am trying to do now – listen to the land.'

It is the call from the indigenous to all of us. Māori often speak of this.[11] Mastery of such listening is the work of a lifetime, conducted in slow time, with observation and contemplation and understanding.

The Limahuli Valley and botanical gardens.

The highest terrace at Limahuli is up near the 100-metre waterfall. These cultivations were gradually abandoned as the processes of disease and appropriation – 'gentleman genocide' – took hold, Kawika explained. 'The ahupua`a isn't large, but it must have had at least 200 families with around 2000–4000 people living here. Now we are trying to restore things, but weeds and other invasive species are a problem – we are still always working at it. There are dunes down by the beach where our ancestors worked to improve on nature by engineering inland fishponds. They developed the terraces, with irrigation water feeding into the ponds, providing nutrition in the form of alga for the herbivorous fish feed. If left to itself, the nitrified water might cause a harmful algal bloom in the ocean. This way of managing natural resources is so much more abundant than it would have been if people were not involved.'

All five native freshwater fish species are found in Limahuli's streams.[12] 'The aquifer is recharged from management of the stream, because springs do dry up – which is exacerbated by changes in forest ecology,' Kawika says. 'There is a lesson here for the global community: humans can enhance

Above: Up near the stream source at Limahuli.

Right and far right: Taro in shallow ponds of water diverted from the freshwater stream. The nutrient-rich water is then returned to nourish crops or fishponds further down the slope towards the sea.

nature, and not simply deplete resources.'

The presence of fish and waterbirds adds to kalo growth – but the animals could result in nutrient pollution in the sea if left unmanaged. Reduction of nutrients to the sea is the other thing. 'So aquaculture adjacent to the taro fields is necessary to complete the equation: if we don't have fishponds to trap the nutrients, ecological problems occur in the ocean.'

High above the gardens the workers even found Polynesian rats. These animals were not honoured in Hawai`i in the way they were in other islands, including Aotearoa. Voyaging Māori prized them, keeping them in wicker-style cages aboard waka when voyaging to new destinations. Here in Hawai`i these rats, as cliff-dwellers, have stayed out of the way of the waves of more recent ship rats and Norwegian rats.

'Once Limahuli was a place of plenty, of fish, birds and produce,' Kawika said. This integrated system tended towards natural balances – but nothing humans do anywhere is without cost to nature. In time the Hawaiian-introduced rats wiped out large-seeded species that were a food source for large, flightless birds; they predated on untold populations of naïve birds, especially ground-nesting seabirds. This happened in Aotearoa, too, and in many places in between. Both the `a`o or Hawaiian shearwater (*Puffinus auricularis newelli*) and the dark-rumped petrel (*Pterodroma sandwichensis*) still nest in burrows here.

Natural forests and the ecosystems and species within them took a serious hit at Limahuli, as they did wherever people turned up across the Pacific. Of the estimated 274 native vascular plant types at Limahuli, 90 are

endemic to Kaua`i, and almost half of those are designated threatened or close to that threshold.[13] But how were Hawaiians, who were so devoted to their land, reduced to such tiny remnants of their cultural and physical life?

Earlier chapters have sketched the organic evolution from kingship and the relationship of `āina groups to place, including to lo`i. The date of early settlement of Hawai`i, including Limahuli, has often been given as 300–600 AD – by Polynesians who sailed from the Marquesas, as opposed to a more aggressive second wave from Tahiti. The original ahupua`a are held to have been constructed by the first wave of settlers.[14] However, Kirch has found that archaeological evidence points to virtually no human presence pre-dating 1000 AD.[15] Māori make similar claims about the existence of earlier inhabitants, but again, sound physical evidence is yet to be found.

Below is a brief sketch of legal changes that impacted with such swiftness on the Hawaiian people, once others began to arrive. The precolonial period is followed by four others (see box).

Under the 1948 Māhele with its constitutional monarchy, the old ahupua`a land divisions were overturned by a board of land commissioners into a system amenable to land sales. Jonathan Osorio was in no doubt as to its effect: 'Historians and economists have concluded that the Mahele, whether a huge fiasco or a devious theft, disinherited the vast majority of the kanaka.'[16] Carlos Andrade saw it as a way 'to address the growing problem of foreign desires for Native land'; the board was 'to investigate, and then to approve or reject, all

FROM THE CUSTOMARY TO CONSTITUTIONAL AND LEGAL – WATER MANAGEMENT

1840	The constitution, formally sought to protect all land under the chiefs and people, with the king as trustee.
1848	The Great Māhele. Despite its Hawaiian name, this haole-inspired process proposed by King Kamehameha III ends traditional ahupua`a land division by codifying land ownership, enabling Euro-American law to prevail.
1897	Annexation by the United States.
1959	Hawai`i officially becomes a state of the US.
1961	The legislature moves to protect groundwater 'for its best utilisation, conservation and protection' and 'to prevent threat of pollution, exhaustion, waste and depletion'.
1978	Article 13, s7 of Constitutional Convention confirms the state as protector of Hawai`i's natural beauty and all natural resources, 'including land, water, air, minerals and energy sources'; enactment of the State Water Code, and the Commission on Water with jurisdiction over water disputes and allocation.
1978	Actual groundwater protections, then mainly on O`ahu, begin to be designated, guided by principles of common law, in which the doctrine of public trust is invoked.
1973–89	Protracted litigation over water rights; further cases continue.
1983	Reppun case: following intense litigation, legal precedent is set; the Board of Water Supply directed by Supreme Court has to reduce its groundwater pumping; ruling that, by reducing flows in Waihe`e Stream, it had damaged taro crops.
1987	State Water Code enacted, setting up the Water Commission. Still insufficient protection of constitutional rights, and issues of ownership, but gradual restoration of some indigenous rights to water.[17]

claims by private individuals, whether Natives or foreigners, to land acquired prior to the creation of the commission. Since the interests of the *mō`ī* [royalty], *ali`i*, and *maka`āinana* in the lands of Hawai`i were so indivisibly entwined, the commission could not at first act on the bulk of the claims …'[18]

Although an equitable division of land was made to the Hawaiian people, soon, under pressures similar to those experienced by Māori in Aotearoa, they sold the bulk of their lands. The old ways succumbed rapidly to the new.[19] For most of the country, as in New Zealand, 'legal process' separated Hawaiians from their traditional lands and from centuries of accumulated knowledge and custom. Hefty surveying and legal fees often stripped out most of the proceeds from land sales.[20] For natives, 'the rule of law' was anything but. To cope with demand, the commission issued fee simple ownership – despite its not yet being part of the Hawaiian legal system. 'With this development,' Andrade wrote, 'the commission blatantly ignored the idea of the *`āina* [land] as the elder sibling to the Hawaiian people conveyed in the *Kumulipo* and the land began its transformation

Waves pummel the lava rocks at Glass Beach, Hanapepe Bay, Kaua`i. *Photo Paulius Dragunas*

into real estate. `*Āina* that had fed the Hawaiian people's bodies, minds and spirits, `*āina* that had been family to the Hawaiian people, had begun to be privatized.'[21]

However, at Limahuli a small group of Hawaiians managed to hang on. Andrade explained: 'Only 14 people here have got land, some obviously had to move, but 39 individuals did buy back that land. For the next 100 years [from the 1850s] ahupua`a was traditionally farmed here, till the 1960s. Here at Hā`ena was the only place where it was farmed for so long continuously by our people. We bought back the land – the process allowed for land claims.'[22] The buy-back enabled this kīpuka to continue to achieve so much against such odds.

In *Hā`ena*, Andrade detailed the traditional ways and their swift breakdown, followed by gradual clawbacks. After Hawai`i became a US territory in 1959 the Wichman family sought protection of traditional sites and water in Hā`ena. As the district was beginning to grow into a holiday and resort area, the county sought to retain its interests in water for domestic supply and its right to make further subdivisions. Eventually a hui (a native group in which the county's interest was included) was formed to become part of the process. This was recognised by the courts as a 'peculiar or special Hawaiian institution': it achieved many of its aims, including upholding its legal interests in water. As a quid pro quo, however, the hui continued to release considerable volumes of water to the county. Even after all they had been through, the Hawaiians still understood the power of 'gift' over 'gain'.

Since most Hawaiian rivers are unnavigable, the old 'ad medium filum aquae' rule, which grants water rights to the landholder bordering either side of a stream out to a fictive midstream line, is not applicable.[23] More importantly, in Hawai`i, 'stream conditions were set that all such lands be subject to the free flowage of water in all streams, drains, swales and ancient `auwai should enjoy water rights established by ancient usage'. Based on historical land use, determined Hawaiian groups were entitled to access such waterways and their `auwai – evident from their maintenance and use. Later, even an absence of traditional use was found not to be a reason for the loss of such rights and, further, the traditional right 'can trump other uses [and demands] if water is in short supply'.[24]

By 2000 the community of Hā`ena had become concerned at the impact of growing numbers of visitors – especially from tourism – on the fish, birds and other wildlife of Hā`ena. In the first of its kind in Hawai`i, an agreement for a community-based subsistence fishing area inside a wider coastal management plan was adopted. The community group, named for the peak above Limahuli, Hui Maka`āinana o Makana, called for regulatory approach. They invoked old konohiki (overseers/managers) and kapu (laws that were exercised like a rāhui) that were recognised and secured between the Hā`ena State Park authority and the community.

Regular meetings are now held within the wider community and with government agencies. This has led to a culturally rooted programme to uphold the 'consumptive needs' of the Hā`ena community while it 'responds to the connection between land and sea and strives to restore the necessary balance of native species'.[25] In such ways, in the incremental manner of Polynesian engagement, the full extent of the old ahupua`a structure, mountains to sea, with all its attendant biodiversity, is being restored as fully as constraints of modern tourism allow. Mindful of ever-increasing human effects on natural life along its once productive coast, Kawika says, 'We became one of the first fully embodied community-based subsistence fishing areas within the state.'

After an absence of 10 years I attended a deeply informed community meeting at Hā`ena. Over those years several more of the community's number had become highly educated in law and science before returning back to their `āina. Among the concerns under discussion were the effects on coastal reef fish of pollution from sunscreen, and of disturbance by too many snorkellers. Since then, the State of Hawai`i has banned the use of up to 80 percent of sunscreens from 2021.[26] Growing tourism numbers on small islands causes myriad such difficulties, and the cost of tourism on environments worldwide is never fully counted.

Even more fundamentally, in a recent controlled study Kawika and others have demonstrated that community-led closures of coastal marine areas 'can result in a significant increase in fish bio-mass' in the face of climate change. The shallow, sheltered backwaters of Hā`ena can still provide a nursery for fish behind high coral cover. But this involves ridge-to-reef management that engages 'spatially explicit' land–sea modelling, which shows that the upgrading of cesspools 'can reduce human impacts on coral reefs in the face of projected climate change impacts'.[27]

Who holds rights to water in Hawai`i?

The fundamentals of modern water rights in Hawai`i were laid out in a December 1997 Hawai`i Supreme Court decision, *McBryde v Robinson*, based on a 100-year-old US Supreme Court-determined principle. Critically, this established 'the public trust doctrine'. Adopted originally in 1899 by the Hawaiian courts, the doctrine asserts that lands held for the people in trust by the state must retain their inalienable status. Navigation, commercial and fishing rights were also upheld by the law to be 'freed from the interference of private parties'. This judgment famously held: 'The State can no more abdicate its trust in property in which the whole people are interested … than it can abdicate its police powers in the administration of government and its administration of the peace.'[28]

In 1973 the Hawaiian Supreme Court ruled that land held under the Great Māhele expressly reserved 'its sovereign prerogative' and that, therefore, any associated water rights awarded were not in any way granting ownership of those rights: 'we believe that by this [sovereign] reservation, a public trust was imposed on all the waters of the kingdom'. This trust included protection of the flow and purity of 'our waters for future generations and to assure that the waters of our land are put to reasonable and beneficial uses'.[29] But rather than automatic justice, these constraints have provoked prolonged battles over water. The Hawaiian Supreme Court, later severely caned by federal courts which, perhaps unsurprisingly, found the decisions lacking in rigour and in defiance of US property law, revisited its decisions twice on this matter. However, effectively those rulings still stand.[30]

Since 1930, the public trust doctrine has been held in common law, designating the state as a trustee over water for the public good.[31] Traditional and customary Hawaiian rights have been given fuller formal recognition in the Constitution Convention in 1978. 'The State … shall conserve and protect Hawaii's natural beauty and all natural resources, including land, water, air … minerals and energy sources [promoting them] in a manner consistent with their conservation and in furtherance of the self-sufficiency of the State.' In effect, water is 'a natural resource held in trust by the State for the benefit of the people'. The McBryde decision also addressed a common law assertion that had been made to the contrary in 1973.[32]

Lawrence Miike saw the amendment as elevating 'the common-law public trust doctrine to a constitutional mandate for the State'.[33] The public trust doctrine can be traced back through British to early Roman law, under Emperor Justinian. This determined that running water, seacoast and air were all public and not for appropriation.[34] The public trust doctrine also has its nature-supporting advocates in Oregon, the US state closest in many ways to Hawai`i. As Jim Lichatowich, a fisher-scientist living near Portland, Oregon, on the Columbia River, wrote: it 'defines natural resources as the common property of all citizens that is held in trust for future generations'.[35]

Would that modern law everywhere had remained this way. Apart from anything else, it surely would have made international responses to climate change an infinitely simpler task.

Miike found that the public trust doctrine 'forms the outer boundaries of permissible government action and therefore the State Water Code does not supplant the protections of the old public trust doctrine'. The legality of such principles has by no means been exhausted in the courts.[36] Miike has summarised Hawai`i's (largely surface) water rights as comprising – until 1973 – three types: riparian, appurtenant and correlative. In New Zealand, of these three, only the English law-derived riparian rights are familiar. These grant landowners whose property is adjacent to water, rights to that water – quaintly, as long as the river is not navigable.

In Hawai`i, riparian rights also grant a landowner access to legal remedy if loss of water (at all flow levels) causes suffering; this has since been amended to the plaintiff having to prove loss. Appurtenant rights relate to taro growing; to that extent they are similar, though not the same, as customary rights for Māori in New Zealand – long-held rights by association and customary use. Correlative rights is a term for water held below ground. There is now, controversially, no special right or protection in New Zealand for artesian or similar types of water.

In Aotearoa, iwi Māori have been moving to gradually strengthen their customary and treaty rights under law. Like almost all New Zealanders, they are dismayed at the rapidly increasing levels of water pollution in many parts of the country, and by the alarming rate of depletion of freshwater levels, notably in Canterbury. And what's worse, between 2008 and 2017 vast quantities of what is, in effect, free water was allocated – invariably to overseas corporations – for bottling in plastic. The granting of these permits was so laissez-faire that it contributed to the National Government's defeat in the 2017 election.[37] It has never been made clear what benefits accrue locally, or nationally, from these often secret agreements that respect neither Māori traditions nor wider local interests. To many New Zealanders, this is the extractive, linear economy as neo-liberalism at its most heedless and reckless.

With Hawai`i's water held in trust, embedded in the constitution as a protection for the rights of people and nature, rights at common law are on a gradual path to being fully harmonised with the constitution. In a highly complicated case – the Waiahole ditch water controversy in O`ahu, which continued through the 1990s – these correlative or underground rights were ruled to be 'contrary to the rule of reasonableness, the basic understanding of usufructuary [responsible user] rights, and the public trust'.[38]

The Alexander and Baldwin case on Maui has been even more protracted, but the case has established that Miike's positive conclusions concerning the public trust doctrine were indeed well founded. The company – Alexander and Baldwin subsidiary East Maui Irrigation Co. – was drawing an average of 700 million litres a day, arguably the biggest uptake of any operation on the archipelago.[39] The sons of missionaries, Alexander and Baldwin built their extensive sugar operations near Kahului in the 1860s. In doing so, they transformed the culture of a people and 27 streams on the island. From the windward east, the wet side of Mt Haleakalā, they built ditches and syphons, commandeering ancient `auwai to their purpose. The water was piped to the flat, dry plains where some 1800ha of sugarcane was irrigated, grown and processed.

Hawaiians seeking reinstatement of their `auwai – their old waterways – and the life that used to grow in them, went to the courts for a fight that took two decades. In 2018 the commission ruled decisively in favour of not only traditional (appurtenant) rights, but also of in-stream values – in particular their natural flows, a range of native fish, the `o`opu, and habitat needs. Finally, water was returned to small farmers, traditional users of water and

Baldwin's sugar factory on Maui, now closed. *Photo supplied*

in-stream life, including `o`opu. Additionally, the decision secured recreational water use in nine locations.[40]

After 130 years, sugar production moved from Maui for economic reasons. Alexander and Baldwin's hulking factory on the plains between the two volcanoes (Kahalewai, 'holder of water' and Haleakalā, 'house of the sun') finally closed its doors in December 2016. Na Moku Aupuni o Ko`olau Hui, a group of Native Hawaiian kalo farmers, fishers and hunters, had litigated for two decades to recover their water rights from a company that, though it has now diversified into agriculture and real estate, remained unwilling to relinquish all legal rights from its vast Victorian water allocation.[41]

The now corporatised Alexander and Baldwin argued it still required 400 million litres a day for its diversification into cattle ranching, food, energy crops and forestry. In June 2018 the Commission on Water Resource Management reached a compromise. It restored water in 17 of 25 contested streams in East Maui while enabling 'reasonable irrigation needs' for the company's 9300 hectares of agricultural lands, its former sugar country. Minimum streamflow standards now applied to all 27 waterways. Camille Kalama, attorney for Native Hawaiian Legal Corporation, said that in doing so, significantly, the courts recognised 'all the public trust purposes including growing kalo, supporting native stream life for gathering purposes, recreation … before allowing any of that water to be diverted for commercial purposes'.[42]

'The return of those natural flowing streams after having been diverted for over 100 years is an unprecedented ruling,' said lawyer Summer Sylva, on behalf of Na Moku Aupuni o Ko`olau Hui. As the largest restoration of water in Hawai`i in 150 years, it was, she said, a 'landmark decision'. Achieving 90 percent habitat recovery for seven streams was 'a huge triumph; and ensuring mauka-to-makai [mountains-to-sea] flows for all previously diverted streams reflects a valuing of the East Maui watershed for the life and traditions it supports – taro farming, fishing, hunting, gathering, recreational use, stewardship and its awe-inspiring beauty'.[43]

The case's significance lies in the commission seeking to give traditional and public stream users as much recognition as those abstracting water. The commission outlined its 'primary trust responsibilities' as 'maintenance of water in their natural states; domestic water uses of the general public, particularly drinking water; Native Hawaiian traditional and cultural gathering; and reservations for water by the Department of Hawaiian Home Lands'.[44]

Restoration of a further seven streams to their natural state, and of good practice (best defined as sustainable use), are high priorities. The commission has pledged to continue monitoring 'in order to identify the need for adaptations to changing circumstances'. Part of the Alexander and Baldwin allocation is to ensure the wind-blown plains are sufficiently cultivated to protect the soils. And, as is so often the case, native Hawaiians were at pains to assure locals that the county water would not be compromised by any court decision in their favour.[45] This is aloha as a way of life, as a principle of law.

The Hawaiians likened their position to that of the Native Americans protesting the construction of oil pipelines across their ancestral lands at the prophetically named Standing Rock.[46] In fact, such parallels now reach across an increasingly legally savvy indigenous Pacific. For Māori, what has long posed an infinite legal maze is beginning to show some daylight. Once part of a cultural monolith, the New Zealand Supreme Court has recognised legal pluralism (other ways of doing law) as distinct from previous established rulings on property rights.[47] As Tamatoa Bambridge points out, in some jurisdictions, aspects of pluralism are taking place effectively within a structure of collective property ownership. These identify family traditional ties as having the ability to uphold the wisdom and applicability of rāhui to help regulate contemporary unreasonable demands on resources. Moreover, this is occurring in the Pacific without invoking the role of the state.[48]

Slowly shifting goalposts in Aotearoa

The overarching guarantees of the original Treaty of Waitangi recognised Māori interests in ways not entirely dissimilar to Hawai`i's formal recognition of water – but only in the broadest terms and, of course, with none of the `auwai provisions of Hawai`i.[49] Many of the water rights were lost by, of all things, the passing of the Coal Mines Act of 1908. This essentially granted the Crown all rights to the beds of rivers that were deemed to be navigable waterways.[50] It was the enactment of treaty rights from 1975 forward and notably, in 1985, the government's legal recognition of historical treaty grievances (dating back to 1840) that have enabled these to be tested in a number of ways. Primarily, the Waitangi Tribunal, with its power to recommend, has been a major catalyst in all aspects of a gradual recognition, over 30 years, that has brought changes that are both legal and profoundly cultural. Judicial activism, sometimes creative treaty settlements and even, in some places, local government engagement on water issues have added to this. Now we are seeing occasionally truly innovative legislation, such as the Te Awa Tupua (Whanganui River Claims Settlement) Act and the Te Urewera Act, which makes Te Urewera a legal entity (not Crown-owned and not a national park), managed by Te Urewera Board with equal membership of Tūhoe and Crown-appointed persons.

The establishment in law of the Whanganui River as a legal entity in its own right has enabled the formal recognition of tribal mana: their connection is bound in to the mauri of the river in all its manifold expressions and potentially, with hard physical restoration backed by financial redress, the mana of the Whanganui from source to sea. The Te Awa Tupua (Whanganui River Claims Settlement) Act 2017 states: 'Te Awa Tupua is an indivisible and living whole, comprising the Whanganui River from the mountains to the sea, incorporating its tributaries and its physical and metaphysical elements.'[51]

The movement to accord rights to nature and uphold them in law is an important movement worldwide.

This concept of the river as having legal personhood is the focus of ongoing international attention. 'One of the more interesting claims about the legal personhood for nature,' writes New Zealand lawyer Catherine Iorns Magallanes, 'is that it is also an example of the protection of – and a victory for – rights of nature … The movement to accord rights to nature and uphold them in law is an important movement worldwide.'[52] This is understood to be the first time a river has been so designated in the world – although a similar idea granting nature rights has been law in Ecuador for some years. The Indian High Court in Uttarakhand granted the Ganges and Yamuna rivers the status of legal personhood, and a court in Colombia in 2016–17 recognised the rights of the Atrato river basin to 'protection, conservation, maintenance and restoration.'[53] As Robert Macfarlane indicates, these matters can be complicated: 'We risk only having respect for things insofar as they resemble human characteristics.'[54] In a later paper, Iorns Magallanes writes: 'Legal personality may well become part of a rights of nature toolbox, but it is not synonymous with rights of nature.' For the four rivers similarly and recently protected across the world, personhood seems to be about responsibility, somewhat distinct from actual rights. What may matter finally, she writes, is collective responsibility. Where violations to that specific environment occur, the custodians have redress in law.[55] Most importantly, the recognition of the relationship between the river and Whanganui iwi will, in time, serve to enhance the recognition of rights to nature. The authors in *Island Rivers* offer an elegant appraisal: 'This approach is a type of modern co-management, through which rights and health of the river are upheld through shared decision-making involving local Māori and other resident communities.'[56]

However, the Crown's insistence – at least until now – that, generally, 'nobody owns the water', is mere pragmatic assertion. Historically this has served to implicitly entrench private water rights through local government permitting of usually long-term, renewable consents to individual farmers, businesses and their own needs. (And local government itself has been heavily complicit in degrading freshwater for its own interests, such as for sewage treatment.) Like ocean fishing quota, these tacit right-to-renew 'ownership' arrangements lack the necessary teeth to incentivise responsible management of a resource such as water. Since the early 1990s a succession of regulation-shy governments has created a number of water issues that are of great complexity and scale. These may now be solved only by addressing the effects of rapid climate change, water flows and temperature and health and biodiversity loss as being parts of an indivisible whole.

Gradual recognition of Māori customary rights in recent decades has been stimulated by a number of Waitangi Tribunal recommendations and several Supreme Court decisions. The Whanganui River settlement legislation, for example, requires authorities to 'act in a manner that is consistent with and gives effect to the principles of the Treaty of Waitangi'.[57]

The 2018 East Coast (North Island) Iwi Leaders Group advanced their 'bottom lines' for reforms to the Resource Management Act. These include the fundamental reminders of respect for water's life-giving qualities, respect for Māori relational values to water, and recognition of the need for greater engagement between local government and iwi and hapū. No longer is the granting of the term 'kaitiakitanga' (guardianship) sufficient – they sought empowerment through recognition of such tools as 'cultural flows' that connect to indigenous notions of sustainability and equitable access to water.[58] Their interests did not differ greatly from those of many other Māori – and some environmental – groups.[59] A supporting view was from Landcare Research, a Crown entity that interfaces with land and water management. Seeking greater clarity in resource management and local government law, it envisages local Māori participation holding equal status with other stakeholders on a collaborative pathway.

In 2019 a government-appointed advisory group of Māori authorities on water produced an uncompromising report, the essence of which was that the Crown has consistently failed to protect freshwater; now iwi must share in such decision-making. In order to effect this, 'a hierarchy of obligations', rooted in treaty obligations, needed to be met to ensure the protection of health and mauri of water; provide for human health; and enable consumptive use. Three treaty-based principles were invoked: iwi–Crown joint decision-making, including mana whakahaere (governance, authority, jurisdiction); kaitiakitanga; and stewardship – together with 'manaakitanga and care and respect in providing for the health of our nation'.

Of course, when the West negotiates in good faith with the indigenous, the defenders of the linear economy must also bring something to the table. Aquatic ecologist and writer Laurel Teirney came up with the idea of 'gifts and gains' as a way of negotiation: the essential dynamic of collaboration is that all parties must be prepared to gift as well as to gain. In this case – and only in this century – the 'gift' from governments at all levels has been their relinquishing of at least some of their assumed power over those they have colonised. Often the most promising way forward for fuller recognition of Māori interests in water is political willingness to seriously reform the law – sometimes using Waitangi Tribunal recommendations. While occasional, mildly radical acts of government courage seem to pass with little of the anticipated disruption of established order, many of these remain still unexplored by government.[60]

A potentially highly contested effort to create a marine conservation strategy in Fiordland – with the full support of minister Marian Hobbs – became a process that successfully engaged interest groups. Tangata whenua, fishers and crayfishers, tourism interests, cetacean ecologists, recreational divers, kayakers and boaties all joined the process. The creation of eight new marine reserves in Fiordland (9529ha) and the strategy and appointment of representative guardians were ratified in the Fiordland (Te Moana o Atawhenua) Marine Management Act 2005. The plan, ensuring the continuation of this integrated approach, continues to be successfully implemented and upheld.[61]

Influential Māori legal academic Jacinta Ruru sums up what ought, for many decades now, to have been blindingly obvious: 'If we want the best environmental outcomes and solutions for our country going forward, and we have some big issues confronting us, we need to be looking to Māori communities to help [develop] the solutions that we are going to devise.'[62] Several Hawaiian leaders have told me that Māori progress continues to inspire their own efforts for recovery of their essential rights.

Kawika Winter, then National Tropical Botanical Garden director, with the author at Limahuli waterfall, 2009.

Limahuli: Hard-graft path back to sustainability

The foundation of today's ahupua`a restoration is the recovery of old knowledge, old ways. As an ethnobotanist, Kawika Winter must straddle both the scientific and the cultural worlds. Out in the garden at Limahuli, he points out an `uala. 'There is a kupuna [elder] who has not seen this particular plant for many years – I'll take some `uala to him tomorrow.' He added, 'The old people who started this place were entirely focused on their culture.'

He observes that Hawaiian health is poor, crime is high, and sugarcane and pineapple – two major historical displacements of land use and diet – have not benefited the health or adaptability of native Hawaiians. Famously, Israel Kamakawiwo`ole, who sang his hopes for a free Hawai`i in his famous angelic vibrato, weighed 343kg when he died at the age of 40. Today, 90 percent of Hawai`i's food is imported, mostly from the United States, 3850km away; the islands usually hold just seven days' supply of fresh produce.

The scope of the ahupua`a as a definition of a unit of land didn't evolve in the Hawai`i islands until the late fifteenth century. Before then, a sufficiency of fish and food from gardens 'within the coastal zone' was taken with no formal definitions of boundaries. There is a suggestion that if it wasn't just a growing population, then an El Niño event may well have triggered the conditions that provoked this development.[63] Certainly, research of glaciers in New Zealand, carried out by the National Institute of Water and Atmospheric Research (NIWA), reveals a 'variable climate' in Aotearoa between 850 and 1850 AD, with cooler spells occurring within every century from the early eleventh until the mid-nineteenth, apart from in the fourteenth century. Cooler summers were experienced in the 1760s, around 1790, and for 20 years between the early 1840s and 1860s. But while these trends over the past millennium appear to align with northern hemisphere events such as the Medieval Warm Period and the Little Ice Age, 'more work is needed to define these climatic intervals more clearly in the New Zealand region'.[64]

One afternoon when I was being instructed in the work of cultivation of taro at Limahuli, an American visitor of similar age to myself approached, wanting to know about an upcoming talk there on archaeology. He turned out to be a property developer from Portland, Oregon. His father in 1970 had bought a place of hundreds of hectares of flats, enclosed on three sides by mountains, with streams, huge scattered boulders and vertical forests, on the boundary of Limahuli. No one else attended the auction, so he purchased this rather formidable landscape for a song.

The son grew to love this land, and later in life, on early retirement, he began a dedicated effort to restore and protect it. After our brief exchange he invited me to accompany him onto his property: he had promised his wife that he would not venture into this broken country littered with mountain rubble and slippery rocks without a companion. The going was difficult because of recently filled streams and wet rocks slick with mud, moss and debris. As we walked he told me about some of the wealthy Hollywood types who had purchased along this coast, and whose interest had driven up property values. A number of Hawaiians were living in shacks on stilts in the same district: some of them were methamphetamine users living deep in the underbelly of Hawaiian culture, and suicide was not uncommon. As we approached his property through the trees we were greeted by several of them, old Hawaiians; they were poor but were most welcoming.

This was not a place he would ever sell, but even in its entirely undeveloped state, his land taxes in 2008 were about $40,000 a year. Needing some income from the property to offset these, he planned to build a small dwelling on the site of an old hale (house), located at the confluence of two tumbling streams. He was also growing orchids, randomly, in clefts in trees in the forest. We climbed up a steep, muddy track to a vantage spot with a vertiginous view across the valley to a stately waterfall, high above the other side of the valley where, he told me, there was a taro plantation from antiquity still growing. It was about 150m above sea level and it seemed inaccessible without climbing gear.

The American is an ardent Democrat. But he was still hesitant when it came to befriending his neighbours at Limahuli. He had been advised by a family who own large swathes of Kaua`i that he should be cautious in allowing anyone onto his land, especially those from Limahuli.

'Kawika is an ethnobotanist,' I venture. 'I know he would be interested in talking with you.'

'But do you think they'd then expect to be able to come onto my land as of right?'

'They're your neighbours, they'd be respectful. I am certain of that.'

He was persuaded, and I became the catalyst for discussion. Since then he and Kawika Winter have formed a partnership, working together on matters such as pest control along their mutual boundary, from the mountains to sea.

On several occasions, he told me, he had heard far-off drumming on his property, 'back in here. One time I went with in a brand new backpack, high quality it was, and I came across the cornerstone of an old hale. I was determined to bring it out. I loaded it into my pack and started walking. But suddenly, both straps sheared off, as if someone had sliced them with a very sharp knife.'

'Don't you think you should return this?' one of his haole friends asked him.

'No, I'm determined to have it.'

So he carried the boulder out, but that night he was visited by the most fearsome dream. He returned the stone to its original location.

The man, who now spends more time at Kaua`i than in Oregon, lives in hope that he will see the dreams of Hawaiian nationalism realised.

Old Hawaiian wisdom is steeped in allusions to weather, cloud patterns, winds and rain, in subtle representations and poetic depictions, as often as not with one eye on the possibility of no rain, no crops and little food. This means that traditional producers of food were often looking northwest: to Kalua, a speck in the ocean, not far west of Kaua`i; to Ni`ihau and Nihoa, a wildlife reserve since 1906, and the French-named Necker, a tall rock in the lonely ocean, 240km west of Nihoa.[65] Nihoa, like Rapanui Easter Island, has only limited seeps of freshwater, and has lost most of its palm forests; old petroglyph face imagery is found on rocks there (and on several other Hawaiian islands, including Big Island).[66] Nihoa was inhabited by early Hawaiians in 1000–1500 AD.[67] Its highly prized red feathers (from the red-tailed tropicbird), loulu palm wood and makiukiu grasses played a part in the economies of both Ni`ihau and Kaua`i. But the idea that Nihoa's western location was in the direction of rain when the Kona winds failed to deliver may have been an even more significant reason for visiting.[68]

As a centre of hope, Limahuli has gradually grown in inspiration on restorative policies. 'We're just working away,' muses Kawika, 'making a difference, at the end of the road on the western-most island, far from politics, hoping that at some point people will pay attention to our work.'

Today, with so many environmental risks and pressures presenting, it is often the local indigenous people who see the signs first. While Honolulu now demands replanting in 90 percent native trees with all new building consents, the botanist in Kawika frets: 'With co-evolution there can be an explosion in biodiversity, but if we're not careful, we can get [genetic] bottlenecks, and soon we're just planting clones.'

The climate has changed dramatically here. The road out to Limahuli, considerably busier than it was in 2008, was washed out and buried in several places by what Kawika called 'the first storm associated with climate change to hit Hawai`i', in April 2018, when 1400mm of rain pelted down in 24 hours. That winding coastal road to Hā`ena 'has been closed for repairs since and won't open fully until next year. They let residents in and out a few times a day through the one-lane sections, but no tourists. Limahuli did relatively okay, although there was significant damage everywhere.'[69]

While still listening to the land and acting strategically and consultatively, what the people of Kaua`i are contending with from the micro to the macro level in their environment is hardly of their making. Colonialism, as it pertains to almost all Pacific Islanders, put them on the receiving end of others' ambitions and weakened their resilience. This, in turn, has made them even more vulnerable to climate change.[70]

If we are to learn something from the wisdom of indigenous culture, as reflected in, for example, its modest kīpuka movement – 'restoration from small, tenacious beginnings' – it is this: as the most successful beneficiaries of heedless economic growth find their own home countries less and less socially and environmentally conducive to a good lifestyle, it is often to Hawai`i or Aotearoa – or similar, seemingly less sullied and more politically stable places – that they retreat. An additional attraction of such places is often what remains of the culture and landscapes of their first peoples.

Photo Aliscia Young

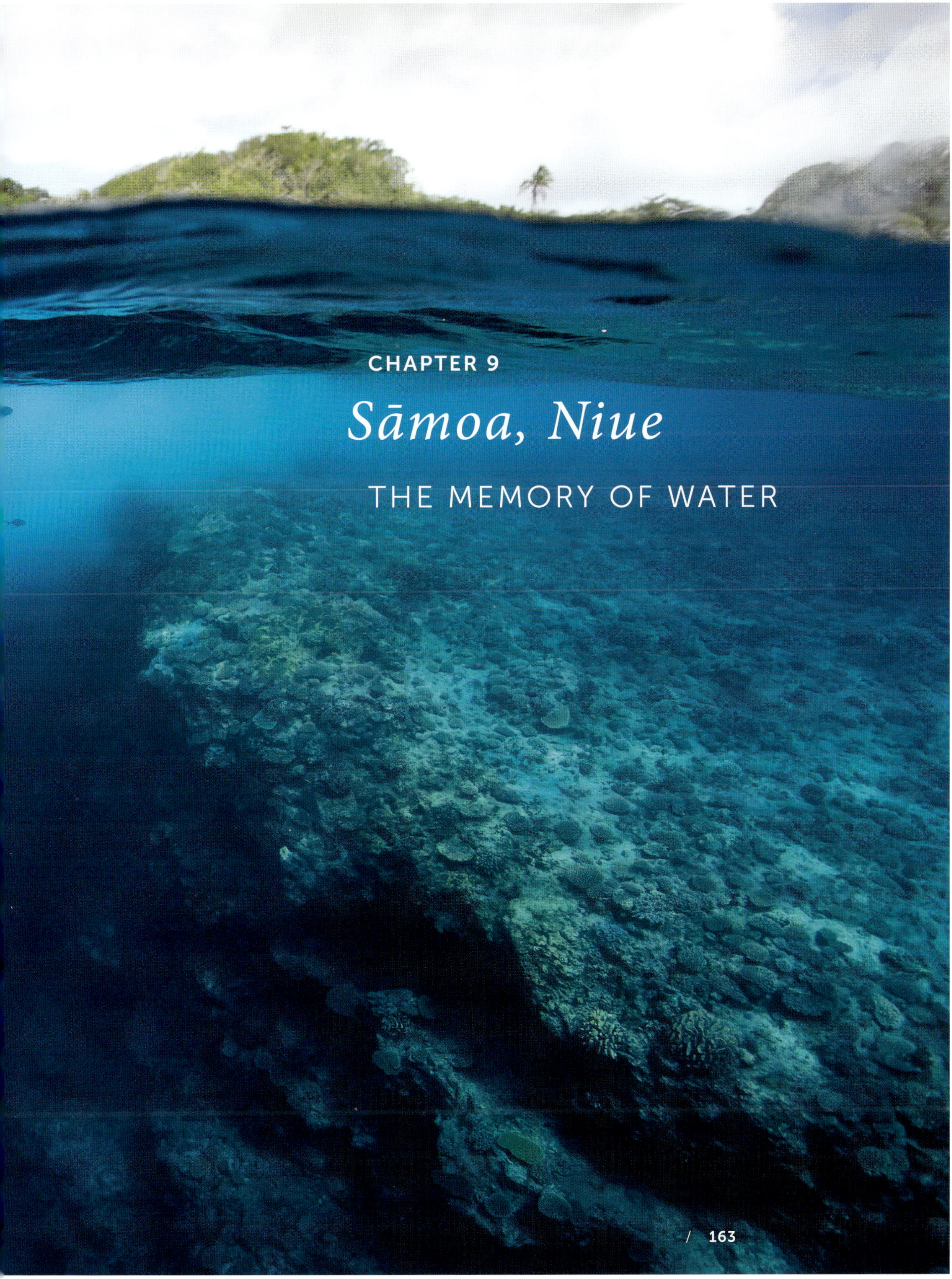

CHAPTER 9

Sāmoa, Niue

THE MEMORY OF WATER

Samoan High Chief Tuimaleali`ifano Si`u and Robert Louis Stevenson at Vailima, 1894.

Wikimedia.org [Commons]

The Mormon women at the reception at R.L. Stevenson's villa in Vailima are discreet, but if you get to chat to them, in hushed Samoan voices they will tell you that there are quiet times when they hear muffled voices. 'Upstairs, by the fireplace. That's where he was when he died, that's where the voices come from.' And Richard Cook, who spent most of his life in Sāmoa where his father held the post of New Zealand high commissioner, recalled hearing children's voices in the grounds behind the original house. The building at Vailima was largely rebuilt by Utah businessmen following the devastation of Cyclone Val in 1991. It now serves as a museum dedicated to the memory of Stevenson, or Tusitala, 'the storyteller', as he was known to the Samoan people. His tomb up on Vaea bears his 'Requiem', which concludes with the well-known lines:

> This be the verse you grave for me:
> Here he lies where he longed to be,
> Home is the sailor, home from sea,
> And the hunter home from the hill.

Stevenson suffered from chronic tuberculosis. After building his 'home for angels'[1] he lived four productive years here with his wife Fanny and some of her family on its 160ha of lawns and gardens claimed from the surrounding forest. He died there from a cerebral haemorrhage, aged 44. One of his works from that time, *A Footnote to History*, draws on the history of the factions and infighting within Sāmoa at the same time as three colonial powers – Germany, the United

States and Britain – were fighting for control over the archipelago.[2]

Writing to his friend and editor Sidney Colvin, Stevenson says, 'Our place is in a deep cleft of Vaea Mountain, some six hundred feet above the sea, embowered in forest, which is our strangling enemy.'[3] Vaea is the remnant stump of the island of Upolu's progression from tectonically active sea floor to old volcanic island.

The name 'Vailima', often said to mean 'five streams', was Stevenson's coinage, not necessarily a local name. But whether he knew it or not, it seems to have had a legendary origin, which we will come to. Sixteen years after Stevenson's death his friend, expat trader H.J. Moors, wrote, somewhat misleadingly: 'It means "five waters" but with the lapse of time the configuration of the country has altered, and you will only find two streams running there.'[4]

Stevenson loved everything about Vailima. Writing soon after he settled on the land, he describes the track he cut as far as a new garden. These observations serve as a useful record of what the forest and its streams consisted of then. In 1890 he again refers to the stream close to the house, 'located in the midst of great silent forests. There is a burn close by, and when we are not talking you can hear the burn.'[5] He noted small creatures that reminded him of hummingbirds 'skirmishing in the wayside flowers'.[6] The track, with the help of workers, continued further on into the forest where, as his map shows, 'the streams conjoin in one'. It is clear from this that there was a main stream that branches into two. Further upstream is a short third branch that is not on his sketch, but is marked on a 1959 Geological Survey map.[7] Stevenson followed the stream to the right behind his house, up to the site of an abrupt gorge fed by an 18m cataract. It was here, an old eyewitness had informed him, that bodies of the vanquished were piled up after a bloody episode of tribal warfare: this was where 'the heat of battle was'. Stevenson noted, 'The

The restored Stevenson villa, Vailima. *Photo Wren Green*

Vaita`e I had explored pretty far up; not yet the other stream, the Vaituliga …'[8] Both of these are tributaries of the Mulivai River, which flows into Apia.

His initial reporting of this rainforest was of his heroic efforts to clear a path through it, ever thwarted by the presence of lianas, saplings and switches. On the Vaituliga track, he adds more botanical detail: 'It is here quite dry … crossed by fallen trees; huge trees overhead in the sun, dripping lianas and tufted with orchids, tree ferns, ferns descending with air roots from the steep banks, great arums … hundreds of bananas …' He finds himself on an old track where the sheer resilience of the regenerative powers of the forest are not wasted on him. Nor, it must be said, is the evidence in his garden of pigs, marauders of the forest.[9]

Today, pigs still ravish the remnants of forests in the hills of both main islands of Western Sāmoa – as they do in parts of the East Coast of the North Island, New Zealand, for example. Pigs are a source of fresh meat for often poor communities, making their eradication to enable forest regeneration an issue of cultural complexity.[10]

Stevenson employs mock-heroism in retelling his adventure to a friend: how he went into battle with the forest, armed with knife, cutlass and spade, loving the physicality of it – though not the forest's own defences, 'poison brush' for example. Close by, woodcutters were hard at work – he thought from the nearby 'bush village' of Tanugamanono. He came upon 'a native coming down with the trunk of a cocoa palm across his shoulder'[11] – a reminder that these were not pristine but occupied forests, with which locals had a long-standing relationship. As M.D. Olson wrote, 'Pacific island landscapes are social landscapes [that exhibit] with little exception the effects of historical patterns of human influence.' Larger trees, for example, were felled for canoes, which were numerous on these islands, the outriggers built to the highest standards. Samoan tradition was renowned for great ocean-going waka or va`atele, capable of carrying two sizeable fishing canoes, rendered out of huge trees.[12] Here, accounts of Laka (Rātā) have him definitely as a canoe-builder – possibly from Fiji – who built two large waka at Fagaloa, Upolu.[13]

About 150 metres from the house along Stevenson's track towards Mt Vaea are twin waterfalls. He found them well filled, 'embowered in the glade of bush' – a term that he offers in Samoan, ulufanua ('a lovely Samoan word, ulu=grove; fanua=land; grove-land'). Later that week Stevenson returned to the subject of the regenerative powers of the forest conspiring against his own efforts; 'the awful silence, the knowledge that all my efforts are only like the performance of an actor, the thing of a moment, and the wood will silently and swiftly heal them up with fresh effervescence'.[14]

On another day, having begun on the same track then turned right into the stem above the fork, away from Mt Vaea, Stevenson writes to Colvin of the falls:

> I have been exploring up the Vaituliga; see your map [included]. It comes down a wonderful fine glen; at least 200 feet of cliffs on either hand, winding like a corkscrew, great forest trees filling it. At the top there ought to be a fine double fall; but the stream evades it by a fault and passes underground. Above the fall it runs (at this season) full and very gaily in a shallow valley, some hundred yards before the head of the glen. Its course is seen full of grasses, like a flooded meadow; that is the sink! beyond the grave of the grasses, the bed lies dry.[15]

Nearby were 'ruinous pig-walls' – signs of an abandoned village. On a later 'recce' above the falls, heading 'N.W. by N., hoping to strike the Vaituliga above the falls', Stevenson was confronted with a dense tangle of forest roots, vines and falling branches. He found the stream and followed it: the water was knee-deep and 'piercing cold'. After 'a crack on the head' from a falling branch he finally made it home.[16]

Stevenson had an eye for truth and a respect for historical facts, but virtually no knowledge of the Samoan language. He was, however, a good listener, and he may have heard the following traditional account of the name Vailima (though he never recounted it). The story concerns a couple from the South Coast, Sina and Soala, journeying across the volcanic central highlands to visit family. They emerged where the short, steeply descending Falefa River discharges into the sea at Falefa Falls, a scenic spot well east of Apia.[17] Their route would later take them through old volcanic country, much of it consisting of dry watercourses.[18] Surprisingly, the pair carried no calabash. Soala became thirsty and weary and, while he rested, Sina descended to find water, which she is said to have carried back in both hands. (Vailima can also mean 'water in the hand'.)[19]

Stevenson, ever alert to accounts of how locals traditionally valued water, refers to the presence of a local guardian as 'our aitu' (spirit). Like the `Aukele legend in Hawai`i, the following story appears to reside in the emphatic Polynesian virtue of generosity (as opposed to stinginess). And as with the `Aukele legend, water is the medium through which the moral is rendered.

> It is a lady, *Aitu fafine*: she lives on the mountain-side; her presence is heralded by the sound of a gust of wind; a sound very common in the high woods; when she catches you, I do not know what happens; but in practice she is avoided, so I suppose she does more than pass the time of day. The great *aitu Saumai-afe* was once a living woman; and became an aitu, no one understands how; she lives in a stream at the well-head, her hair is red, she appears as a lovely young lady, her bust particularly admired, to handsome young men; these die, her love being fatal; – as a handsome youth she has been known to court damsels with the like result, but this is very rare; as an old crone she goes about and asks for water, and woe to them who are uncivil! *Saumai-afe* means literally, 'Come here a thousand!' A good name for a lady of her manners. My *aitu fafine* does not seem to be in the same line of business. It is unsafe to be a handsome youth in Samoa; a young man died from her favours last month – so we said on this side of the island; on the other, where he died, it was not so certain. I, for one, blame it on Madam *Saumai-afe* without hesitation.[20]

In traditional Samoan culture, aitu are the spiritual upholders of mana, connecting the people to their land: 'only aitu possess mana, and people's fortune is tied to their relations with their gods'.[21] This relationship worked in a more seamless fashion in the pre-Christian era, suggests Olson, when the ocean and forest were simultaneously harvested and revered; typically, lines between conservation and utility were not as they are now. The modern order has set up a tension between 'the degree of direct, consumptive use and substitutability between resources, as opposed to values and practices promoting the maintenance and protection of specific natural resource flows or ecological processes.' He adds – speaking for much of the Pacific – 'In this sense, the current pattern of nature transformation in Samoa, the decline of native forests and living coral reefs, appears as an extension of material-economic practices devoid of the more symbolic-religious aspects of Samoan relations.'[22]

About 60 percent of land in Western Samoa remains in forest – much of it on steep country. The rainforest includes Savai`i's extensive cloud forests – cloud-shrouded, moist at canopy level and richly diverse.[23] Together these are home to hundreds of species of flowering plants – 30 percent of which are endemic to the archipelago – and over 200 ferns. Many of the

The falls at Vailima. *Photo Wren Green*

35 native bird species here are vital pollinators and seedbearers of the cloud forest, and the pe`a or Samoan flying fox (*Pteropus samoensis*) holds the status of keystone species in the rainforest.[24] In 1999 much of the forest on both islands was scarred by cyclone damage.[25]

Modern science is revealing that all forests are 'cloud forests' inasmuch as they interact with the atmosphere in ways that create 'giant rivers of water in the air – rivers that form clouds and create rainfall hundreds or even thousands of miles away'. Consequently deforestation in the large remaining reserves on the planet – in the Congo Basin, Southeast Asia and the ever-threatened Amazon – 'will put water and crops at risk for breadbaskets halfway round the world in parts of the US, India and China'.[26]

Ecologist Wren Green has recollections of Upolu 65 years after Stevenson lived at Vailima. As a youngster in the mid-1950s he went to live in Western Sāmoa where his father was a civil engineer in the Public Works, his mother a teacher at Apia Primary School. Wren recalled: 'Primary school was over at 1pm and we had two options for play – snorkelling over the reefs or biking up to Vailima. We lived about a mile away and would head over to our "secret pool".' He recalled the setting:

> Beside the grand house the expansive lawn dipped into a small valley where it surrendered to deep tropical forest before rising, unbroken, up the green sides of Mt Vaea. A large stream ran through this valley draining some of the forest headwaters of Upolu. We loved playing in the dense shade of tall trees where water rushed over the stream bed of black lava to plunge over a two-metre waterfall into a small deep pool. It was the perfect tropical swimming hole, open any day we wanted, all year. No tickets required.

Decades later he returned during a Samoan dry season:

> It was a shattering disappointment. The stream was a defeated trickle, caused by the removal of the forest trees in the headwaters for their timber to meet local demands. The 'secret pool' was filled with rocks and sediment washed down from the scarred hills above. The place felt forever diminished. Perhaps, I thought, the stream would flow strongly again when the rains returned, but no brave kids would ever jump shouting from the top of the waterfall into the now rocky puddle below – only in my memories.[27]

Gavin Bell, a fellow countryman and aficionado of Stevenson who literally followed in the Tusitala's footsteps, made similar observations. His 1994 analysis of what Stevenson knew and what Bell discovered reveals the profound changes that have taken place in the streams and forests of Stevenson's Samoan haunts.[28]

Despite international pleas for decades to halt damaging patterns of forest extraction, some of the changes undergone since the 1950s may now be irreversible. Rainfall then ranged between 250cm and, in the ranges, 760cm per year, with an overall average under 254cm. Rainfall measures seem approximately the same today.[29] However, post-war the combination of rapid national population increase – on a par with Africa's, but mitigated by migration – and rising living standards, including increased milling for timber-framed housing, have affected the hydrological cycles that serve such favoured swimming holes.[30] And these impacts now operate within the overarching effects of climate change.

Cyclones on the prowl

Cyclonic activity presents another grave threat to Sāmoa's ability to conserve its native biodiversity and freshwater.[31] South of the equator, these offspring of violent upwellings have been sired by warm subequatorial seas throughout much of the Holocene. Bringing lacerating winds and torrential rain, cyclones regularly affect the islands of the tropics from November through to March: they affect Polynesian islands of the South Pacific, as well as the Melanesian archipelagos of Vanuatu and the Solomons to the west. The Fiji group, too, are regularly on their hit list. To the far north, Hong Kong, Japan and the Philippines contend with seasonal typhoons. Over the past 20 years Sāmoa and its neighbours, Tonga and Tokelau, have all been seriously damaged by cyclones, causing loss of life and, at their most destructive, a toll of damage to villages and infrastructure estimated at up to US$300m.[32] This is big money for small places.

When the Apia tropical cyclone hit on 15 March 1889 it was, in many ways, the 'perfect storm'. The cyclone swooped on Apia Harbour with gusts of up to 160kph, exacerbated by the lack of a protecting reef and with the northern end of the harbour exposed to the open sea. Seven naval ships were anchored in the harbour, representing the three colonial powers that were vying for dominion over Sāmoa. Four were destroyed; two more were picked up and thrown on the beach but later repaired; and just one, the larger HMS *Calliope* with its powerful, newer engines, narrowly escaped unscathed by steaming into the lesser perils of the open sea. The death toll was estimated at between 140 and 200.[33]

It was 100 years before another cyclone of similar magnitude and destruction occurred. Cyclone Val, which hit in December 1991,[34] generated winds of up to 240kph and 15-metre waves. Sixty-five percent of residences in American Sāmoa were destroyed, and the cost of damage in Western Sāmoa and American Sāmoa combined was estimated at US$300 million.

American botanist and ecologist Paul Alan Cox witnessed the devastation caused in American Sāmoa by Cyclone Ofa, which struck two years before Val. He recorded his observations in *Nafanua: Saving the Samoan rainforest*.[35] Cox had worked to raise US$85,000 towards the protection of Nafanua forest, home to a rapidly diminishing population of pe`a. Cox recounts an example of the modernist dilemma faced by elders throughout the Pacific, when his village of Falealupo was faced with the need to remove a 'pe`a forest' in order to pay for the cost of building a new school: in 10 years, milling had already taken 80 percent of the large island of Savai`i's lowland forest.[36]

New Zealand meteorologist Erick Brenstrum explains that, correctly speaking, only ex-tropical cyclones (ETCs) visit New Zealand. Once tropical cyclones have strayed beyond the tropics, they tend to gather speed – but not always.[37] They grow considerably larger in cloud shield but are less intense. The strongest and most damaging wind is confined to a relatively narrow belt around the cyclone's eye-wall (of cumulonimbus cloud), says Brenstrum. 'Once they become ex tropical cyclones the maximum wind is generally much less but the area covered by gale- or storm-force wind is much larger – possibly a hundred times larger or more.' These deep lows cause elevated sea level through a process of storm surge, he says. Most of the deaths worldwide in tropical cyclones are caused by storm surge.[38]

Cyclone Bola struck Vanuatu and Fiji in March 1988, leaving devastation in its wake. Heading south towards New Zealand as an ETC, it dropped over 900mm of rain in three days behind Tolaga Bay on the East Coast, forcing warm moist air over vulnerable 'greasy-back' hills, exacerbating the rains; one area had 'more than 514mm in a single day'.[39] Three lives were lost. Elsewhere in New Zealand, Bola caused damage and destruction to roads, rail, bridges, stopbanks, farms and horticulture, as well as 300 homes, at an estimated cost of NZ$160m.[40] As a landmark weather event, Bola is still referenced regularly in New Zealand; Maori Marsden noted that its presence in New Zealand was a reminder from nature of the follies of heedless pursuit of 'a fast buck'.[41]

Bola marked the beginning of a new paradigm for the South Pacific in climate change discussion that coincides with the formation of the UN Intergovernmental Panel on Climate Change (IPCC). Since then we have seen patterns of intensifying climate change, including those carried by cyclones that break out of their geographical bandwidth, spilling into the cooler oceans around New Zealand. Complex in origin as well as in manifestation, these events often arise in conjunction with extremely high tides.[42]

Is some of this awareness simply a matter of perception? Advances in communication technology and forecasting have given us a heightened awareness of weather patterns and behaviour. This is also shaped by our shared experience of such occurrences, as today we are better informed about the events and cultures of our Polynesian and Melanesian neighbours. Brenstrum, a scholar of weather history, cautions that many momentous and quite different weather events in New Zealand's past can actually be traced to ETCs.[43] However, what science can tell us today is that, though cyclones may not necessarily increase in number, they will be more powerful than previously.

According to an official government report that is restrained in its findings and limited in its conclusions, intense rainfall events are, with two exceptions, expected to increase over the entirety of New Zealand as global temperatures inevitably rise over the next century. In the South Island's southwest region, with its already globally significant rainfalls, rises are estimated by NIWA to be as much as 20–30 percent. At the same time, at about one quarter of sites monitored, 'soils have been getting drier'. NIWA also finds dry days will increase over most of the country.[44]

The report contains little discussion of cyclonic-related events between 1972 and 2016. It found the frequency and magnitude of extreme wind had decreased across one third of sites – including notoriously windy Wellington – and increased at only two, New Plymouth and Queenstown. However, a subsequent report from NIWA signals that at three degrees of global warming the water cycle accelerates, but 'the recharge of atmospheric moisture takes a bit longer'. This means that rainfall patterns will be more extreme: 'more heavy rainfall, more dry days'. NIWA states explicitly that 'more extreme rainfall, one in 100 year events, is expected to show even larger increases'. Its most telling point, in the context of this chapter, is that current climate change global modelling lacks sufficient resolution to provide clarity on cyclonic events of the future.[45]

CYCLONE VAL

Cyclone Val struck just after New Zealand ecologists Geoff Park and Rod Hay had completed, but not written up, the lowland forest survey that they were commissioned to undertake in light of the 'unprecedented clearance' of forests under way in Western Sāmoa. They noted that few of the forests they had selected for conservation had escaped Val undamaged, with the prospect of decades of recovery. Several of the sites selected became 'critical refugia for birds and flying foxes'. Damage to some villages' crops and buildings was so extensive that it imperilled their capacity for the sustainable relationship that could ensure their survival and that of the natural forest.[46]

Lasting over a week, Cyclone Val destroyed many houses on the Samoan islands of Upolu and Savai'i.
MFAT photo, Archives New Zealand, R23578793 AAEG 7398 W4835 7

Conservation 'is about more than endangered animals and plants', Park and Hay reported. It must also reflect the primary factors from which landscape and ecosystems emerged; 'that is as much a matter of cyclones as it is volcanic activity creating new lava surfaces'. Although the giant buttressed trees, the ferns and the birds of rainforests are 'a romantic sight', they are also an expression of the cycles of growth, destruction and renewal, 'of which cyclones are integral'.[47]

After that century of cyclone-free life, the authors suggested, Sāmoa's extractive timber industry seemed have a licence to mill, almost heedless of the vulnerabilities on which those forests were founded. Hence the necessity of retaining natural forests in national parks, under conservation agreements and similar protections – notwithstanding that 'protection' is no protection in the face of a cyclone, and that the damage caused to forests by cyclones is exacerbated by the irruption of invasive plants, many of recent provenance, that impede and choke natural plant regeneration. The situation worsens when two cyclones descend in close succession: 'The cyclone of December 1991 heightens rather than lowers the importance of protection for the range of ecosystems that, damaged or otherwise, still survive.'[48] All this, of course, has implications for the availability and potability of freshwater.

Biodiversity, culture and survival

As early as 1854 a survey of the endemic plants in the Samoan archipelago gave a total of around 3000, with a further 100 that the Polynesians had brought and adapted for their use.[49] Since the Rio Convention on Biological Diversity was signed in 1991, Sāmoa (along with other signatory nations) has been appraised for its ecological diversity, its protection of threatened species and efforts to regenerate native ecosystems. Since World War 2 the popular tourist sites at Mt Vaea, the small Stevenson memorial reserve and the Samoan National Botanical Garden at Vailima have been recognised, gazetted and fully protected. Restoration work has been more recent.[50]

By the early 1980s it was realised that a growing population and higher housing standards would deplete most productive forests within 15 years. Considerable efforts to replant this rapidly diminishing resource then began on Savai`i and Upolu, to ensure sufficient plantation forest to provide more durable, suitable timber for housing.[51]

Nu`ufou Petaia has raised the question: 'Should we recapture traditional science?' He concluded that, as a pathway to sustainability, it was not only possible but essential. Through his research he found that, in the late 1980s, 70 percent of Samoan households were still 'agriculturally active', with 50 percent producing for home consumption. The impact of urban life, especially on nutrition, was taking its toll, as were natural disasters such as hurricanes. However, traditional Samoan agriculture was strong, with most farmers not wishing to change those patterns; and the same was true of traditional fishers. The strength of tradition and its superiority overall to non-traditional techniques was identified as a preferable mode, even by business people, he wrote.[52]

The retention of the proven old alongside the supplementary new applies in forestry, too. As the effects of climate change on small Pacific islands become more apparent, Sāmoa has had to recognise that the often dominant stands of exotic, sometimes invasive trees will have a role to play in mitigation and adaptation, at least for the interim, especially in watershed management. Consequently, the country is adopting a more integrated management approach. Sāmoa, the Biodiversity Convention report suggests, needs to work in conjunction with rehabilitation and regeneration of old forest and – essentially – community education.[53]

To this end, nurseries have been set up for watershed rehabilitation, upland forests and agro-forestry, including medicinal plants. Since 2011 the United Nations Development Programme has invested US$4.9m in 14 pilot villages for increasing forest, coastal and community resilience in the face of the effects of a warming planet. Nearly 17,000 people are involved in this scheme. From this approach the Samoan government has built a wider action plan to respond to climate change with further tree-planting, consequential better water supply and quality, improved soil retention and related policies.[54] Of course, in such mountains-to-sea programmes, mangroves – those extraordinary guardians of the naturally fecund intertidal zone – also play an essential role; in Sāmoa they support 29 different species of fish, crustaceans and shellfish at different life stages.[55] And they are an effective and inexpensive bulwark against the storm surges and king tides that are becoming all too common in warmer waters. Unfortunately, though, mangroves are too often sacrificed to the culture of marinas and sandy beaches.

With one eye on climate change, Sāmoa's former head of state, Tui Atua Tupua Tamasese Ta`isi Efi, has spoken of a need for a new respect for nature. He emphasises the necessity of reinstating customary approaches to nature and freshwater, under a 'trusteeship system' rather than water ownership – guardianship rather than property rights. This upholding of 'the old commons' is essential to the plans of many other islanders, including i-Kiribati, whose lands have been retained by the people.

Such an approach, Tupua Tamasese suggests, would involve revivifying age-old proven concepts of accepted practice and boundaries. In particular, he selected the following: tuā`oi (boundaries; jurisdiction), tulāfono (laws; customary lore), agānu`u (general cultural principles and custom laws), agāifanua (cultural principles and custom laws relating specifically to a village or district or family), matāfaioi (designated work and/or responsibility), tofā sa`ili (the perennial search for wisdom) and vā tapuia (relationships that are sacred or spiritual).[56]

In her recognition of the wisdom of the ancestors, Karen Lupe has prioritised the renewal of the traditional honouring of food requiring 'a whole new mind-set, a different kind of ego, that doesn't objectify the Other'. The power of indigenous teaching to revitalise, she argues, could enable Samoans to renew their culture and care for their communities, 'whether the community has wings, fins, roots or paws'.[57] Given the most recent trends in re-examination of Western food choices – towards a kinder, more regenerative approach to agriculture generally – her audience for this message is potentially a much wider one.

We can see, then, that Tanugamanono, the forest village Stevenson came upon near Vailima, was no aberration. Like other Polynesians, Samoans lived in practical and spiritual relationship with forest and shore in what T.D. Salesa calls 'an already living and ordered universe', until the new arrivals came. He argued that far from being overwhelmed by the newcomers, devout Samoans took in Christianity in the manner of 'a confluence of rivers, a meeting of the waters'[58] – which sums up much of what underlies this chapter.

Given the longevity of forest culture in those islands, one wonders at Tāne's place in the scheme of things. And while Lata still stands proud in the Samoan pantheon, with the lessons he learnt and then taught Polynesians about respect for the mauri of the forest, Kāne seems not as evident in the surviving traditions of Sāmoa as in other parts of Polynesia.[59]

M.D. Olson suggests that, before contact, an ethic more utilitarian than conservation seems to have served the Samoans for the 3500 years they have occupied their homeland. But these lived-in forests, traditionally harvested for food, medicine, canoes and firewood, served a much smaller island population and were subject therefore to what we would call a much lighter ecological footprint than that of today's demands.[60] This was a people who, with their intergenerational reach, knew where their next big tree was coming from. To think otherwise of so enduring a culture of land and sea is unthinkable. Further, this is a jurisdiction, like the other traditional island cultures discussed, in which land–water–forest–sea coexist in an unbroken arc of life and meaning, differentiated through the intersecting powers of the gods.

Epeli Hau`ofa, in his seminal essay on identity and Oceania as a people defined by the sea, drew inspiration from the words of respected elder of oceanography, Sylvia Earle:

> The sea shapes the character of this planet, governs weather and climate, stabilises moisture that falls back on the land, replenishing Earth's fresh water to rivers, lakes, streams – and us. Every breath we take is possible because of the life-filled life-giving sea; oxygen is generated there, carbon dioxide absorbed. Both in terms of the sheer mass of living things and genetic diversity, that's where the action is.
>
> Rain forests and other terrestrial systems are important too, of course, but without the living

ocean there would be no life on land. Most of Earth's living space, the biosphere, is ocean – about 97%. And not so coincidentally 97% of Earth's water is Ocean.[61]

Earle, an insightful leader in her field for decades, is making an ecological statement about the seamless interconnectivity between all life across ocean and land that an indigenous and modern scholar of Oceania clearly felt entirely at home with.

Whatever the future survival strategies for water and forest in Samoa, there are two points that the policy writers on climate change will be bearing in mind. The first is that one tropical cyclone can take out years of forest growth in entire valleys, even whole atolls or small islands, by crushing a forest's synergies for new life, freshwater and sustainability. Given the extraordinary powers of renewal of tropical vegetation, however, there is all the more reason to be planting resilience wherever plants will take – now.

The second point is that as the planet's interconnected systems, be they hidden or obvious, seem to be spinning into bouts of chaos, there is no knowing where cyclones, typhoons, hurricanes and other such disturbances will track their trajectories of destruction. If the science and compassion of Sylvia Earle won't cut it now, then this increasing 'democratisation' of the effects of our future weather is surely an added reason for serious, concerted measures … today.

Niue: The Rock and a hard place

Niue, or 'The Rock', sits in a triangle with its near Polynesia neighbours, Sāmoa and Tonga. Ovoid in shape, it is one of the largest raised coral atolls in the world, about 18km across and rising steeply from the ocean floor to stand at some 60m above sea level.[62]

Parts of its coast are impenetrable; and, as James Cook discovered, its warriors had a reputation to match. Three times Cook was denied access to the island, and his men managed only a brief botanical foray: hence his name for it, 'Savage Island'. In the best of Polynesian theatre, this idea was promoted by the Niueans, who still recount using the stem of the red fe`i banana tree to stain their teeth, like bloody cannibals.[63] Today self-governing in free association with New Zealand, Niue has a population of around 1700; almost 31,000 ethnic Niueans are resident in New Zealand, and around 5000 in Australia.[64] Niue has been described as 'an upstanding mass of infertile coral with little surface water': all freshwater found there is from rain, filtered through limestone into chasms and caves that discharge into the sea. Bores enable the artesian water to

The wreck of the German gunship *Adler* in Apia Harbor after a hurricane in 1889.
US Naval Historical Center

Aerial view of the Niue coastline. *Photo Richard Sidey*

sustain Niue's population. Frequent droughts have shaped 'the cultural, economic, social and religious life', meaning the people of Niue 'treasured rain when it came'.[65]

Humpback whales visit the island with their calves between July and October, coming in close to the beaches; Niueans believe it is the freshwater streams and springs entering the ocean that attract them. They have stories as to the whales' purpose in these visits: one says that the newborn humpbacks bathe their eyes in freshwater. Whales take in freshwater only through their consumption of fish, so scientists are keen to explore further this apparent attraction to the freshwater off Niue.[66]

The role forest plays in all of this is recognised in Niue and Sāmoa, as well as Fiji and Vanuatu, by an intensive programme designed to build capacity in conservation management that will assist in reducing the degradation of forest and land.[67]

Similar steps may be under way to reduce the use of paraquat, the herbicide of choice through much of the verdant Pacific for decades. Twenty-eight times more toxic than the potentially harmful glyphosate, paraquat is now banned in many other places, including Europe. While some organic gardening is practised on Niue, paraquat is sold there, as it is on a number of islands, where it has tended to

Before the coronavirus pandemic, increasing numbers of tourists were visiting Niue each year, drawn in large part to snorkelling in the deep blue sea amidst the alluring coral. Warming seas are causing coral bleaching all over the Pacific. *Photo Aliscia Young*

A humpback whale calf. *Photo Aliscia Young*

replace the machete as a weedkiller.

Currently, paraquat is not listed among the internationally ranked, euphemistically named 'persistent organic pollutants' (POPs), a highly toxic band of agrochemicals. It is, however, one of those organic compounds that persist in water, in runoff, and move freely with water in porous soils after rain, building up in layers of film. Known as bio-accumulators, POPs can be taken up by soils; most may be absorbed through skin and blubber. Over time, in mammals – including cetaceans and humans – exposed to it, paraquat affects the endocrine system, exposing the foetus to these pollutants. POPs are also expressed in milk. In a word, paraquat is a neurotoxin, rated 6.1 toxicity – the maximum for acute (meaning 'immediate') – by New Zealand's Environmental Protection Authority. Recently it has been banned by the Fiji government.[68]

Given paraquat's ubiquity, the porous nature of Niue's volcanic soils, and the possible exposure of humpbacks, not to mention humans, to this chemical from groundwater discharge, suitable control measures seem perilously long overdue.[69]

Photo Aliscia Young

CHAPTER 10

Rapanui Easter Island

RATS, FORESTS AND GOOD WATER

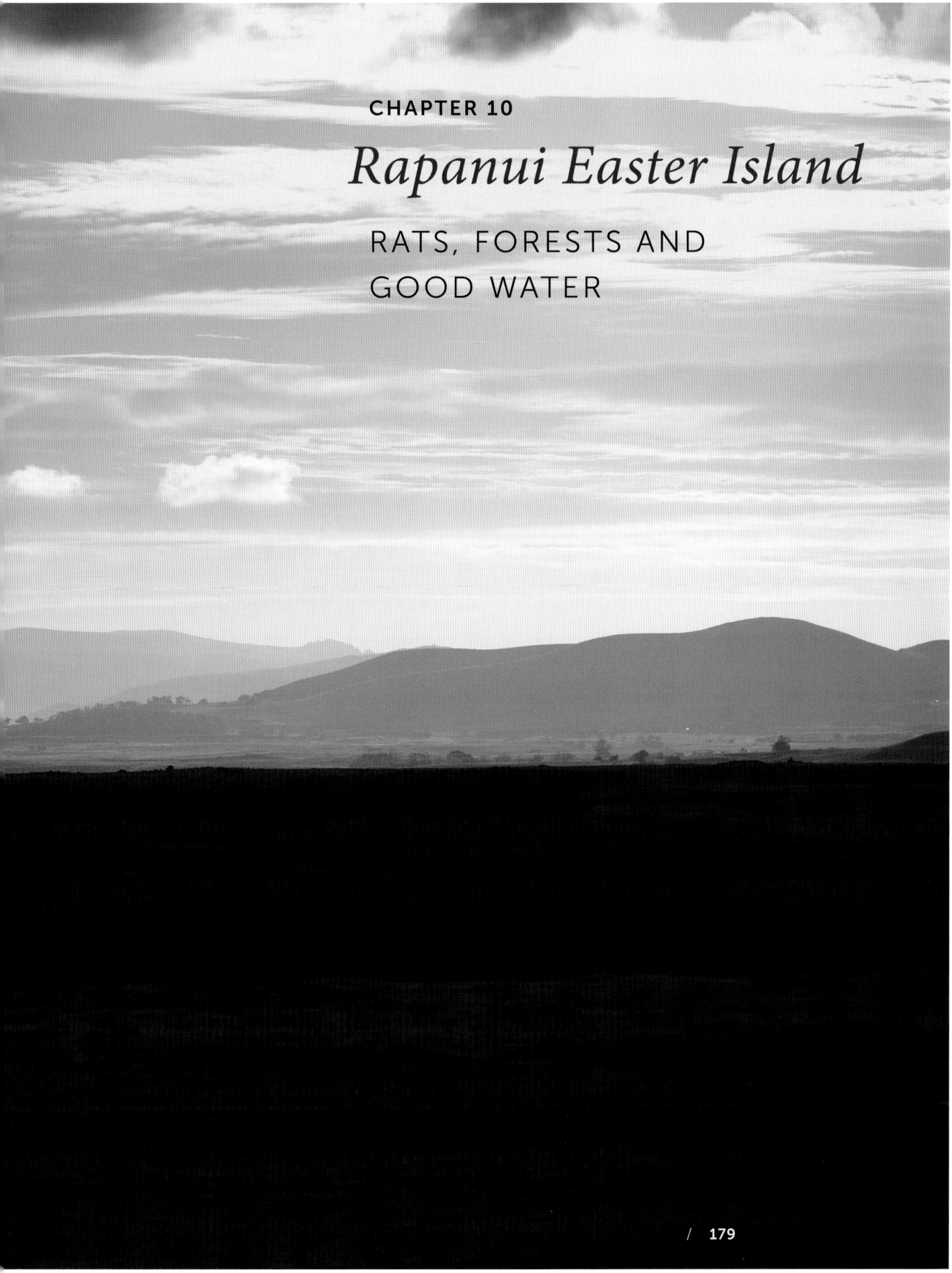

In spring on Rapanui Easter Island the nights are filled with the chirruping of insects. As Aliscia Young has observed: 'The crickets sparkle and twinkle with layers of sounds, all with their rhythms and beats, producing an amazing collective ensemble that somehow seems to speak for the whole island.'[1] Perhaps these insects have language, too, for the vanished ecosystems once dominated by millions of giant Easter Island palm trees (*Paschalococos disperta*) that became extinct on the island after some seven centuries of human habitation.[2] And the lost animal species: six land birds, lizards, several sea mammals, and untold insects and other creatures that were part of the once undisturbed island's ecological pulse.[3] Easter Island today is a dry, wind-blown Oceanic place of isolation and palpable absences, often paradoxical and interconnected.

The explanation for the absence of forest and land birds was deftly tied together by philologist and linguist Steven Fischer, of the University of Auckland, in his observation of another haunting absence: of the culture that raised and then, seemingly just as perplexingly, demolished many of the famed 14-tonne, four-metre-high mo`ai statues: 'It was like the last palm to be felled on the island. With the deafening thud, an entire way of life suddenly came to an end. Ironically, it coincided with the beginning of the reign of East Island's great `ariki mau – "the last man to fill the post of ariki with its original dignity".'[4]

Of all the absences thrown up by Easter Island, it is its remoteness – a telling absence of neighbours – that reigns eternally. Rapanui is 3000km east of its nearest Polynesian neighbour, the Pitcairn Islands, and 3200km west of Chile, which governs the territory. It was James Cook who 'joined the dots' while moored in this solitude, and freely acknowledged the skill and intrepid nature of Polynesian sailing. This was the seasoned navigator's second voyage into the Pacific. Like his Dutch predecessor Abel Tasman, Cook marvelled that 'the same Nation should have

More than 900 enormous mo`ai statues erected on Rapanui (Easter Island) between the years 1250 and 1500 had been toppled by the mid-19th century, but many have been restored, and others remain visible. The statue above, right, in Rano Raraku quarry, has a petroglyph representing the arrival of a European ship carved on its chest. *Photos Aliscia Young*

spread themselves over all the isles in this Vast Ocean from New Zealand to this Island which is almost a fourth part of the circumference of the Globe'.[5] (Both places were colonised by Polynesians within generations of each other in the thirteenth century AD.) Cook went on, in his third voyage, to chart the apex far to the north of the Polynesian triangle: Hawai`i.

Cook's was not the first ship, however, to call at this 19 x 12-km, three-cornered island, shaped rather like the navigator's own tricorne hat. That distinction goes to Jacob Roggeveen, another Dutch explorer who, in 1722, named it Easter Island (the expedition arrived there on Easter Sunday). A number of visiting ships recorded that locals scooped up brackish water to drink from its rocky shore. French anthropologist Alfred Métraux later described more than a 'handful' of water sources: 'Along the shore, where the coast is low, natives dug reservoirs (*puna*) which impounded rain water and perhaps some fresh water springs.'[6] The water was taken from 'a kind of basin, separated from the sea by a wall, where fresh water mixes with salt water. Previously the natives drank that water …'[7]

Certainly, Cook landed water for his ship in Hangaroa Bay on the southwest coast, above the southern tip of the island, but he found it too brackish for drinking. However, his crew took on at least enough water to lift their spirits. They then caught the trade winds that took them swiftly back to the Marquesas.[8] Nevertheless, the impression left in the ship's wake that this was an impoverished island was not accurate.[9] Joseph Gilbert, master of HMS *Resolution*, spoke of a dry, porous and impoverished soil, with no trees; apart from potatoes and plantains 'in the valleys', he had found no 'kind of vegetables useful for refreshment. Water is exceedingly scarce and indifferent throughout the island.'[10] And John Linton Palmer, who visited the island in 1868, believed he could read the old pathways of former streams in the rocks.[11]

The availability and potability of water for its original inhabitants – and for passing sailing ships – continued to exercise the curiosity of many voyagers – as, of course, did the mo`ai. All who anchored off Rapanui's rather daunting coast seemed perplexed by the culture they had intruded on, and not least by their own observations of the water supply. Métraux observed, 'Little salt, if any, was used to flavour the food, and even today the natives do without it. It has been noted that they do not need salt, since they drink brackish water and absorb air impregnated with salt.'[12]

Alexander Adams, who visited in King Kamehameha's brig *Kaahumana*, recorded what he thought he saw, thereby spawning a persistent fallacy:

> Perouse says he had seen some of these people drink salt water like the Albatross of Cape Horn, though his officers discovered a spring of a less saline nature. This fact I may corroborate by one of our party having witnessed a native stooping down on the rocks and slaking his thirst from the water of the great Pacific Ocean.[13]

In short order, Cook's exploration party traversed some 32 kilometres overland on the island: 'The only relief was one spring of what would have been good sweet water if the natives had not combined drinking with bathing.' They found the island 'treeless and infertile', with some sugarcane, taro, yams, turmeric and a few gourds but no coconuts.[14] This absence of trees gave the Polynesian-wide word for wood, tree or timber, rākau, a new meaning of 'riches' or 'wealth' in old Rapanui language – 'a meaning attached to the word nowhere else in Polynesia', as Fischer notes.[15] (Similarly, the exclusively Hawaiian coinage of the word kānāwai, meaning 'law', is based on the word for water.) There had once been extensive forest on Rapanui but, according to Bahn and Flenley, the last trees were felled in about 1660, probably for firewood.[16]

Almost a century later Palmer found that, contrary to Cook's experience, the water of a sulphur spring close to Terano Kau, although it 'had a distinctly mineral taste', was quite palatable and 'in sufficient quantity to serve our large and thirsty party. The rocks in most of the gullies are evidently stream-worn, but now not the smallest brook exists. The soil … seems moist enough not to require particular irrigation.'[17]

Palmer, a naval surgeon on the HMS *Topaze*, wrote his observations of water on the island in a detailed Royal Navy casebook in 1868: 'On the road from Otuiti [Rano Raraku volcano] are many pools of small size but the natives warned us not to drink of them.' He recorded that, while journeying, they chewed sugarcane 'to appease their thirst'.

> At meal times they use salt water as seasoning with their vegetables, and this must have led to the belief that they used it alone, from the absence of fresh water. At Otuiti I was told distinctly that there was no water, except that in the pools of the crater.*
>
> **Ina vai – ina ina*: no water at all, none at all – a pleasing notice to us, when parched with thirst, and at sundown![18]

The mystery of the availability of freshwater endured for many years to come – only intensified by the island's remoteness, and compounded by a suspicion that in the 52 years between the first and second contacts of the eighteenth century, something tragic but imponderable had occurred. From his brief stay in 1722 Roggeveen did not report toppled statues; in 1774 Cook's

Photo Aliscia Young

companion, the naturalist Johann Forster, most certainly did. Roggeveen saw the island as 'almost destitute of wood'; yet, he noted, 'The size and goodness of their potatoes, yams, sugar canes, etc. are proofs of great fertility and strong vegetation.'[19] Roggeveen's visit, however, was for just one day.[20]

If the assiduous Captain Cook had not been unwell during his stay, it is likely that both mysteries – the existence of freshwater and of the island's palm forests – would have been interrogated and more fully explained, if not solved.[21] John Flenley, the first to publish evidence that Easter Island was once forested, reminds us of that.[22] Certainly the apparently towering mystery as to the genesis and later toppling of many of those 900 statues – excavated, executed, transported and erected by a Palaeolithic population of fewer than 3000 – might have been better understood earlier.[23] In those missing 52 years between Roggeveen and Cook, when the Rapanui scripture tablets might have been translated, many cultural clues were lost. 'Cook encountered a different land,' wrote Fischer, whose research into those tablets offers extraordinary insights into the culture. 'It was obvious that something terrible had happened in the interim, probably a war.'[24] Fischer also brought his skills to the task of 'glyph-cracking' Rapanui's hitherto inaccessible proto-language – the rongorongo 'code', enscribed on polished tablets of wood.[25]

All signs pointed to a Rapanui in rapid social and more gradual, but profound, ecological decline. And underneath that, the inevitable effects of long- and short-period climatic shifts.

Cook's missed opportunity has given rise to a body of other research over the last few decades, with evidence to explain such haunting remnants to a wondering world – not least from modern physical sciences. In recent decades, ecologist and palaeoclimatologist John

Flenley and veteran archaeologists, including Terry Hunt, Carl Lipo, Patrick Kirch and Easter Island's Edmundo Edwards, have offered increasingly nuanced and comprehensive explanations.

At latitude 27.9° south, Rapanui is a 3000m volcano, rising 510m out of the sea, that formed the island a mere 400,000 years ago.[26] The island is without rivers or permanent streams on its weathered and porous surface, and with variable rainfall and droughts.[27] The only significant reservoirs of potable water, called rano, were trapped in the old craters of Terevaka, the volcanic ancestor of the island.[28] Irrigation was not feasible, and with the available rainfall it was not possible to grow taro. Standard cultigens such as coconut, kava and breadfruit – a staple of its neighbour to the west, Mangareva – were beyond their southern range on Rapanui. Hunt and Lipo, drawing on the DNA analysis work of archaeologist David Addison, point out that: 'As options for establishing the mainstay of irrigated taro were dashed the colonists of Rapa Nui must have watched in dismay as critical crops such as breadfruit simply died.'[29]

Lack of groundcover, combined with porous surfaces, has been problematical on Rapanui for centuries. One observer wrote: 'In spite of the morning's downpour, the grass isn't even damp and neither is the ground. How can this land dry out so rapidly, with the earth turning to dust within several hours of rain …?'[30] John Linton Palmer noted a 'lack of minimum surface water – high temperature and humidity, high evaporation and porous soil.'[31] A scoria quarry on Rapanui is named Puna Pau, 'exhausted springs' – a reminder that sources of freshwater once existed on the island.[32] Yet Flenley and Bahn ascertained that

Opposite and above: Rano Kau, a freshwater crater lake, is one of three bodies of fresh water on Rapanui Easter Island. *Photos Aliscia Young*

'almost everywhere on the island, from the steepest parts of the cliffs and the youngest lava surfaces, there is soil sufficient in quantity and quality to support the growth of trees'.[33] Palmer observed: 'The ground is so fertile, that a few days' work suffices for keeping any family in subsistence for the entire year.' He also noted: 'The whole of the island has formerly been under cultivation, and rahui-stones are met with in every direction.'[34]

Part of the explanation for productive fertility in such unpromising ground, say Hunt and Lipo, seems to lie in the practice of 'lithic mulching' – the use of buried rock in gardens across the island. This technique, used in many old cultures, can have the effect of regulating the temperature night and day, deflecting wind and introducing minerals to the soil.[35] But Fischer noted that 'the removal of forest cover had increased local temperature and humidity, causing greater erosion and leaching of the soil. As the evaporation rate increased, the soil became more porous. Drainage worsened; water was poorly retained …'[36]

There are three main lakes or swamps in three craters that hold rainwater: Rano Kau (on the most southern edge), Rano Aroi (the crater near the centre of Terevaka) and Rano Raraku in the eastern corner. These were once reliable water sources, but for many years now 'none of the fresh water is of value – only reeds grow there'.[37] Rano Kau is lined with the endemic toromiro (*Sophora toromiro*, a cousin to the kōwhai), the predominant plant, and hau hau (*Triumfetta semitriloba*), a shrub.[38] The crater lakes were never the most convenient source of water, and were also highly mineralised.[39] Rano Raraku was once a quarry to the mo`ai; today it is home to nesting tropicbirds.[40] However, seepage through lava tubes and caves yielded water beyond the craters – some as springs[41] and some trapped and held in stone basins that had been 'pecked out' by the Rapanui.[42] Old placenames with vai (water) or puna (spring) are often the site of a village with a form of well close by. 'The ruins of ancient settlements are always thick around water holes,' wrote Métraux.[43] He listed the most famous of them: Ahu Te Peu, Puna Marengo and Vai Tara Kai Ua; the latter was deep, 'covered with big blocks of stones so that only a small, narrow passage remains, just wide enough for one person. Behind Anakena [bay] abundant water is found.'[44] Palmer formed the impression of 'many deep holes' in the volcanoes:

> in those of the Terano Kau these are fully 25 feet [7.6m] deep, and I have tasted it pure and fresh from many places near the shore. At Winipoo [Vinapu, in the southern 'hook' of the island], not only is there a subterranean reservoir (to which a tunnel leads from the face of the cliff) but on the very sea beach the natives have made a cistern to catch the water which distils from a little runnel. I did not see that the natives had sunk any wells.[45]

Vinapu was a water source for at least 80 habitations and related cultivations[46] – and another place marked by monuments. Nearby Vaihu was the site of an early ahu, or sacred mound.[47] Each of these sites is marked by mo`ai hewn from volcanic tuff, today facing outwards towards the ocean.[48]

There were other sites for drinking water. In the western corner of the island, Vai a Heva, on which a human face was carved, was an outlet for water,[49] and 'well-watered ravines' were to be found, including one near Vaitea, in mid-island.[50]

Among other water-related petroglyphs on the island is a stone named Pu o Hiro (Hiro's trumpet). To Edmundo Edwards, the ethnographic and archaeological evidence indicates this is Hiro, god of fertility, 'with two natural cavities that produce a sound similar to that of a [blown] conch shell. The holes are surrounded by petroglyphs of vulvas since according to Rapanui oral tradition, fertility spread over the earth when the holes were sounded.'[51] Hiro is definitely an expression of a culture that well understood drought: as the god of rain, he was called on to 'ensure rainfall or prevent a drought'. This was especially the case with the planting of the Rapanui staple, kūmara (sweet potato).[52]

The mo`ai in their original state seem to have both marked and addressed freshwater. They wore red scoria headdresses (noted but not explained by the 1786 La Pérouse voyage, which recorded much of island culture and geography).[53] Today the headdresses of the mo`ai are 'small stones covered by calcareous algae … found lying near many ceremonial platforms and have been found in [two] excavations. Ethnographic sources indicate that these stones were used in ceremonies to promote rainfall and increase fertility of sweet potatoes.'[54]

On Rapanui, Tangaroa is still god of seas and all oceanic life. As in Hawai`i, Tangaroa and Tāne are the water-finders, the pair who conjure up the 'streams, springs, pools and heavy showers'. Interestingly, both Polynesian homes share extensive lava tubes and underground networks of water – the kind of geology that invites the notion of conjuring up water from the ground. Both these gods also govern the Rapanui fortunes of farming, 'and were honoured with prayers and offerings during the different stages of crop cultivation, especially when planting and harvesting'.[55]

Matariki, the rising of Pleiades, was as important to Rapanui as to Hawaiians and

Māori, with a lineage back to the observances of the early Lapita people. Matariki was observed in autumn, with festivals and the bringing in of yam and kūmara harvests for feasting. In parts of the southern hemisphere, such as Rapanui, it also marked the end of the rāhui, that other pan-Polynesian practice.[56]

Given such observances, the question inevitably raised by scientists is, what happened to all the trees? Carl Skottsberg is a thoroughgoing Swedish botanist who recorded Rapanui's botany quite late, during World War 1. Both legend and Skottsberg are clear that a sufficiency and small variety of tree cover at an earlier time in the history of Rapanui had kept the islanders in wood from native plants such as toromiro for canoes, paddles, house framing and extensive carving. In addition, they cultivated plants they had introduced to the island, such as paper mulberry.[57]

It is difficult to explain how a phenomenal 16 million palm trees were 'chopped down' within 700–800 years. One possibility, raised in a joint paper in 2004, argues strongly that 'the overuse of palm trees was the major reason for their destruction'.[58] Christopher Stevenson and his colleagues noted that, together with evidence of a growing demand for firewood, the demise of these Easter Island palms might well have been based on the lack of suitable drinking water. This scarcity may have driven the Rapanui to 'harvest palm sap', as is practised in Chile with a similar palm: 'The cut trunk of a full-grown palm provides the amount of fresh, drinkable sap needed each day by one person for a period of more than a year.' The hundreds of litres each palm provided could have served as a highly nutritious drink, 'replacing the scarce freshwater sources on Rapa Nui'.[59] As the last of the natural woodlands were cleared, such a practice would have been another contributor to the breakdown of the centuries-old practice of sustainable land use, they argued. This in itself would have brought profound change to life on the island, possibly exacerbated by 'a second wave of occupation from outside or new external contacts of the already established people of Rapa Nui'.[60]

It can be argued that this hypothesis is literally too 'clear-cut'. The effects of wind and storm damage, starting at the forest edges, may also have been a factor here. The trade winds at this latitude are driven largely from the east and are affected by many vectors, not least the powerful currents driving out of the Southern Ocean and the cold Humboldt Current moving northwards up the coast of South America. The easterly trade winds are prevalent from November to February.[61] Of course, once an established canopy is cleared, a strong wind can do a lot of damage.

The authors concluded that the inevitable and intensifying conflict over resources may have led to 'the abandonment of a peaceful and well-organised megalithic culture'.[62] By 1722 the forest, if not already gone, was severely diminished. However, this does not explain the failure of the palms to be sustained naturally from seed throw. For Hunt and Lipo, one question remained: was it introduced rats, thoughtless firing or tree-felling that was the Rapanui people's greater contribution to the island's demise?[63] And underlying this are the fundamental, long-period climatic shifts of the Pacific.

The work of Flenley and Bahn anticipated such possible reasons for Easter Island's ecological demise in the early 1990s:

> It is theoretically possible that a drought made things worse, but clearly the islanders brought disaster upon themselves by gradually destroying a crucial resource, the palm, and unwittingly preventing its regeneration through having imported rats. A delicate balance was put under strain and finally upset. The environmental degradation was irreversible.[64]

Drought struck again across much of the planet from 500 to 1200 AD. However, the effects were not all negative. This was a time of extended climatic stability. It saw the development of irrigation and terracing, with high food production, long-distance navigation and new settlements on many Pacific Islands, notably in Eastern Polynesia – including Rapanui. A group of climate scientists argue that, from 1200 AD, wetter weather filled Rapanui's lakes, creating a cooler climate and stormier conditions until about 1570. This was marked by a densification of the palm forests at Rano Aroi, but also what may have been deliberate firing, which reduced forest at Rano Raraku. Until 1500 these conditions underpinned prosperity and the cultural flowering of the Rapanui people, embodied most evidently in the tablets and monumental mo`ai. Valenti Rull and his team argue that the palm forests were virtually removed, very swiftly, from both sites between 1570 and 1620: 'This was the "coup de grace" of the deforestation of these basins and, likely, of the whole island.'[65]

The questions of agency in regard to the ecological and cultural decline on Rapanui have, with ongoing research, become more and more entwined. Techniques for core sampling sediments in peatlands have proved to be a highly accurate tool in many Pacific islands: they can identify not only the pollen of old forests, but also the arrival of Polynesians to new islands. The core samples extract the original layers of hard seeds and palm nuts with teeth marks throughout the Pacific that – in these hitherto mammal-free lands – are proof positive of the time of arrival of the Polynesian rat or kiore.[66]

Rats were brought on waka as a source of protein to new, unfamiliar islands. Once established they ravaged the flora and mammal-naïve fauna, especially ground-nesting creatures such as seabirds.[67] Core sampling by a team at Ordy Pond on the Ewa Plain, O`ahu, showed telltale layers of charcoal but, unexpectedly, these occurred at a later date than indications of the presence of rats.[68] Some forests here too – palm and woody shrub – disappeared within just 100 years. Hunt and Lipo commented, 'What Athens proposed was quite radical: the Polynesian rat (*Rattus exulans*) had caused the rapid demise of the forest.'[69]

But more detailed research took scientists to a slightly different, more subtle interpretation of the sequence on Rapanui. While the Polynesian rats had something to answer for, sampling of sedimentation from Rapanui's crater lakes revealed that it was not until the introduction of the much larger ship rat and Norwegian rat (*Rattus rattus* and *Rattus norvegicus*) that changes possibly already under way became irreversible.[70] So it was exotic predators – ship-borne, not waka-borne – that administered the coup de grâce to an already stressed island ecology: the obliteration of nesting seabirds, the end of palm regeneration and, with these, the collapse of the culture.[71]

Such lessons are indicated on many islands, large and small, throughout the Pacific. Wherever it fetched up, the impact of the kiore on ground-nesting birds and other natives has been variable – and often considerable. On islands where birds have not co-evolved with rats, the effect of their introduction has 'out-weighed all other causes of extermination of island birds'.[72] Those on Rapanui are no exception. In cooler conditions, and when starving, kiore will abandon their vegetarian diet to prey on smaller animals such as chicks and lizards and, in Aotearoa, young tuatara.[73] As always, though, it is the relationship between climate and forest, forest and water, water and soil, and the over-exploitation by human communities of these matrices, that is fundamental. Many islands where kiore were introduced by Polynesians and, later, the larger rat species arrived as stowaways on foreign shipping, now have extermination programmes in place, usually targeting the larger rat species.[74]

We can't yet say definitively exactly what brought such abrupt changes to the ecology and society on Rapanui. But when it comes to cultural decline, by 2011 Fischer, with his historical linguistic reach, and Hunt and Lipo were in no doubt. The Easter Islanders continued to survive for hundreds of years in their now highly modified environment, only to be savagely compromised and defeated by outsiders: 'Through a series of disastrous encounters with foreign visitors, the Rapanui population had collapsed, rebounded, collapsed again, and then recovered to a degree, only to be ravished in slave raids.'[75]

We who inhabit the islands across the Pacific have learnt with ever greater clarity of the intertwining vectors that were in play to bring down the forests, the mo`ai and the civilisation that shone so brightly from a small, highly inventive and industrious Polynesian community. Climate and water are both implicated in this narrative, as are, now most certainly, *Rattus exulans*, *Rattus rattus* and *Rattus norvegicus*.

This chapter, then, comes down to an understanding of the relationship between rats and good water on Pacific islands – large and, especially, small. The modelling for climate change since the 1980s continues to be shown, over and over, to be, if anything, far too conservative. Meantime, the role of forest as both carbon sink and 'curator' of water quality and availability assumes ever greater significance. If native forests are to play their full role in natural soil and water conservation, the birds, lizards and insects that co-evolved with them need to be actively protected. The intrinsic and extrinsic reasons for this are inseparable – the role of fauna in seed dispersal alone justifies their protection. That is why today, on archipelagos such as Aotearoa and Hawai`i, there is such an emphasis on eliminating the exotic predators, as well as exotic invasive weeds.[76] For such work to be fully realised in democracies, a shift to a deeper understanding by populations and their politicians becomes imperative. Like measures of climate change itself, some policies are simply too urgent to be at the mercy of populism and other hubristic manifestations of politics.

It is only quite recently, however, that scientific concerns are being raised about the effects of cats, dogs, ants, rats and introduced hawks on Rapanui – in particular with regard to a relatively recent colony of nesting red-tailed tropicbirds at Rano Raraku.[77]

The regeneration process: a nesting red-tailed tropicbird. *Photo Aliscia Young*

Right: A kea, one of the world's most intelligent birds, supervises a stoat trap, Mt Aspiring National Park.

Opposite, top: Tiritiri Matangi Island in the Hauraki Gulf has for decades been the focus of considerable regeneration effort and species transfer.

Photo Aliscia Young

Opposite, middle and bottom: Pipit and wandering albatross are both an important focus of conservation efforts on South Georgia that use techniques pioneered in New Zealand.

Photos Richard Sidey

PREDATOR-FREE NEW ZEALAND

The goal for a predator-free Aotearoa by 2050 was formulated by the Royal Forest and Bird Protection Society in 2014:[78] no mice, rats, stoats, weasels, possums or feral cats by 2050. It may sound grandiose but, in fact, the task has been under way for 60 years – since 1960, when a group of schoolchildren, with Forest and Bird backing, cleared tiny Maria Island of rodents.[79] Today, 171 of the world's 1123 predator-free islands (15 percent) are within Aotearoa's territorial waters. And the size of the islands where pests have been successfully eradicated keeps growing: the 450,000ha Auckland Island, 465km south of Aotearoa, is the next target.[80]

Conservation volunteers give thousands of hours every year to helping regenerate or replant natives. This activity is now often carried out in tandem with predator trapping, which is regarded as a necessary additional obligation to the birdlife the forests engender.[81] The Wellington suburb of Crofton Downs, edged by the Wilton Bush Reserve, is an example of an urban pest-free zone maintained by vigilant trapping. The success of this 'nature neighbourhood watch programme' has inspired many other such zones.

New Zealand's growing expertise in sophisticated trapping and better targeted aerial use of 1080 poison is an enabler of healthy re-wilding, both on and offshore. New Zealand has also pioneered 'mainland islands' – special areas of biodiversity set aside on the mainland for regeneration of plant and birdlife within predator-proof fences, often near urban centres. The first of these was a Forest and Bird initiative at Karori, south Wellington, in an old city water supply catchment area, now 'an island in a city' called Zealandia.[82] Today there are 20 such mainland islands, publicly developed and run, and another five under government administration.[83]

The techniques for exclusion zones have been exported to islands beyond New Zealand such as the Seychelles and, biggest yet, South Georgia in the southern Atlantic Ocean. And, after careful planning, in 2016 the remote subantarctic island of Antipodes underwent a successful extermination campaign of millions of mice, which were killing albatross chicks.[84]

From Kiribati, there is nothing over the horizon but more sea and tiny atolls.
Photo Richard Sidey

CHAPTER 11

Kiribati

LIFE ON AN OCEAN WAVE

To understand that nothing, but nothing stands between life in Kiribati and millions of square miles of warming, potentially destructive seas is a giddying experience. However, observing myself as a visitor from much larger islands, I also found my mind quickly internalised a comforting facsimile of higher ground that it 'knew' lay somewhere just over the horizon. Sadly, that's where most of the world seems to have parked this skinny wishbone of separated and conjoined atolls and their more scattered associates, with their slim pickings and brave leaders.

The long, draped coral necklace floating in the equatorial heat that is Kiribati 'Central', Tarawa, has just one high point: Eita. This bump rises halfway along the causeway that joins the northern atolls of Tabiteuea south to historic Betio, where in November 1943 thousands of US Marines and Japanese perished fighting in WW2. That 'hill' is just three metres high. No land is more prominent than this for almost the entirety of the Kiribati nation.

Betio is at one of the wider points in the series of islands. It is densely populated, even in comparison with Hong Kong or Tokyo, and packs around 10 people per household. In 40 years the i-Kiribati population has doubled to almost 120,000, with most of the increase concentrated in South Tarawa. In many places these islands are not a lot wider than the two-lane highway with its newly built southern seawall. With space so limited, all the telltale spikes, starting with sea-level rise, that graph the concerns the entire world is soon to face are here, already exposed and raw. The numerous darting dogs, the chickens, children, houses, the waste of all kinds and even the shoreline seem to jostle for a place in the shade.

Meteorologist and climatologist Nakibae Teuatabo observes that the shorter the distance across an island the greater the variability in weather. Indeed, smaller islands, especially those to the south, are generally more vulnerable to water shortages. Extreme weather events, particularly severe drought punctuated by more frequent, intense rainfall, have become something of a trend in the past decade. Teuatabo says that the El Niño cycle in the Southern Oscillation weather pattern that intensifies the stormier and wetter conditions, usually from the northeast, seems to be occurring at ever closer intervals. Cyclonic behaviour is also on the increase with these El Niño events. 'The observation is factual,' he says, 'but the explanation is missing.' Overall, droughts may become less frequent, with extreme rainfall days likely to increase.[1] The average height above sea level of islands in the Kiribati group is under two metres, and the encroachment of the sea is relentless – as was eloquently explained in the documentary *Anote's Ark*.[2] And in addition to rising sea levels, saltwater incursion into wells and the nation's tiny allotment of arable lands is threatening freshwater supply and crops.

The Micronesian–Polynesian inhabitants across Kiribati stand to lose more than any other people from continued international tardiness and inaction on climate change: advancing seas, loss of freshwater, loss of crops, loss of the very land beneath their feet, if not their culture and life itself. Predicted rising temperatures also bring greater risk of water-borne diseases, such as cholera, as do the effects of ocean acidification now showing up in the data.[3]

On Tarawa, when these unwelcome events occur together with all the usual downsides of modernity, there is no escape. Plastic, sanitary, medical and human waste – not to mention a legacy of toxic residues – and human proximity to freshwater all detract darkly from traditional scenes of reef and tuna fishing, swaying coconut palms, gardens and the welcome shade

Kanton Island (also known as Canton Island or Abariringa) is the largest of the Phoenix Islands, one of the three island groups that make up the Republic of Kiribati. *Photo Richard Sidey*

of thatched beach houses. Fish migratory patterns are changing, according to the fishers. Fish has always been a staple of the i-Kiribati diet, but many of the nation's fish stocks have been poached out in the vast Kiribati exclusive economic zone by commercial fishers from China, Taiwan and Korea.

In late January 2018 Kiribati was hit by an unseasonal tropical storm, Fehi, causing extensive damage on the island of Tarawa (and it went on to invade low-lying coastal communities in New Zealand). 'There's a gap between the seawall and the causeway,' toxic-waste engineer John O'Grady recounts. 'The sea was coming through it and over the causeway. People's homes were affected, along all of those causeways … People are scared.'

This wasn't the first time. Sea-level rise has averaged between 1mm and 4mm a year since 1998. By far the most urgent account of rising seas threatening any nation at the Paris 2015 United Nations Climate Change Conference was that from the Kiribati delegation. Then president Anote Tong spoke for his and many other people who have lived for thousands of years on the low-lying coral formations of the tropical mid-Pacific, where two islands have already been 'wiped out' and villages have had to be relocated. Tong, who graduated from Canterbury University and the London School of Economics, spoke of his initial anger as he realised the indifference of the carbon-emitting nations towards the plight of his low-carbon nation. In an allusion to past events, he likened Kiribati's current predicament to slavery: 'no one spoke about slavery for a long time,' he said, enigmatically, before moving on.

Population pressures, warfare and shortages of food and water gave impetus to migration. Popular traditional accounts that disinherit younger sons from land rights, such as the pan-Pacific Māui story, are all part of the 'movement narrative'.[4] R.L. Stevenson saw it in this way: 'Over the whole extent of the South Seas, from one tropic to another, we find traces of a bygone state of over-population, when the resources of even a tropical soil were taxed, and even the improvident Polynesian trembled for the future.'[5] He saw that through the islands, 'abortion and infanticide prevailed. On coral atolls, where the danger was most plainly obvious, these were enforced by law and sanctioned by punishment.'[6] Stevenson thought Darwin's theory of coral islands being inundated by sea-level rise might have been a factor in moving for 'a people of sea-rovers, emigrants from a crowded country, to strike upon and settle island after island, and [who] as time went on to multiply exceedingly in their new seats. In either case the end must be the same; sooner or later it must grow apparent that the crew are too numerous, and that famine is at hand.'[7]

Kiribati, a transliteration of its colonial name, the Gilberts, is part of Micronesia to the north of Polynesia, spreading far across Oceania. One cannot speak of Kiribati, with its 32 atolls and its reef islands population of about 107,000, without thinking of Tuvalu (Ellice Islands), population 10,000.[8] The island groups of Kiribati, which include Christmas Atoll in the Line Islands and the once phosphate-rich Banaba (Ocean Island), lie within an area of ocean equivalent to the size of the continental United States.[9] The people of Tuvalu occupy a more compact array of nine atolls that retain strong Polynesian influences, predominantly of ancient Sāmoa and its northern neighbour, Tonga.[10] Together with the Marshall Islands and Vanuatu to the northwest, and the Maldives in the Indian Ocean, Kiribati and Vanuatu are at the frontline of climate vulnerability.[11] So far, only the Solomons, with the inundation of five small islands already, can count their losses territorially.

For more than a century, until independence, Kiribati formed the Gilbert and, with Tuvalu, the Ellice Islands British Protectorate.[12] With nominally over five million square kilometres of exclusive economic zone – most of which was in modern Kiribati's 'paper' jurisdiction – the two became separate, tiny nations in 1979. Before independence the protectorate stretched well beyond either side of the International Dateline – to the west and, including the furthest of the islands of Kiribati, more than 40 degrees longitude to the east.[13] In 1994 the government 'bent' that section of the dateline eastwards to keep the i-Kiribati island population within one time zone.[14]

Perhaps no one has greater moral claim to securing the increasingly elusive 1.5°C cap on global warming than these people. As a backstop, Tong directed Kiribati's purchase of 20 square kilometres of land in Fiji – more as a signal investment for future contingencies than necessarily as a bolthole for his people. Meanwhile, he argued, other options such as land reclamation remained a priority. Making his international appeal to conscience, he argued for 'migration with dignity', if and when it must happen.[15] The current Kiribati government, under Taneti Maamau, has renewed the plan for its people to stay, eschewing the worst aspects of modernity and re-embracing old ways combined with other sustainable practices. The first phase of this Kiribati Adaptation Programme (KAP) began in 2005.

The plan exemplifies the spirit of i-Kiribati to push back – through advocacy internationally and through action, often supported by aid, domestically. The government is committed

Opposite: Manra Island, in the Phoenix chain, contains evidence of prehistoric habitation. In the 19th century guano was mined there, and later copra was produced. The atoll has been uninhabited since 1963. *Photo Richard Sidey*

Opposite, below: Another low horizon: Orona Island in the Phoenix Island chain. *Photo Richard Sidey*

In many parts of Kiribati today, islands are just holding their own above sea level. *Photo Richard Sidey*

to upholding the ancestrally owned lands and practices of its people. It sees a return to old ways as being directly connected to survival through self-sufficiency. Director of Agriculture Kinaai Kairo explains that response to climate change is being built into all government policy: 'In the early 1990s we were more than 80 percent self-sufficient in many aspects of agriculture.' She sees a direct connection between rising fuel costs and the demands of cooking traditional foods (two hours for taro) that, through substitution by convenience food, has had serious consequences. Not only is there more waste, but the effects on health of a diet of processed food – especially the increase in heart disease and diabetes – are a burden on Tarawa's tiny hospital and medical services. And as part of the return to viable traditions, agrichemicals were banned in Kiribati in 2005.[16]

Taro, that great standby of the Pacific, is a healthy and sustainable food option – but only where water is in good supply. And pandanus, used for thatching, is also a traditional food with roots that are rich in vitamins, says Kairo, 'a miracle plant that is even a cleanser of freshwater'. On the northern, wetter islands of Kiribati a giant swamp taro, babai or bwabwai (*Cyrtosperma merkusii*) is a naturally occurring staple, dug deep from freshwater. In the southern islands, where water is often scarce, this taro can be grown as a traditional crop. However, Kairo says, 'at times high salinity is more common because of incursion due to rising seas' – and that is a real hazard for babai. Coconut oil from copra, another traditional product of Kiribati, is today a growing extractive industry; and it would be even more financially beneficial if coconut sugar could be refined in large quantities. Kairo notes that while coconut trees on Kiribati mature quickly, in just six years, economically efficient harvesting remains an issue.

However, soils are poor and sparse on Kiribati, and only about 2.5 percent is cultivable. 'Forests are not as productive here as Vanuatu, and land is poor compared with Fiji,' says Director of Environment and Conservation Nenenteiti Teariki-Ruatu. Until 2020 there was a long-term Taiwanese programme in place, trialling vegetable production in the high-heat, low-moisture, thin soil areas of the atolls. The Taiwanese described crop-growing in Kiribati as 'like growing crops on the moon,' says Teariki-Ruatu. Alkalinity and porosity constitute further problems. Cucumber, pumpkin and taro are being produced, assisted by scientific composting in agricultural pits and other natural soil enrichment using products such as liquid compost from fish waste.

In the past, islanders sometimes mixed brackish water with freshwater to make it go further. Taina Temakei, who works in water policy, is aware of small freshwater wells on isolated northern atolls that are held back privately by families for emergencies. She cited a small one she knew of, only a few metres deep, right on the seashore. 'It was not a freshwater lens, 'she said. 'We tested it for salinity: there was none. You could scoop clear water from it. It was probably on compacted, bedrock coral.' Seawall protection is probably needed to prevent overtopping of such wells, she suggests.

A new, relatively inexpensive technology – the solar-powered desalinisation plant – seems

ANIMAL–VEGETABLE AND THE RISK OF SALT

In corners of the Pacific the giant taro or Cyrtosperma has been identified in folktales as also having a life as a cetacean – a porpoise. 'A Cyrtosperma transforms itself into a porpoise in the ocean and back into a plant on land. Human or divine beings, living or dead, become porpoises and travel from land above the ocean to a land below it where porpoises (and other fish) and people and Cyrtosperma (and other plants) exist under conditions like that on land above the ocean,' explained Katherine Luomala. In the myth of the Arorae islanders, the plant-intolerable waters can force the metamorphosis: 'the Cyrtosperms become cetaceans to tolerate salty water as they swim to a new home'.[17]

If cultivated for 10 years or more on the Kiribati island of Arorae, this special food may weigh in at 80kg, and 2m in length. Given its size and its cultivation in large pits that access the freshwater lens, and understanding that the realms of sea (marawa), sky (karawawa) and land (tarawa) are seamlessly connected, this tradition, grounded in fable, is hardly surprising. The giant taro is also 'a critical food source to these tropical reef dwellers'.[18] Since each plant needs to float in a muddy medium, in which small fish may also be found, it is ever at risk from an increase in the brine balance. Therefore, 'The [i-Kiribati] are expressive of gratitude for how supernatural kin provided a feast for hungry natives or their haughty guests.' This, she thought, reflected lives marked by 'famine-producing droughts and destructive westerly storms'. In the vulnerable freshwater lenses, crop failure in such conditions and after such sustained cultivation could be a harsh reality.

to offer a panacea for some of the limitations noted above, especially on the water-short smaller inhabited atolls, where droughts can easily trigger salt incursion into water supplies. These portable units have been installed on Tamana and Arorae in the southern islands, and at least three in the northern islands. Each unit costs about AUD$10,000, and is supplied by a donor. Technical maintenance is an ongoing problem in the remote Pacific. On Tarawa are two reservoirs that can barely supply the population, with rationing now down to two hours' use (500 litres a household) for just two days a week, explains Itienang Timona, a water and sanitation engineer. 'The government spends 9 million AUD a year on diesel for pumping water to South Tarawa,' he says. 'We are now looking at a desalinisation plant for supply to South Tarawa. The plant will offset the extra energy demand with solar.'

Most dramatically, under government supervision a million trees have been planted by volunteers to protect land, freshwater and a way of life. The faithful mangrove, amphibious defender of the atolls' integrity, is the major species planted. Ironically, in Fiji, as the country seeks to boost its tourism economy, chunks of the coastline between Lautoka and Nadi have been cleared of mangroves; new shoots battle against cutting and herbicidal sprays among the rocks of marinas.

Although i-Kiribati are not ones to cry victim, frontline climate change is just the latest indignity visited on Micronesia by outsiders in modern times. Apart from the effects of thermonuclear testing, whose malignant legacy from the Cold War out on the far edge of Micronesia remains toxic, in the nineteenth century some of these islands were regular stops on routes for blackbirding – a euphemism for enslavement; hence Tong's allusions to slavery and war. A number of these islands, including Tamana in southern Kiribati, were a calling place for the depredations of the slavers, beginning with New South Wales farmer Benjamin Boyd in 1847.[19] Then, in the twentieth century, Nauru and Banaba populations were largely dispatched elsewhere while their lands were systematically quarried by British phosphate interests. A lingering industrial legacy of loss – of extensive quarrying and of extensive asbestos-clad buildings – remains, making British Phosphate 'the gift that keeps on giving'. In this century Christmas Island and Nauru became the Australian government's internment camps for the 'boat people' out of Asia. The issue of climate change (and its refugees), until now defiantly ignored by successive Australian governments, in the years ahead may well become the elephant in their crowded waiting room.

The occupants of these reefs and atolls of usually tiny bits of land in the vastness of tropical Micronesia have often found nature decidedly challenging. Before tank storage of roof water 'there was a total dependence on wells sunk to the depth of the narrow freshwater lens,' writes historian Barrie Macdonald.[20]

Rainfall across the Tuvalu group is graduated from north to south. Butaritari and Makin to the north have the highest and most reliable supply of rainwater; and this gradually diminishes through to the southern islands, which receive the lowest. According to Lawrence, 'Drought probably explains [why] the Phoenix and Northern Line islands, including Christmas Island, did not support permanent settlement.'[21] Differing soil types 'reflect different water-holding capacities of the soils'. The hardest of underlying cemented coral material is Te Ba: it is said to contain water that is of greater purity and less salty than may be found under other, softer cemented substrate materials on Tamana.[22]

CORAL GROWTH AND THE FRESHWATER LENS

Corals dwell tribally on the bones of their ancestors: the new ones literally stand on the shoulders of the biophysical accomplishment that precedes them.

Freshwater almost never exists at the ground surface level on coral atolls and reefs. With rare exceptions, rainfall and land height are insufficient to enable freshwater to 'actually flow out from the edges of the island into the beach,' says Roger Lawrence, who researched social and economic life in Tamana in the early 1970s. In 1973 the population of Tamana was 1392. The atoll, a 1 x 5km reef, is the closest of the Kiribati group to Tuvalu. The average rainfall was just over one metre a year.[23] Historical and official reports indicate the island was subject to drought of varying extent in every decade forward from the 1870s, and even in moderate drought, trees died.[24] However, 'at no time either in recent history or oral tradition did the freshwater lens beneath the island fail or become too saline to drink'.[25]

A fringing coral reef is formed from coral growth around an emerging volcano. As a product of activity along tectonic plate boundaries, the volcano will eventually subside and be slowly pulled away, leaving the reef encircling a lagoon originally shaped by the volcano's caldera. Strings of atolls therefore mark areas of tectonic movement – active or inactive.

Lawrence explained that 12,000 years ago Tamana was 'a plug of coral debris extending well above sea level, and, as coral cannot grow much above the sea, subject to weathering'. As the seas rose after the last glaciation this 'drowned plug' slowly rebuilt towards the new sea level, eventually exposing the reef at low tide. Wave-driven coralline algae and coral debris attached to the reef and grew outwards as a platform. On Tamana the reef, hardened by the effects of air, is shaped by a tug o' war between natural build-up and knock-down wave and freshwater action.

Rainwater is held above sea level in the accumulated ridges of coral. Since this is where the new growth is, the edges of islands such as Tamana, especially to windward, are higher than the centre. Particulate coral just above sea level has the ability to retain a film of freshwater through capillary action, which slows the mixing of incoming fresh rain with seawater: this is what is referred to as a freshwater lens. The coral can hold freshwater at a height above sea level, moving up and down on the tide.

The differing densities of fresh and saltwater enable freshwater to sit atop seawater – water salinity increases with depth. But not all coral islands have freshwater lenses: their existence depends on certain essentials, such as the porosity of the island surface and the rapidity with which rainfall can soak in. Clearly, in times of high rainfall the lens is thicker than in times of drought.

Plants – and inhabitants who dig wells through the coral rubble – depend on such lenses for survival, especially in times of drought.[26]

So inherent was drought to the islanders' way of life that they named three divisions for their entire island, according to viability – a specific indication of the availability of reliable water and soil quality for growing crops in hard times. The area with ridges of stony ground thrown up on the eastern coastal edges by gravel-bearing storms was called aba ni mate, 'land of the dead'. The other two divisions were described as aba ni maiu, 'land of the living' and tetabo, 'the place of staying alive'. This was knowledge developed from at least 2000 years of apparently continuous habitation.[27] The Tamana classification wisdom enabled them to visualise, comprehend and name what for modernity showed up only in aerial photography.

Even in times of moderate drought, palms struggled to set fruit in aba ni mate, whereas aba ni maiu, with its better soil, sustained many palms and fruiting pandanus trees, replanted following severe droughts in the mid-1930s. It was only here that palms could self-establish more prolifically in the dark humus that supports a number of different plants. While palms tend not to thrive in water-heavy soils, in years of moderate rainfall and drought in 'the land of the living', Lawrence wrote, 'these lands are of greatest importance as drought refuges'.[28] Critically, the low-lying tetabo held a lens of water close to the surface, sometimes evident in wet conditions; here, 'it's a real jungle, nuts germinate where they fall and the palm forest is a mixture of different-aged trees'. Even in a drought, palm trees could flourish. 'Most importantly,' says Lawrence, 'ownership rights were suspended – there was no notion of superior rights in a drought.'[29]

It was in the 'land of the living' and the 'place of staying alive' that Tamanans survived during a notoriously challenging drought of the 1870s that affected the entire Pacific. When no palm fruit formed, 'the decimated Tamana population kept itself alive on fish from the sea and toddy [highly nutritious sap] cut from palms in the *tetabo* lands'.[30]

Some land favoured healthy palms, while neighbouring patches did not. Lawrence explained: 'I didn't realise what I was looking at was drought response and replanting history in the "land of the living". These were natural-born ecologists and their naming was so evocative.'[31]

At the heart of the 1873–76 drought was a story Lawrence was unable to fully authenticate: such accounts are exceptional because there are few survivors. He was told of a mother, Nei Miringining, a 'free woman' with no husband and therefore no one to cut toddy for the family. She supported her children through the extremities of the drought by feeding them thin slices of flesh cut from her arm. The day the drought broke, she died.

But there is another, longer story from that same drought that belongs at the heart of all endeavour recounted in this book. The story is from Banaba (meaning 'hollow land'), the raised coral reef island right out on the westernmost edge of Kiribati, east of Nauru,[32] with a population today of about 400. To Banabans their island was 'the navel of the world'. Their founding myth was of Auriaria, a giant of the tree people who, significantly, took the pandanus tree from Tabiteua, northwest of Tamana, to Sāmoa, where he planted it before returning north to Banaba.[33]

In the annals of climate change in Oceania, there is no more graphic eyewitness account of the devastating effects of a prolonged El Niño event on island people than that which occurred here in Banaba. At 81m it is the highest piece of land in the entire Kiribati configuration, but it harbours neither stream nor freshwater lens.[34] As a raised atoll, pushed upwards by its 'parent' submarine peak, it is subject to underground weathering effects that over millennia have cut caves in the calcium carbonate of the coral debris. It shares its particular geological history with Nauru and Christmas Island. Through much of the twentieth century, it was a so-called protectorate and seat of British rule, in thrall to the British Phosphate Commission. Much of its population was shunted largely to Fiji during that period of intense extraction.[35]

Banaba was a prominence, a subterranean peak crowned in coral in an archipelago of flat reefs and atolls. Though never more than five metres above sea level, its relative elevation was due to phenomenal deposits of guano. This was from a time when bird populations in the

Given half a chance, nature will seek to rebuild life, even on Tarawa's remnants from the bloody Pacific War. *Photo Richard Sidey*

uninhabited Pacific rose above land in circling clouds of unimaginable numbers: to voyagers the appearance of birds that were closely associated with the shore, such as frigatebirds, fairy terns and some nobbies and boobies, was a welcome navigational aid to locating nearby land.[36] In fact, the frigatebird is said to have been carried by Polynesians on journeys in Kiribati and Tuvalu and then released to fly back to the point of departure as a messenger.[37]

This was when the deep seas off the reefs carried an even greater biomass of fish life. Over millennia the guano, in places lying nearly 20m deep, turned an atoll into a grey hump in the ocean.[38] The abundance of seabird life may also lend weight to the reputation that the Banabans – essentially a Polynesian people out of Sāmoa, but with a Melanesian backstory – were regarded as the finest fishers in the Pacific.[39] Their revered Tinirau, widely esteemed as a guardian of fish, is a son of the great god of the sea, Tangaroa. A dancer and lover of exceptional prowess throughout the greater Samoan range of Polynesia, Tinirau, whose name figures as far south as New Zealand's Whanganui River, is a Tuvaluan favourite.[40]

When the British Phosphate Commission mined guano for superphosphate from Banaba and Nauru and exported it to Australian and New Zealand farms it enabled more intensive farming of sheep and cows in the Antipodes for decades. Historian Damon Salesa, of Samoan heritage, sums up the mutual costs from another Pacific source of guano: 'Nauru produced almost all of the phosphate that proved key to the domestic New Zealand agricultural revolution – the 'grasslands revolution' of the 1920s and 1930s … This was dependent on an exploitative colonialism aimed not just at possessing the land, but also uplifting it and taking it away. In a strange irony these phosphates were then mixed with the soil of farms, themselves obtained in earlier iterations of New Zealand colonialism.'[41]

During the 'protectorate' period, Britain, Australia and New Zealand extracted 80 percent of Nauru's surface area, and on Banaba the effects of mining were similarly significant. There is a deeper irony. The atolls' contribution to short-term improvement in agriculture in Australasia has increased their vulnerability to climate change in unforeseen ways. Methane (CH_4) emissions from sheep and cattle make up one third of New Zealand's carbon emissions.[42] That issue has only been intensified by more recent importation of extract from Indonesian palm kernel – so often grown in areas where tropical rainforest has been razed – to supplement herd feed. And over some 30 years, cumulative phosphorous levels from several sources became a major pollutant in New Zealand rivers and lakes.[43]

So, in the brief course of European expansionist history, these violations of both the land and the long-settled peoples of Oceania have been compounding. Now, the fate of the low-lying islands seems almost inevitable: perhaps not even coral growth can keep the atolls abreast of storm surges and rising sea levels. The inhabitants will be deprived not simply of their liberty to live as they choose, but of their culturally underwritten ability to stand on their own ancestral land. For them, life and culture may never be the same. Diversity – in language, ways of seeing and doing – would be lost to the world.

Arthur Grimble, British resident commissioner of the Gilbert and Ellice Islands from 1926, offers a definitive and disturbing account of famine and drought in all its horror on a remote outlier of the Pacific. If ever there was a graphic account of a prime reason for traditional Oceanic migration, then it was this. Nothing accounts for the respect for and reverencing of water throughout the region more than does famine. More poignantly, Grimble's account offers a jolt as to how it is that freshwater figures as a primary and often transcendental place in the islanders' culture.

After serving as a junior functionary to the Gilberts earlier in the twentieth century, Grimble returned in 1926 to preside over the Gilbert and Ellice Islands. (While he grew close to the people, when the islanders protested in 1927 at the inequity of a new deal proposed for phosphate extraction, it was plain that Grimble's deepest loyalty remained to commerce and Empire.)[44] His record of a serious and growing water shortage, leading to drought, was taken from the account of Eri, an aged survivor who had been a native magistrate at Banaba. Today we know that it occurred during what we now refer to as an El Niño event of the mid-1870s.[45] (We already know how it affected Tamana.) This struck at the very heart of the island, whose only link to the outside world was the occasional passing ship.

Grimble explains that in years of plenty Banaba was tropic in vegetation, 'beautifully bowered … in the dark green of forests, the starry white of lilies, the flung foam of scarlet and crimson petals'.[46] Macdonald, who speaks of a 'narrow freshwater lens' two to three metres below the surface, says that only a few basic plants – taro, bananas, pawpaw and yams – were able to be grown, together with babai, the giant taro.[47]

Drought came in seven-year cycles, marked like a gauge in rings on the trunk of every coconut palm left standing for more than 10 years.[48] That nearby Tuvaluans were familiar with hardship may be seen in the tradition on Vaitapu, the group's largest island, where couples were allowed only two children; and on Nukufetau, just one.[49] However, this El Niño event continued to deepen, affecting many places in a broad sweep from northern China to northeastern Brazil.[50] For Banabans it was a calamity. Grimble chronicles the drought's grim progress:

> There were no more flowers anywhere after two rainless months. After six, the pawpaws and guavas, the custard apples and soursops were dead, the mangoes and wild almonds dying. After twelve, half of the island's coconut palms stood headless, while those that lived on, their leaves burned rusty black, had been fruitless for many weeks. Then, even the mighty, deep-rooted forest of calophyllum trees [tall, straight and flowering] that covered the island's middle was stripped of its leaves. Our two thousand acres [80ha] of phosphate and coral rock, left naked to the sun-glare, lay flinging back the savage heat in a white-hot column to heaven.[51]

He describes the tantalising presence of soaring 'black battalions' of cloud standing offshore, filled with the promise of rain:

> Almost, yet not near enough: in the last moment, you saw it waver, halt in the middle, torn apart there by the uprushing column of furnace-hot air. You watched its sundered halves pass by, spilling their torrents into the sea a few hundred yards from the coast on either hand, while, between them, under a sullen grey but rainless sky, the stricken land thirsted on.[52]

By then the Banabans had run out of the usual brackish water available deep in the water lens of the island's coral flat, on sea-borne sand deposits, or 'sand pans', as Grimble calls them. The men headed their canoes out into offshore rainstorms in the hope of collecting runoff, channelling rain with palm mats into wooden bowls. There were also grottos, the bangabanga of Banaba: old-formation coral weathered into a system of caves through which freshwater flowed, lodging in depressions.[53] As a descendant – and a woman – Katerina Teaiwa was taken to the bangabanga in a three-hour journey crawling through water-filled, extremely confined spaces bedecked with stalactites. 'The water was refreshing, very cold, but I did not shiver … I knew I was in a special place with so much significance, a realm that had once been exclusively for women and the only source of water for countless generations of Banabans. It was exhilarating.'[54] Customary sanctions back then prevented visits to the caves by individuals, but in times of stress, parties of women were permitted to find their way into these places to collect water in rationed numbers of coconut shells. Village councillors strictly enforced the rations. More water would be discovered deeper in the bangabanga, but not until much later when technology arrived with the phosphate miners.[55]

Such was the horror of the event of the 1870s, none of the islanders spoke of it to Grimble until 1930, after they had known him for 16 years. However, according to the growth rings on the palm trees of that time, there were none older than 60 years – back to the time of the famine. Eri, the old magistrate, referred to it as the curse of Naka; he recalled how the drought went on for a year, so that each family was limited to a daily ration of one coconut shell of water; this measure kept them going for another year, at which stage even the fish that were usually around the island had gone much further out to sea.

Gradually, whole families died; the bangabanga waterways dried up and survivors were reduced to sucking seaweed and moisture from the eyes of flying fish. Eri had been engaged for some time, but when he visited the family home of his fiancée, he found her parents dead and his beloved lying gravely ill; she died in his arms. Eri's account is steeped in pathos: it ends with him and his mother lying down at the edge of the sea on mats, for relief, and Eri awakening to find his mother floating – drowned while she slept.[56] Soon after, a trading vessel from New Zealand arrived and took Eri, his father and a few other survivors to O`ahu, Hawai`i. 'There we lived until my father died, six years later, and then I returned to this place, because I owned no land anywhere else.'[57]

Harry Maude, drawing from other accounts, suggests that during the drought a number of whalers and cargo boats picked up distressed Banabans and took them to other inhabited islands. He documents cases of 24 people being taken to Kosrae in the Federated States of Micronesia, and others to New Caledonia, Fiji, Sāmoa, Tahiti, Hawai`i and the Caroline Islands.[58]

But home is where the heart is, and nowhere more so than in the Pacific: even after such a tragedy, people still returned to Banaba. And now, 150 years later, with an inexorable climatic phenomenon in train, cultural ties remain just as strong.

Scientists from the University of Auckland have studied the physical effects of climate change on the Tuvalu island chain. In their study of the Tuvalu nation's string of 101 reef islands and nine atolls, Paul Kench and his team have shown that the disruptive effects of climate change in islands immediately southeast of Tamana, Banaba and Nauru may not be simply a case of atolls and reefs disappearing beneath the waves of rising sea levels. The situation could be far more nuanced than this. Remote sensing data collected over a 40-year period, when 'sea level has risen at twice the global average', showed that eight atolls increased in land area (by 73.5ha). Just one, Nanumea, recorded a loss of land of 0.01 percent. In the case of reef islands, the pattern was variable: 74 percent increased in size and 27 percent decreased in size. The total land gain was 80.7ha, but this varied hugely from island to island: it was the tiny islands of less than 0.5ha that showed loss of land of greater than 50 percent.[59]

The authors concluded that far from being a reason for inaction, such trends should allow for more considered, gradual and culturally sensitive adaptations for communities of the Pacific who are facing the effects of climate change. Some of this may involve shifting to better-placed islands within 'the neighbourhood'. They argue, too, that the international community, which has ordained so much of the islanders' recent destiny, needs to make room for the kinds of choices those who may need to move will face. Funding is essential, 'including the development of opportunities for appropriate economic growth'.[60] Once again, in the midst of this great perturbation, science reminds us that there are subtleties within the apparent chaos.

Other research, however, shows a picture for the dwellers of reefs and atolls that reinforces the bleaker predictions of Anote Tong. Another recent paper on 29 atolls in the Marshall Islands sums that up in its title: 'Most atolls will be uninhabitable by the mid-21st century because of sea-level rise exacerbating wave-driven flooding'. The authors hold that the anticipated 1.6m rise in sea level would be devastating for the low-lying territories of the Pacific in less than five generations. More frequent 'wave over-wash' would cause a saturation of the ground surface, threatening infrastructure. Equally crucially, the natural freshwater lens, already projected to be less frequently replenished by rain, would be damaged by its inability to recover between wash-over events. This trend, anticipated within a lifetime for some places, would occur as 'the freshwater lens floats increasingly higher relative to ground surface'.[61]

Whether in the Pacific, Atlantic or Indian ocean, those larger islands that stand high enough for their freshwater to withstand such

A flying fish rapidly flicks its tail to gain speed while skimming the ocean's surface. *Photo Richard Sidey*

impacts are still vulnerable to over-washing waves. This is because in most cases critical infrastructure – towns, airports, hospitals and so on – is 'located within a few metres of current sea level'. This means that island homes as far apart as the Cooks and the northwest Hawaiian islands, the Gilberts and the Line islands of Kiribati, and even west to the Seychelles and the Spratly Islands in the South China Sea are vulnerable.[62]

Roger Lawrence, writing nearly 50 years ago, argued that the prognosis for future perturbations of climate on Tamana was 'that the population will be less well able to cope with future vagaries in climate and could need periodic support if droughts of several years' duration occur'.[63] While drought will be an issue, the global warming paradox is that the cycles of weather perturbations now are also delivering record rains across Micronesia in its entirety and many other parts of the planet besides.[64]

But as always, remote, exotic and easily dismissed as they may be, Kiribati and its neighbours are us, writ small. Everything they must now contend with – rising air, waterway and sea temperatures, fish shifts, water shortages, crop failures, lack of space for people and their waste – are issues that the rest of the globe will also face, sooner rather than later.

TIDAL WRACK'S LARGESSE

Gardening was essential to the survival of inhabitants of the atolls and islands of the Pacific. Some of the techniques for crop growing were innate, some were adaptation; and nature itself also threw up ideas for new arrivals. English missionary S.J. Whitmee visited a number of islands on his 'missionary cruise in the South Pacific' in the 1860s, including some in the Tokelau group where, on one atoll, he found a population of 136, 'having been greatly diminished by the Peruvian slavers'. He described 'gardens' that occurred naturally in the tidal wrack:

a series of sandy mounds running parallel with the coast, and varying from 10 to 50 feet [3–15m] in breadth. Some had cocoanut-palms and pandanus already growing on them and producing fruit; on others the vegetation was of a more recent growth; while others were bare, or with only a few of the pandanus fruit which by some accident had been cast upon them, sprouting and giving promise for the future. The outer mound was fully 50 feet across it, and had been washed up during heavy weather at the beginning of the present year.[65]

On Funafuti (Tuvalu) there were other signs of naturally flourishing sandbanks, where 'besides the cocoanut Palm – which is almost sure to be found wherever a sandbank raises its head above low water mark – and the pandanus, we found a few breadfruit trees and bananas, with two species of the edible arum or taro ...' The effort in growing plants such as taro impressed him:

They dig large trenches, like wide moats along the centre of the islands. Some of these are from 100 to 200 yards [90–180m] across them and from 8 to 10 feet [2.4–3m] deep. To carry out the sand from these trenches must have been the work of generations. These low levels are moist, and on them they make as much soil as is possible by throwing in decayed wool and leaves, and here they plant everything which requires social care. Nothing edible but the Cocoanut and the pandanus grows on the upper sand. We brought two cases of useful plants from Samoa, and distributed them amongst the islands we visited, in order to add to the temporal, as well as the spiritual well-being of the people.[66]

Nukunonu, Tokelau, where the sea currently washes within a few metres of the island's infrastructure.
Photo Richard Sidey

CHAPTER 12

And in the Northwest

THE PACIFIC AS ECOSYSTEM

The forests of the Pacific Northwest – in this case Alaska – have similarities and differences to those of Aotearoa. *Photo Aliscia Young*

There are signs, however tentative and fragile, of a re-awakening to the wisdom of the past, a wisdom rooted in place.

– James Lichatowich, *Salmon, People and Place*[1]

Nothing brings home the symmetries between the State of Oregon and the nation of New Zealand – at diagonal poles of the Pacific Ring of Fire – more than the mirror-imaging lifecycles of their two most emblematic native fish: the Pacific salmon and the longfin eel. Both places are home to indigenous peoples who underwent abrupt colonisation within a few years of each other. Each place wears the signs of active volcanism and earthquakes, in a temperate climate that sustains – apart from fine wine and boutique beer – a critical rarity: their rainforests (in Oregon's case, technically, these are 'seasonal rainforests', in Aotearoa 'temperate').[2] There are also striking physical similarities in the big rivers and chains of dunes and dune lakes along parts of their western coastal flanks.

As a state within the 'Pacific Northwest' (Oregon to Alaska), Oregon has forests that are predominantly western hemlock, with Sitka spruce in the fog-prone coastal district and Douglas fir dominant further inland.[3] The forests of Aotearoa are more diverse: 80 percent endemic, they are internationally recognised as one of the world's biodiversity hotspots. The dominant trees are giant podocarps, notably tōtara, rimu, mataī, miro and 'wet-footed' kahikatea.[4] The Aotearoa rainforest grows in the moist and temperate lowlands where it shows characteristics of tropical rainforest such as buttressing for stability in damp soil.

In both places, wherever there is water – from brackish water at the forest's edge, to peat wetlands, river systems or mountain streams – top predators exist: the salmon in Oregon and the eel in Aotearoa.[5] There are two main species of eel in Aotearoa and seven species of salmon in the Pacific Northwest.[6] The shortfin eel (*Anguilla australis*) is shared with Australia and New Caledonia; the longfin eel (*Anguilla dieffenbachii*) is endemic to Aotearoa.[7] Māori have multiple names for various species of eel; the common name is tuna.[8]

The bulk of Aotearoa's other freshwater species derive from the 25 galaxiidae species.[9] Constrained by earth's equatorial barrier like so many freshwater fish, including salmon, the galaxiids therefore are of 'southern circumpolar' sources arising from 'South American freshwater fish fauna'.[10] The equator upholds the distinctiveness between the two hemispheres.

The lifecycle of the Pacific salmon and the longfin eel are in curious ways polar opposites. From diametrically opposed parts of the Pacific, each embarks on a journey that is epic and transformative – usually at the onset of autumn in the respective hemisphere.[11] Both are ancient species, with a lineage of 10–20 million years. Evolving through the cycling of the planet out of the Pleistocene into the Holocene, Pacific salmon and the three Australasian species of eel carry the trace of even older lineages, back 40 million years – that is, through the Oligocene era to the Eocene, in the *Anguilla* eel's case originating in Indonesian waters.[12] From the Pleistocene era the modern expression of these two very different fish has evolved in step with the dynamic formation of rivers in their freshwater habitat. Neither is present in the territory of the other, but each has evolved as the land has changed, through massive upheaval – tectonic plate shifts, volcanism, erosion and glaciation – to coalesce into today's landscape ecology of forest and rivers.[13]

Though some salmon were left landlocked in the process, the five different species of Pacific salmon in genus *Oncorhynchus* separated out some six million years ago from their ancestor and each other.[14] Of the seven varieties, only steelhead and sea-run cutthroat trout survive their arduous task of spawning;[15] the others die soon after they reproduce – 'a finely-tuned and effective evolutionary strategy that … seeks to maximise the transfer of genetic material from one generation to the next'.[16] Their anadromous lifecycle, defined by a freshwater juvenile phase, a maturing adult phase in the ocean followed by a return to their natal stream to procreate and die, is evolution's way of optimising access to their food requirements. Their return, by smell, to their natal streams involves hundreds of kilometres of travel that, once in freshwater, is in a state of 'fasting'. The returning females shapeshift the gravels of upriver beds to create a redd (nest) in which to spawn. The males battle for mates and fertilise eggs before, depleted and exhausted, this congress ends in death. Covering a range of 32–70 degrees latitude North on the coastal margins of North Asia and the Pacific Northwest,[17] salmon have been exposed to a recent human history that is out of step with their ancient biological cycles. Despite massive loss of original habitat, highly engineered rivers and pollution, the prospect for renewal, based on greater respect for those natural patterns, still remains.

The shortfin and longfin eels, on the contrary, live in a catadromous cycle – equally epic, but a reverse journey to that of the salmon. It begins with the eels spawning in an as yet undiscovered site in a 10-km-deep trench off Tonga, an island group 2300km away from New Zealand. In these tropical seas, each longfin female produces up to 20 million eggs.

From the fertilised eggs the eels evolve in four stages. First the larvae drift on currents that carry them towards their homeland. From there they move gradually into estuaries where, miraculously, they transform into slender glass eels: in flows of teeming numbers the glass eels peak at the height of the whitebait

There are seven salmon species in the Pacific Northwest. *Photo Richard Sidey*

Millan Ruka handling a longfin eel.

Where river health is not in question, neither is the health of eels such as these pictured on a kahikatea-lined tributary of the Waiohine River.
Photo Jason Burton

fishing season, but they are distinctive from the galaxiids. Then, as greyish elvers they migrate far upriver: they are capable of climbing 20–30m waterfalls (tuna riki). High-country tuna take an average of 25–30 years, sometimes much longer, to grow to maturity in undisturbed inland waterways. Feeding on native fish such as galaxiids and bullies, the longfin can mature into one of the largest freshwater eels in the world, weighing well over 25kg; some may grow to 40kg.[18] Then hormones trigger changes in their body, sending them – like the spawning salmon, foodless – on their great return journey to their source: in this stage Māori call them tuna heke.[19] After mating, they die.

Just as salmonoids (trout or salmon) never evolved in Aotearoa, neither, apparently, did the eel evolve in the Pacific Northwest. The eel

FROM THE STARS: GALAXIIDS

It is the native fishes of the *Galaxiid* genus that come closer to the salmon cycle in New Zealand. Most of these, apart from a few rarities, are diadromous (their lifecycle contains both a river and a marine phase, but without quite the odyssey of salmon or eels). The fry of these species, together with young bullies and smelt, make up the fabled whitebait catch. Would that those who harvest and consume them might show similar enthusiasm for the survival of these species![20] Despite the loss of habitat and decline in many places, these galaxiids are still unprotected. They breed in freshwater; the juveniles head downstream to the ocean for maturation into adulthood, then they return to repeat the cycle. There is no 'death spawning' by galaxiids.

New Zealand's native grayling (*Prototroctes oxyrhynchus*) – extinct now for about a century, probably as a casualty of trout introductions – was not related to its homonym in the northern hemisphere.[21] Southern waters also sustain a 'sucker-mouthed fish', the piharau or lamprey (*Geotria australis*) – genetically unrelated to the northern lamprey – which spends its adult life in the ocean before returning to rivers to spawn.[22] In the Pacific Northwest the ecological contribution of the Pacific lamprey (*Entrosphenus tridentatus*) lies in its oxygenation of creeks in its bed-nesting activity and enhancement of nutrient availability for other species, especially salmon. It seems entirely possible that, just as science is recognising its significance, it is heading, as Native Americans say, for endangerment.[23] In Aotearoa this may not yet be the case – certainly not in Central Otago – although traditional takes of piharau on the Whanganui River are today but a shadow of those of a century ago, when they swarmed in abundance, and were a highly prized food of manaakitanga (hospitality). The lamprey coexist with eels and other natives such as the *Galaxias* species and are found in many parts of the southern Pacific, including Australia, the Chatham Islands, and Chile and Argentina in South America.[24]

has long rejected much of the central Volcanic Plateau of the North Island from its range; and it may well have never evolved in the regions of fire and ice of the Pacific Northwest for similar reasons.[25]

The life patterns of the Pacific salmon and the longfin eel have meant that not only did they evolve into keystone species, but they became keystone cultural species for the indigenous Native American and the Māori who occupied their ecosystems.[26] These patterns are indicative of so many other such rhythms and pulses across the Pacific – including those of the whales, bartailed godwits, and short- and long-tailed cuckoos – that have also found a place in the migration stories of Polynesia. They are linked to seasonal and weather shifts, currents, tides, and the passage of weeds and insects, other fish and marine mammals and birds, seeds and spore across the Pacific. Each strand is a study entire unto itself and all may be spun into a golden narrative of relationship to the life that underpins sustainability. Many strands are broken or scarred by those who, in past centuries, occupied, rejected and often overwhelmed – seldom comprehending that they were breaking into a carefully evolved compact between culture and nature, where culture mediates perception and understanding.

So critical were the returning salmon to the entire ecosystems of the glacier-scavenged upland country that their influence extended well beyond the reaches of their water habitat deep into the life of the watershed: 'returning salmon substantially enrich carbon and nitrogen cycles in the vicinity of spawning areas'.[27]

Thanks to the popularity of books such as Tony Taylor's *Fishing the River of Time*, set in British Columbia,[28] the unanticipated collateral benefits of the salmon's anadromous cycle are now more widely understood. This animal is a dynamic replenisher of the interior river

The bald eagle is a top predator of salmon country in the Pacific Northwest. *Photo Richard Sidey*

systems, bringing many ocean-derived minerals to the upper reaches and streamlets. At least 22 species of mammals and birds benefit from salmon-food. Magnesium and phosphorus, for example, are vital to the diets of bears as they prepare for another winter of hibernation. In those cold, wet uplands, the discarded remains of untidy bear meals, spread by the opportunistic raven, eagle and wolf, further contribute to land ecosystems within about 500 metres of the river.[29] These animals deliver the ocean-borne nutrients into the forests of the watershed, where their dispersal may enhance the stature of the riparian spruce, cedar, willow and hemlock forests by as much as threefold. And when these huge trees die, their collapse into streams 'sculpts pools in the streambed, providing deep water for large salmon species and calm eddies in which juveniles can mature'. In other words, the salmon are the keystone species for the ecosystems of the

The rich minerals in the salmon consumed by the black bear are ultimately returned to the forest.
Photo Richard Sidey

entire northwestern region. They bring life to many other species and, equally, are dependent for their continued vitality on the feedback mechanisms of the watershed, such as the fallen trees, to which they have contributed. Without them, over time, nothing that exists naturally there will remain as it was.[30]

As far as we know, this is where the ecological narratives of the eel and salmon lifecycles part company. The line we can draw is between a continental-edge ecology and an island one. Since the Tongan site for eel mating is as yet poorly defined, we could speculate that the nutrients from their bodies do contribute to other life on the deep ocean floor, on slow release. But in those dark depths, who knows? Whereas the 'salmon forest' of the watershed in the northwestern Pacific benefits from the disposal of salmon flesh, we would never refer in New Zealand to 'eel forest' or 'galaxiid forest'. But Aotearoa's forests may have been similarly dependent on other animal interventions, especially of birds and, in particular, petrels, whose presence across the prehistoric Pacific was on a scale unimaginable today. Burrowing petrels were crucial in the dispersal of phosphate across pre-human forests. This again is a natural, inestimable agent of dispersal of nutrients on land from the 'oceanic foodchain'.[31] Today, as pasture, these rich lowlands of the old forest lands require massive doses of expensive imported phosphates that, with intensified dairying this century, have created one of the big environmental challenges Aotearoa faces. Our waterways and lakes are especially vulnerable.[32]

Salmon were not the only species with a history rooted in the water cycle that caught the eye of the first human explorers who crossed from Asia during the last glaciation 'along the edge of mile-high glaciers to the ice-free areas of the Pacific Northwest'.[33] There they met a vast range and abundance of species. The traditions of the Gitga`at First Nation people of the Northwest, for example, hold memory of that time in their deep reverencing of the Moksgm`ol (Kermode or spirit bears). These creatures may still be found only with the assistance of native 'bear whisperers' in Canada's Great Bear Rainforest. The traditions explain that these light vanilla-coloured animals arise from a request of the creator, Raven, to Black Bear. They were to ensure that one in every 10 cubs would be white, as a reminder of a time when the world was covered in ice.[34] For many thousands of

In the traditions of the Gitga`at First Nation people of the Northwest, Moksgm`ol or spirit bears can only be found with the assistance of native 'bear whisperers'. *Photo Richard Sidey*

Moksgm`ol (Kermode or spirit bears) in Canada's Great Bear Rainforest. *Photo Richard Sidey*

years the First Nation people, made up of four culturally distinct groups out of Asia, joined to share the salmon with the bears and other creatures of the catchment.[35]

Those First Nation people who discovered the Pacific Northwest founded a culture rooted in respect for and unshakeable connection with nature. Salmon and bear, among others, became intimately in and of their culture. They acquired the wisdom to put the salmon at the heart of their relationship with the land. They learnt the ways of salmon and how to catch and prepare them for food. Theirs was an attitude grounded in humility, gratitude and respect towards nature in the context of the place they occupied.

Robin Kimmerer, a member of the Potawatomi Nation, is a distinguished professor of environmental and forest biology at the State University of New York. She is a leader in the movement that seeks to recover community as it works to recover the land and old ways. She advances the practice of receiving from nature, but never taking – nothing but 'an honourable harvest'.[36] So too, as Jim Lichatowich reminds us, 'Northwest Indians melded the concept of the gift and the biology of the Pacific salmon into a sustainable economy.'[37]

Similarly, for many Māori the eel represented the cultural keystone species. The taking, curing and giving of the bounty from big-river, eel-rich foodways like the Whanganui River was as embedded in their culture as the tuna was embodied in the waiora (health) of the river.[38] Māori Marsden explained this principle of manaaki, giving, as a gift of Tāne. It was Tāne who, in the first whare wānanga, set up the principle of mauri manaaki. Out of this came the tikanga of tuku rangatira (noblesse oblige). One of three associated principles, manaaki (to bestow a blessing) is engrained in the utu of reciprocity: 'The presence of visitors was equivalent to the bestowal of a blessing upon the hosts. On the part of the hosts, they bestowed a blessing upon the guests by giving them the best of their provisions in the hākari (banquet) and hospitality provided. This was a reciprocal relationship which could be extended by the exchange of gifts.'[39]

An elegant North American expression of such reciprocity is the potlach, one of the most famous of gift cultures, which addressed the fish relationally as the salmon people. For thousands of years, potlach was central to the northwestern tribes, the Kwakiutl, Tlingit and Haida among others.[40] Lichatowich describes it as deriving from nature, based on 'the age-old concept of the gift and a belief system that treated all parts of the earth – plants, animals, rocks – as equal members of a community'. Tribes who embraced it understood their debt to nature and the dangers of personal accumulation: an essential aspect of gratitude in all relationships.[41] This, with its echoes of Polynesian traditions, is completely at odds with the market economy: 'The gift economy of the native northwesterners was fundamentally in harmony with nature's own economy.'[42] More specifically, the potlach is founded in myth and story, as Lewis Hyde explains it:

> It was the Indian belief that all animals lived as they themselves lived – in tribes – and that the salmon, in particular, dwelt in a huge lodge beneath the sea. According to this mythology, the salmon go about in human form while they are at home in their lodge, but once a year they change their bodies into fish bodies, dress themselves in robes of salmon skin, swim to the mouths of the rivers, and voluntarily sacrifice themselves that their land brothers may have food for the winter.[43]

The first salmon of the season to show on a local river was caught and ritually welcomed by a priest as a visiting chief. Speeches were made and songs of welcome were sung before everyone was given a piece of the fish to eat. Finally – and this is what makes it truly a gift cycle – the bones of the first fish were returned as an intact skeleton to the sea. The belief was that they would reassemble; 'the fish would then revive, return to its home, and revert to its human form'[44] – and that respect shown ensured the cycle would continue for another year. As Hyde writes, 'Organic increase is at issue, gift exchange preserves the increase; the gift grows because living things grow.'[45] It is this that distinguishes those who live by and of their land from those who see nature as an infinite resource for exploitation.

Jim Lichatowich, fisher, scientist and long-time advocate for natural salmon flows. *'Angling Oral History', Montana State University*

It is also true that the myths and practices the First Nations developed were not the result of innate kindness to nature, but grew from experience, including disastrous early encounters with their new lands. While causation is still debated, in North America, 73 percent of the megafauna was extinct within a short time of their arrival.[46] Māori caused the extinction of a quarter of the bird population they encountered after settlement in Aotearoa,

Lake Papaitonga is an extensive dune lake and once the home of the Muaūpoko in the Horowhenua, on the Tasman coast. *iStock.com/MollyNZ*

most dramatically the country's megafauna, the moa.[47]

But both pioneer groups came – slowly and, being human, perhaps clumsily – to a deep understanding of nature. Geoff Park offers insights into how the most fertile ecosystems on the planet – wetlands – were a representation for Māori of reciprocity at work in nature. Lake Papaitonga, where Māori had an island fort, was one of his favourites, surrounded by a poignant remnant of what was once the extensive forest wetlands of the Horowhenua Plains – one of many natural homes for eels in Aotearoa:

> Tāne, manifest in the tall trees of the dry sand ridges, is locked in a reciprocal embrace with the swampy arms of the lake, feminine in its flows. Embued with the female mauri, or life force, Niwareka, daughter of the union between Papatūānuku and Rangi, made possible the plants like harakeke, nīkau and kiekie that bind the people to the land – and gave the people the skills to weave them.[48]

Unfortunately, the usual combination of 'guns, germs and steel'[49] devastated Native American and Māori in tragic numbers. With them went rights to land and standing in ways that defy the rule of law, not to mention common decency. So, as the 'neo-Europeans'[50] moved in with their technology and their attitude of 'improvement', those massive migratory and regenerative dances performed by the two keystone species took hit after hit from either side of the Pacific – as did the old cultures who had built their lives, their societies, around and with them.

As early as the mid-eighteenth century, it was the eastern colonists of America who watched 'their' Atlantic salmon fisheries go into rapid, continuous decline. And despite what should have been a chastening experience there, the same attitude continued to compromise healthy wild salmon flows in much of the exploited Northwest more than a century later, notably between 1829 and 1929.[51]

It is a story retold countlessly across the planet.

The first impacts were those of canneries and lumberjacks, with no thought for past or future. They took all that they could, by whatever means:[52] the settlers in the Northwest fished, netted, racked and canned the wild salmon flows to endangerment. And in New Zealand for much of the twentieth century, acclimatisation societies, seeing nothing in the eel but a predator of their imported trout fingerlings, were legally empowered to extirpate them, together with the equally 'culpable' native shags – actions that were literally red in tooth and claw.[53]

The assaults in both countries also involved deforestation on a swift and ecologically destructive scale, thereby removing an element of river habitat that was vital for salmon and eel alike. In Oregon much of the clear-felling took place more intensively after the waterways had been overfished – followed by crude exploitation of fish, often to the point of extermination. Foresters penetrated this country, often thwarting flows and denaturing entire catchments. Today it is old-growth forest in the coastal Oregon range that 'contains some of the most unfragmented old forests remaining on the West Coast of the United States'.[54]

From the time the first neo-Europeans entered the west, drawn by the vast and millable 'timber lots', the old ways and many of the old inhabitants of this land found their number was up. The new culture that never relented came yet again with its uncomprehending extraction, depletion and flow disruptions. Lichatowich has established 'the characteristics of the natural and industrial economies' (see table below): it may be summed up in his poetic observation that the salmon's plight was nothing more than

CHARACTERISTICS OF THE NATURAL AND INDUSTRIAL ECONOMIES[55]

(with a focus on rivers and native fish)

Natural economy	**Industrial economy**
Watersheds and water define organisation	Political hierarchies
Fish and habitat co-evolving relationship, defined by biophysical limits	Conflict and competition, often over boundaries & ownership
Complex, sometimes invisible, still poorly understood by science	Infrastructure visible, recognisable, understood
Driven by solar energy	Driven by fossil fuels and profits
Favours dispersed production, e.g. varieties of species	Favours centralised production; factories/hatcheries, monoculture
Productivity maintained by internal processes of the ecosystem, e.g. nutrient transfers	Fertility maintained through use of externalities such as fertiliser, pesticides and hybrids
Circular, renewable, production-enabling	Linear, extractive, production-oriented
No waste, all is recycled	Resource-demanding, creates waste, toxins
Watersheds are complexities of connected habitats	Watersheds divided into discrete parcels of production, logging, irrigation etc
Rivers are complex, interconnected mazes of habitat	Rivers are simplified and modified for human use

'lack of ecological companionship'.[56] In the first column of the table, 'Natural economy', the watershed is supreme, all its elements contributing to a diverse, 'renewable', 'recycled' ecosystem with the capacity to 'internalise' nutrient capture and transportation. On the left-hand side, 'Industrial economy' refers to 'political hierarchies' of development that are monocultural, external inputs solely for the maintenance of production. These have proved to be intrusive and wasteful. In the first category rivers are seen as complex, connected mazes of habitats; in the industrial category they are 'restructured, simplified and controlled to meet [short-term] human needs'.[57]

Permaculture farmer-philosopher and engineer Gary Williams of Ōtaki, Aotearoa, finds resolution between the linear and the circular in the universal exchanges that timelessly underpin natural life in the energy of the spiral.[58] This is represented in Māori culture as the koru, often said to be the unfurling frond of the silver fern. Lichatowich has emphasised how the relationship between salmon, people and place was broken down by the new culture: 'Euro-American fishermen moved from the rivers to the ocean where they harvested salmon in mixed aggregates of stocks from many different rivers and regions.' Then, as 'the relationship between salmon, people and place was strained to breaking point' the neo-Europeans moved on, or turned to fish factories (a term used for hatcheries); these they 'managed in ways that contributed to the separation of salmon from place'.[59]

There is no easy fix. Some US$17.9 billion has been spent on salmon regeneration in the Columbia River system alone since 1984. This has contributed nothing to the wild fishery, says Lichatowich from his Oregon home with views to Mount St Helens, overlooking the broad back of this mighty river.[60] Over a lunch of his own canned salmon this reflective man made the point that engages conservationists across the planet: 'Different world views have different social norms.'

He had an early apprehension of this while on a trip in 1992 to the Olympic National Park, northwest Washington, with foresters. Lichatowich had eyes only for the resplendent old-growth forest, contributing to healthy river systems with their good salmon flows. However, 'All the foresters saw was waste – these trees that would make fantastic timber.' Through the great New Zealand forestry debates of the 1970s and 80s, especially, I interviewed senior European-trained foresters who, when they saw the magnificent trees of Aotearoa's natural forests, were itching to get out their chainsaws. One senior official protested sincerely, 'But in Europe we manage our forests in 300-year cycles.'

In a recent shared paper, Lichatowich stated: 'The extirpation of wild Pacific salmon in forty percent of their historical range in the Pacific Northwest and the extensive listing of Pacific salmon stocks and under the federal Endangered Species Act (ESA) are a strong signal that the current salmon management paradigm has failed.' Government agencies, his paper argues, 'must act as trustees of the wild salmon and protect them consistent with the long-standing public trust doctrine. Their obligation is to maintain the wild salmon legacy in good health for citizen beneficiaries of present and future generations.'[61]

Then there were the wonders of hydroelectricity. This was an imposition that again took place with the shallow optimism of modernism, without regard to old patterns. The ever-ready US Army Corps, which created 'Cadillac Desert' in the American West, from 1945 threw up 60 hydro dams across the rivers of the greater Columbia River watershed.[62] The benefits of hydroelectricity to humans are undeniable, but unaccounted-for salmon flows diminished as crucial habitat contracted, their ancient cycle halted by the towering imperviousness of the dams. Few who tried won against the relentless atavism and skulduggery of the US Army Corps of Engineers. One – old-fashioned Republican Richard Wilson, who defended the Eel River, Round Valley, Mendocino County in the 1960s and 70s – once remarked to me, 'The developers always have more money, often better lawyers and time on their side.' After preventing the planned colossus of a dam, Wilson went on to become, for many years, a respected director of the California Department of Forestry and Fire Protection.[63] Northwestern salmon takes – commercial, offshore, in-stream and recreational – have persisted in spite of the evidence of declining yields. Like the accumulation of toxins, however, species decline up until a sudden breaking point can be deceptively gradual.

That very same process took place in many parts of Aotearoa. Major rivers of suitable head were dammed and diverted, with the result that the essential relationship between forest, topsoil retention and water was weakened by the twin fiats of farming and forestry, as well as mining, not to mention growing urban demands. And often, those who were last to receive the benefits of 'improvement' – notably electricity – were the indigenous people who had had to make the greatest sacrifice. For Ngāti Hikairo – to cite but one of many examples – the harvest of kōaro, their mahinga kai, especially on Lake Rotoaira, was effectively wiped out overnight.[64] Again, a vital element of nature was taken, not received – and without consultation. Loss of the biomass of Aotearoa's longtime traditional fisheries has not been documented to anything like the same extent as that of the salmon fisheries of the Northwest. However, loss of habitat for eels in the Waikato catchment alone, since colonisation and exacerbated by the five hydroelectric dams on the river, is estimated at around 84 percent.[65]

While the deployment of artificial fish passageways has long been discussed, sometimes tried and occasionally executed, lack of committed resources and of testing has often left them as wasting assets, with dams inflicting heavy ecological casualties. On the Rangitaiki River, the Matahina Dam has been the site of manual transfer of elvers and, later, mature migrating eels across the dam in both directions for 25 years.[66] The technique, largely the work of volunteers, has been picked up at a number of other New Zealand dams, notably the Waitaki Dam in North Otago, in a cooperative effort between the operating company, Fish & Game New Zealand (the angler umbrella group) and local iwi.[67] While it's far better than nothing, this is still very much 'after-the-event' mitigation. Only in 2018 has a reforming New Zealand government legislated for better access for migrating native fish. It is a first step, with a comprehensive reform of water use and protection under way.

With wild salmon flows now limited in the Pacific Northwest, the web of life to which they are central and the people whose entire culture was once shaped within that web have been deeply compromised. It seemed that the ingenuity of the invaders knew no limits. Over 150 years, buoyed by the exuberance of American entrepreneurship, the salmon runs went into dizzying decline.[68] Wild salmon throughout the region today are represented in just 40 percent of their original runs, nearly all of them significantly diminished.[69]

The answer, as Lichatowich has consistently advanced in his many publications since the 1970s, is that hatcheries are not habitat. They have, he asserts, added almost nothing to the vitality, plurality or numbers of wild salmon. On the contrary – and this experience is by no means limited to America – the diseases of the captive stock have sullied and weakened what remains of the wild populations.

Enabling salmon habitat to revert to its former state is key to the approach adopted by rancher-conservationist Johnny Sundstrom, of the Siuslaw Basin. He is the noted negotiator between users, fishers and ecologists through the Northwest over the protection and regeneration of its salmon rivers.[70]

Restoration of suitable habitat is also the only hope that Mike Joy can hold out for New Zealand's declining numbers of tuna and other freshwater native fish. Without the ecological activism necessary to secure such protection, he predicts that 'most if not all' of our entire range 'will be extinct by 2050'.[71] His fear for the longfin is that, while populations still appear to be standing up well in a number of rivers, there is evidence that they could go the way of America's passenger pigeon: here today, in numbers estimated at six million birds; gone tomorrow.[72] 'You have all these breeding eels heading for the same spot, out in the depths of the Pacific,' Joy says. 'There's got to be sufficient numbers – they've got to be able to find each other, but the crowd is too small.' Besides, our rivers now are warming, another reason for more riparian forest, given the sensitivity of all our native fish to small changes in temperature.[73]

This predicament is amplified in fishing for eel, says Don Jellyman, emeritus researcher for NIWA: 'Historically, longfin eels have been our most frequently encountered freshwater fish, but today it's becoming increasingly difficult to find large females.' He finds that, 'commonly, fishing takes out all the females'.[74] For entirely different reasons, this is how one of the world's few territorial river ducks, the native whio (blue duck), came perilously close to extinction. The ground-nesting females die off because they are the ones most vulnerable to introduced predators such as stoats and feral cats. Happily, strenuous efforts in habitat-based science, management and investment have painstakingly lifted the whio's status to Threatened – Nationally Vulnerable.[75]

It's a similar story for the Columbia River and its two major tributaries, the Willamette flowing through Portland, its valley enfolding most of the population of Oregon, down to Eugene; and, further east, the Snake River. Historically they supported runs of salmon reliably estimated at between 10 and 16 million, including five species of Pacific salmon and the steelhead trout.[76] Today, these salmon runs number around 1.5 million, 80 percent of which are hatchery-based; wild salmon are just a small percentage of this. It comes down to what Lichatowich continues to demonstrate is a failure of management systems to take into account 'their ecological and life history needs'.

So broken is the system that Native Americans may be prosecuted for fishing without a licence. In Washington State – and there alone – this has been corrected, following determined litigation. Today its First Nation people have a right to 50 percent of the fishery.[77] In Aotearoa a similar arrangement has been in place under rulings pertaining to all tribes under the Treaty of Waitangi, but only in relation to ocean fisheries. Māori rights to values in freshwater, including fish, remain largely unresolved.

Apparently enacted in the best interests of First Nations, a series of treaties were signed between 1850–55 in Oregon, but the enscribed protection of the respective indigenes in their land and forests and fisheries proved to be hollow promises, and the treaties have proved largely chimerical.[78] Consequently, Native American views seem to be, at best, still operating more at the 'gift' than 'gain' level.

The Treaty of Waitangi, New Zealand's single, multitribal (though not universal) 'high-

Thriving, healthy tuna at a river's edge, Aotearoa. *iStock.com/LazingBee*

level' agreement with the Crown, was signed in 1840, but it was not until 1975 that the Treaty was formally recognised by the New Zealand Parliament. Only since the 1980s have Māori been able to make retrospective claims on the Crown for limited compensation for wholesale loss of their rights. Some considerable progress and positive cultural shifts have taken place since then; however, yawning injustices remain. The copious flora and fauna report, based on a case for the loss of mātauranga Māori (evolving Māori wisdom, bound into culture and identity) has until recently sat silently unaddressed by governments for a decade.[79]

So, a mere 150 years of disconnection and 'ownership' into a 20 million-year narrative have brought regenerative stocks of many wild salmon and longfin eels to the point where they are officially endangered. This hasn't stopped commercial fishing of eels in Aotearoa, and the remedy being discussed by officials is to farm them. It is to be hoped that other approaches will also be taken, given the expensive lessons from Pacific Northwest salmon. There is no good reason why some of our rivers, perhaps the Whanganui's main stem, could not, as habitat improves eventually under the Treaty settlement reparations, become eel/lamprey preserves.[80] Commercial farming is also a line of least resistance to the interests of the independently funded and serviced Fish & Game New Zealand. The irony is that the exotic trout and salmon that give this body statutory focus enjoy a legal protection not yet extended in a meaningful way to any living native fish.

The Native American Student and Community Center at Portland State University is as much about recovering and upholding culture as it is about formal study. One senses here a quiet

Forest fires in northern California turned its skies red in the summer of 2018.
Photo Bruce Warrington

persistence and strength of purpose in the face of profound cultural and ecological dislocation. The attitude seems to be that so much has been lost, yes – and that there is so much to do. Prompted by my questions, concerns ranged from ecosystem recovery and urbanisation's effects to hydroelectric dams. There is also the toxic history of their rivers, particularly from the postwar era, that provoked Rachel Carson's *Silent Spring*; and oil and climate change are top of mind, too.

It was here I met Christopher Rempel, who affiliates to the Chinookan, Kalapuya and Klamath native peoples. As we spoke the California wind drift carrying the smoke of a string of horrendous fires had reached as far north as the Klamath Basin, edging into southern Oregon. Rivers without forest, in Jim Lichatowich's aphorism, are salmon without rivers.

Christopher is aware that a number of hydro dams in the Northwest have reached the end of their life. Usually rendered redundant by siltation and age, some dams have been dismantled in the Klamath Basin, on California's border with Oregon. Rempel said: 'A lot of companies have resisted, but there is evidence that it is safe to do it. But there are no incentives, however. And it is the same with old pipelines. On paper they can nominally still be used, but the companies never shut them down entirely because they are resistant to the cost of dismantling.'

Since reclamation of the vast 73,000ha wetlands in the Klamath Basin began in 1906, a fractious history has coloured water-rights

Klamath Basin, in its lower reaches a place of inspired cooperation between farmers, First Nation people and conservationists in wetland/wildlife management. It is on the Pacific flyway, but climate change is impacting its water supply. *iStock.com/Samson*

contests between farmer–irrigators and Native Americans. The city of Klamath Falls in southern Oregon also has interests in extracting geothermal water for heat from aquifers in the basin.[81] The wetlands, home to their usual rich diversity of benthic invertebrates and extensive bird and fish life, including coho and chinook salmon in the river and protected sucker fish in the lake, have become 'walking wetlands'. This management system has involved 'draining wetlands to create farmland, and flooding farmland to create wetlands', then swapping their roles. This idea from Ron Cole, then manager of the Klamath Basin National Wildlife Refuge, entailed considerable negotiation. An agreement was secured in 2009 by collaboration between the Klamath people and the farmers who rely on the water for their production.[82] A seemingly impossible conflict between preservation of nature and its exploitation was resolved largely through ingenuity and flexibility of attitude on both sides. Crucially, this intriguing collaboration 'seems to work as long as there is sufficient water,' Rempel observed, but 'it can be fairly tense … As water supplies are reduced, we say "shut it off for the salmon".'

The wetlands, which had been reduced by 25 percent through reclamation, have now been increased by an additional 1600 hectares. Potato production in the area is up by about 25 percent. Fertiliser and pesticides inputs – and costs – have dropped by the same percentage, making organic farming possible in some places. Wildlife has increased by an estimated 50–70 percent.[83] Recent droughts, however,

A waterfall on the McCloud River near Mt Shasta, Northern California.
iStock.com/CampPhoto

have threatened the trade-off arrangements.

Many other First Nations are still awaiting a time when their relationship to the salmon might be restored by a company that's prepared to make sacrifices, or to cut their losses responsibly, in order to enable salmon to run their natal streams again in numbers that allow a return to customary harvest. The modern cult that grants shareholders primacy over all other interests is a major barrier to change.

One such tribe has been the patient Winnemem Wintu on the McCloud River, Northern California. Their sacred relationship with the chinook was severed by the 183-metre-high Shasta hydroelectric dam raised between 1938 and 1945: their wild salmon population, deeply bonded into their sense of place, history and culture, was effectively wiped out. Imagine their incredulous delight this century when they learnt that chinook originally out of this very same river were still extant and running wild – in a number of the braided rivers in the South Island of New Zealand. This benefit was an unintended consequence of early twentieth-century acclimatisation policy.

The South Island chinook population is descended from ova collected, packed in sphagnum moss on ice, and shipped across the Pacific by the New Zealand Marine Department to their purpose-built hatchery in Kurow, South Canterbury.[84] Although efforts to achieve populations that might support salmon canneries were eventually abandoned, sufficient salmon 'took' to their new environment. These are the forebears of today's sporting salmon fishery. Further research confirming their

The enormous Shasta dam, which has imperilled the traditional fishery of the Winnemem Wintu nation. *iStock.com/Ranger Ron*

genetic origin is still in process, but Ngāi Tahu of Te Waipounamu, who have been actively supporting the tribe, remain optimistic. Attempts by the guardians of the river to dismantle the hydro dams on the McCloud, however, have been met by the federal administration with plans to raise the dam even higher, with further loss of sacred sites.[85]

As the world knows from the resistance in 2016 at the aptly named Standing Rock, oil pipelines carried across natural waterways are another ongoing issue over tribal and environmental rights. Rempel fears that the outlet proposed for oil shipped out of Washington State, which may have been defeated, will now shift to Coos Bay in southern Oregon. Apart from the usual risks, such as pollution, installation would necessitate the dredging of Coos Bay, where the tankers would be loaded before their journey across the Pacific to China.

For indigenous peoples, who are often closer to the action, fossil-fuel emissions and their multiple denaturing effects loom large. 'Two degrees C warming will simply make it intolerable for salmon,' Rempel says. Under normal circumstances nature provides an equilibrium, a stasis model, that enables the fish to move between ocean and into riverwater, but 'as the freshwater becomes warmer, it interferes with that state'. Evidence from the Columbia River Inter-Tribal Fish Commission, formed by the four Columbia River Treaty tribes in 1977 for the comprehensive restoration of migrating salmon, shows this is happening. Just upriver from Oregon on the Columbia River are the

Fog rises over the Tongass National Forest in southeast Alaska. *Photo Richard Sidey*

Yakama, one of the four tribes that make up the inter-tribal commission. 'We've seen the tell-tale signs of climate change for several years now,' they report. 'It will affect each and every living thing, including us. Everything depends on water for life – our foods, our salmon, our roots and berries. Many tribes will be affected.'[86]

Toxins: What lurks beneath

Legacy issues, buried deep in the gravels of some rivers of the Pacific Northwest, will resist best efforts to restore them. For instance the Columbia River, which is 15m deep in places, has long supported a succession of inland seaports and shipbuilding and dismantling activities. Shipwrecking is often code for a graveyard of some of the worst toxins known to man. Consequently, this river system hosts a series of sites the Environmental Protection Agency has designated for clean-up. Accumulations of PCPs, DDT, lead, arsenic, mercury and heavy metals are set to shadow in-stream restoration for many years to come. The worst of these is a remnant of the much-quarried Ross Island on the Willamette River at Portland: now a park, it remains a toxic site of daunting proportions.[87]

Chris Rempel worries that 'nature is being poisoned faster than we can compensate. When companies can still come in and make messes, what is the point? Near Mount St Helens there has been some good clean-up work on the Columbia, but new companies are coming in to create more mining, more waste.' Twenty years after designation, these 'Superfund sites' have still not been addressed. The issues are systemic. 'Clearly, we can't restore fully, but it's ridiculous to keep allowing it to happen,' he says. 'Modern consumer life creates waste naturally, whether plastic or e-waste, fossil fuels or nuclear production.' He is speaking of a closed economy.

When Jim Lichatowich says, 'The obligation of public trust is to pass on to future generations an environment that is as close to what it was and not just what was left over by industry and even hatcheries,' the ecologist and the First Nation spokesperson are not far apart. Government agencies, Lichatowich and co-authors write, have an 'obligation to maintain the wild salmon legacy in good health for citizen beneficiaries of present and future generations.'[88] In fact, the argument that public trust and Treaty obligations have been betrayed by successive governments, to the detriment of both habitat and long-sustained indigenous fisheries, has been to the fore in Aotearoa for over 20 years. The situation just may be about to change, though.[89] A collection of essays edited by water ecologist Mike Joy summarises both these issues for Aotearoa: the book offers a collective of sustainable/profitable alternative thinking along already-tested pathways.[90]

In both countries it comes down to this. In Oregon:

> The lack of fish is not the problem. It is a symptom of the real problem, which is the degraded state of the salmon's sustaining ecosystems. A conceptual foundation that defines the problem in terms of symptoms will have difficulty identifying an appropriate solution.[91]

And in Aotearoa:

> Key to progressing in the right direction is keeping people with vested interests as far away as possible from decision-making positions; thinking long-term; and finally, choosing the change that maximises multiple gains by ensuring problems are not considered in isolation.[92]

For both Oregon and Aotearoa it is a given that genuinely indigenous voices will deepen informed decision-making. Joy's plea to eschew special-interest groups has merit, and does not mean the exclusion of tangata whenua, nor of collaboratively derived recommendations. Recently tested, extensive collaboration has foundered, however, on the then government's 'failure to respect the consensus' that might have been achieved in protection of New Zealand rivers.[93]

Oregon's natural areas are part of the great coastal temperate rainforests 'stretching from the redwoods of California to the vast stands of spruce and hemlock on Kodiak Island, Alaska'.[94] Common through the coastal region of Oregon are forests of hemlock and maple, together with Douglas fir (or Oregon pine, *Pseudotsuga menziesii*). The latter is the timber that gave its name to the state. (Douglas fir is, today, the second most-planted exotic tree in New Zealand; and a similar tree, macrocarpa (*Cupressus macrocarpa*) – the iconic exotic of old New Zealand farm landscapes – is a survivor from small native stands in Monterey, California.)[95] Oregon's southern border with California is at 42 degrees north: it is thanks to the effects of benign ocean currents that its climate is normally warmer than might be expected.

Twenty years ago, science reported that in the coastal temperate rainforest – essentially coastal redwood forest between San Francisco Bay and the south Oregon coast – 'ground fires are common in summer and fall, but generally they have little impact on dominant trees. Catastrophic fire occurs along the coast only every 100 years or more.' However, in the lush seasonal rainforest from southern Oregon to central Vancouver Island, periodic fire was also important to forest life.[96] Today, however, as climate change starts to kick in, textbooks are being rewritten.

Take the year 2017. In August, Oregon's southwest forests were riven by fires in series and simultaneous, taking out about 59,000ha. These were historically unprecedented. The fire up in the Columbia Gorge took months to put out. Further fires in October were even more persistent and widespread; and the following year more than 170,000ha of forest was severely affected by fire.[97] Tragically, a fire in California wine country, further south, saw the loss of 47 lives and 2500 businesses in 2017; another blazed concurrently in Orange County. Fire struck again in December in the Santa Ana region to the north of Los Angeles, with destruction of 10,000 structures and property. From a population of 370,000, 250,000 were at least temporarily displaced. Heavy rain brought on new forest growth in the fall and this, followed by water shortages – much in the north induced by horticulture on an industrial scale in the Sacramento region – exacerbated the risk of further fires. And in the summer of that same year there were fires further north in British Columbia that took out 900,000ha of forest, its greatest fire in recorded history.[98] The following year was worse. Between July and November 2018, what came to be called the Mendocino Complex Fire set all-time records in its range (about 185,750ha), duration and number of fatalities (104); the costs are estimated at US$3.5 billion.

In all, 12 of the 15 largest fires ever in California have occurred since 2003, each attributed by the state government to the effects of climate change – with these trends expected to increase in range by 77 percent this century. In 2020 the story continues. From a salmon perspective, rivers without forests are rivers without salmon.

In the Pearl District of land-locked Portland, a biggish city in the American Northwest, I came upon a class of children in the rain tending a public garden of about half a hectare of open space. The city towers ranged on all sides of this space, a reticulated rainwater holding-pond where engineering had enabled the restoration of water-edge plants and trees. It was like a scene from my village at the top of Te Waipounamu, where schoolchildren have added to the beauty of a recently re-established wetland by planting a variety of eco-sourced natives in a low-lying park, a buffer to insistent sea-level rising on a shore just a few hundred metres away.

The Willamette flows close by Portland. Ecology as utility, design as nature. It was still raining – the salmon were just starting to run down in the Siuslaw River – and anoraked pupils weeded in that random way of 8- to 10-year-olds.

These were pupils from a central urban school nearby, its premises another refurbished building in the now hipster neighbourhood of repurposed old industrial/commercial buildings: many are now apartments, or outfitted as offices, restaurants, art galleries and accommodation for agencies like the Ecotrust Building/National Capital Center complex. Such places are reminders that Portland has long been a leader among those cities and other centres working to integrate their water engineering into the wider weave of policies and initiatives that will better serve citizens, and land and water environments.

On the east side of the city is Crystal Springs Creek – cut off from its parent Johnson Creek for 80 years by earthworks and culverts. Since 2014 it has been reinstated and restored sufficiently for coho salmon and cutthroat trout to return. At 3.8km long, with its old-growth riparian forests granting shade and food to cleaner water, it has become a home for birds and other wildlife within city limits.[99] 'I'm realising,' says Portland City environmental services senior engineer Tim Kurtz, who shows me some of his agency's handiwork, 'that the little pieces we still have here of original nature are more important than I understood in my training … We really do view the plants as part of our infrastructure,' he continues. And with the added advantages of improved habitat for birds and other animals, and enhanced nature for humans to enjoy, 'It comes down to a value judgement as to how much does this really mean for you as a citizen – how much paving? For us it is managing an ever-growing amount of impervious surfaces.'

From my hotel room in the Pearl District, close by the river, I gazed onto a rooftop atrium garden of trees, grasses and a mix of absorbent and hard surfaces. Similar efforts are evident across the CBD.

The Native American Student and Community Center at Portland State University is housed in a striking architectural statement that invokes both modernity and tradition, with a rooftop garden of lushness and fecundity and views of large trees all round. The city of Portland is intent on reducing runoff and mitigating carbon emissions, and rooftop gardens on larger buildings are key to this. 'They not only reduce stormwater,' says urban ecologist Marion Dresner, 'they can cool air and reduce energy use in a commercial or domestic building.' Ecoroofs are now mandatory on many new buildings in Portland. There is a lot of this happening, and not just in the US: Paris announced recently that it would require all new buildings to have either solar panels on the roof or an absorbent cover of green growth; and China is now energetically planting vertical gardens on high-rises as one response to climate change.[100]

In Portland's centre and suburbs the happy union of engineering and ecology is evident. Not that Portland doesn't have its share of roads and pavements, but many have been transformed into what Dresner refers to as bioswales – sometimes just simple plantings of rushes on a verge. The idea is to keep surfaces – whether at ground level or on high-rises – as green and permeable as possible, and to give trees a chance everywhere in the city. In many parts, where space allows, concrete or steel-edged soak zones have been established. Open parking lots, supermarkets and zones of disconnected downpipes all lend themselves to planters, swales and rain gardens.

Of course, with leadership, motivated citizens can also make changes, sometimes by entire 'green streets' that, in some cases, cut stormwater flows by up to 85 percent and pollution runoff by even more. Entire neighbourhoods are turning hard surfaces into greenscape edges on pavements and roads, and claiming the stormwater utility discount from the Clean River Rewards 'treebate program': 2000 green streets and 350 streams are being cared for under neighbourhood adoption schemes.

Nothing is simple, though. The more berms and roads quarantined – in reeds, grasses or in trees and hedgerows – the less space there is for other demands. The more public parks, the less rates are available to fund reforms. And the more humane and friendly a city's social policies are, the more likely it is the interstate homeless will fetch up there – as has happened.

Dresner sees a big role here for more citizen science, which she refers to as garden ecology: 'I believe in science … and the need to pay attention to what is going on in a garden. People have been sold a bill of goods about the importance of natives – exotics too can often serve the needs of birds very well. We should be making decisions based on first-hand experience.' She cites Bob Schaffer who, writing for 40 years for Columbia Broadcasting System, the *LA Times*, *Boston Globe* and others, advocated for an understanding of science as essential to democracy. 'We could use him and his platform today,' Dresner observes.

Even in the 1920s, however, Portland was decidedly reactionary, with a mayor who was known to be a Ku Klux Klan member. By 1930 its defining feature, the Willamette River, was pronounced biologically dead.[101] Between 1967 and 1975 Robert McCall, a long-time TV reporter with a bent for environmental stories, was governor for two terms. The city was already growing rapidly, its population made up of the kind of folk the world over

The Native American Student and Community Center at Portland State University: urban regeneration, cultural regeneration, hard surfaces softened by nature.

who might just as easily protect a fishing environment as empty their trash in it. McCall valued wildlife, open spaces, access to beaches – and he achieved strict boundaries to contain city limits. Dresner recalls that there was active consultation with the public, who came along with him. McCall was succeeded by Robert Straub, who from the 1960s had been influential as a senator and then state treasurer with progressive environmental and social policy.[102] It was Straub who led the addition of 4500 hectares of park and recreation in the now thriving Willamette River Greenway area.[103]

In the 1960s Portland State University pioneered pedestrian access and park design around its library buildings in particular, in a relatively tight space; the project was led by renowned modernist landscape architect Lawrence Halprin. He adopted a holistic approach, based on awareness of individual and community needs. He designed open space sequences contained by buildings that brought nature, water and pedestrian conductivity together in a way that now seems prescient. At the university, Halprin 'unpaved the way', connecting soft surface corridors between buildings, using big trees and the creation of parklands.

The Willamette River (the name derives from the Native American name Wallamet) is the thirteenth largest river in the US, and drains 12 percent of the entire state of Oregon. Rising in the mountains south of Eugene, it flows 300 kilometres atypically north to join the even mightier Columbia River at Portland.[104] The river has shaped the steep parts and the economic character of the inland city port that is Portland. Reforms of the city intensified in about 1992, when the US Environmental Protection Agency (EPA) required the City of Portland, among others, to separate out its sewerage system from its stormwater. Most cities in the world still do not do this, so when heavy rains fall, human waste usually gathers in harbours, rivers and along beaches. We call it civilisation, but it's Victorian in concept, from when populations were significantly smaller and humanity's slops routinely were discharged into rivers and harbours. In Aotearoa, all major cities run treatment systems for sewage that discharge into long-outfall pipelines into the ocean; or into rivers. Many small towns, however, still lack effective treatment. Separation of stormwater and treated sewage remains an issue for many places.[105]

A United States EPA report in 2008 indicated that nationwide repair and refit of sewerage and stormwater systems to allow for the impacts of climate change would cost at least $298 billion.[106] 'Our first task,' says Kurtz, 'was to separate as much stormwater as we could from the sewage.' Greywater can now be directed into green environments for direct absorption in green soak areas, or held in special storage chambers 7 metres in diameter. Stored water can refresh groundwater supplies while reducing summer urban heat, with benefits for local climate. 'We often needed to use both as there is no physical tolerance to manage all the volume required,' he said. And as we have seen, there are also green garden programmes in schools.[107]

Portland began this work in 1993, carefully marshalling supportive policies, regulations and remedies as it evolved. It has been world-leading in terms of its integrative management – always the hard part. The overall result, at a time when many cities are struggling with more frequent intense downpours, is that 20 percent of stormwater is now diverted from drains that used to overflow into the sewerage system.

With incentives from the council, 56,000 residents disconnected their downpipes from the system between 1992 and 2011, removing more than 4.5 billion litres of stormwater a year. In tandem with this, 3000 sumps and sedimentation traps that remain in the combined sewer system now collect residential runoff that is allowed to seep into groundwater.[108] 'In some places it is possible to manage nearly 100 percent of the runoff,' says Kurtz. 'It depends on the soil.' Essentially, he says, it comes down to three things – volume, cleanliness and flow control – in ways that protect soil, surface water and, just as essential, the tiny invertebrates that live in clean water.

As severe episodes of rain increase in temperate zones, the city's parks have taken on additional value. Kurtz tells me that a single mature tree with a 9m crown can absorb 2650 litres of rain annually. Encouraging the planting and retention of trees with incentives and policies aimed at achieving a more livable, beautiful city beside an enhanced Willamette River, involves a matrix of related actions. These go well beyond wrangling stormwater runoff from impervious to porous surfaces. It means mobilising citizens in mass transport systems and self-propulsion. Posters on buses proclaim cycling's benefits; bicycle racks are mounted on the streetcars and the city's bus service is frequent and comprehensive. A new cycle and pedestrian-only bridge spans the Willamette, and there are many cycleways. Kurtz says sometimes these hard surfaces compete with the limited space for absorbed runoff. In some parts, however, the tracks between the trams have been planted out in small, tenacious grasses and other groundcover.

As early as 1993 Portland also became the first city in the United States to formally head down a carbon-neutral pathway. Its climate action plan (CAP) set up a strategy for the greater district; the aim is 'to achieve a 40 percent reduction in carbon emissions by 2030 and an 80 percent reduction by 2050 compared to 1990 levels.'[109] In New Zealand it took three successive governments from 1993 until 2018 to develop climate change targets that meant what they said. These – currently still exempting farming's contribution – were announced in 2019 by the Labour–Green–New Zealand First coalition that took power in late 2017. Both New Zealand and Oregon share a major problem of reconciling sound environmental management on the one hand with significant social inequality on the other.

After nearly 30 years Kurtz sees challenges, not least in balancing policies and costs with a housing shortage in a fast-growing city. 'We want lower-cost housing, but there are impacts and the city has to manage. Equally, restoring beautiful places in the city may attract the underbelly of the city – hyperdemic needles and waste.' As for the Willamette River itself, it had been defiled for so long by its citizens that 'no one wanted to go there, it was no longer our lifeblood,' Kurtz noted. The separation into two distinct discharges cost US$1.4 billion at the time; and a number of other river cities, including Seattle, Milwaukee and Chicago, made the same move.

A fundamental for Portland was to reduce the runoff with all its contaminants and residues – everything from automobile toxins and the ubiquitous plastic shards and particles to dog faecal material. Too much of any of this means gradually accumulating toxins in the river, which is already carrying a dead load of PCBs and other dioxins. Natural filters of bioswale can double as ecological restoration to counter such runoff. Dresner was concerned, however, that the original swales were established without much underlying research, and that forests and other ways for holding water are still the priority. But they were part of a suite of options, determined by engineers, landscape architects, planners and citizens.

The approach is to integrate the fundamentals of water management – 'hydrology, habitat, water quality and biological communities,' as Kurtz says. According to some

TIMBER INDUSTRIES' CHEMICAL LEGACY

The dioxin-based herbicide 245T (used to make Agent Orange) was deployed in forestry practices in Oregon and in New Zealand, and dioxin was used as an anti-fungal agent in milled timber. New Zealand has addressed the most chronic cases, but many old timber mill sites still exist, and buried, leaching drums of the stuff lurk in rural countryside.[110]

In 1987 the *Oregonian* reported that dioxin in its most toxic form, 2378-TCDD, had been isolated in EPA samples, one of which was on the Siuslaw River near Eugene and one, at a high 4.5 parts per trillion, 'in two fish caught near the Southern Pacific Railroad Bridge in Portland Harbour'. The latter is in the old industrial heart of Portland, close to the EPA-designated 'Superfund site' of former heavy petrochemical industries and pulp and paper-making, both implicated in the PCP residues still in the river and available to send dioxin up the food chain.[111]

sources, Portland is among the top 10 cities in the world for environmental progressiveness. But the city can't afford to rest on its laurels. 'We still have plenty of work to do – expanding our stormwater, focused metrics to encompass all the benefits city-wide, understanding the impacts of climate change and how that changes the math, and balancing costs between and stable funding for construction (capital) and long-term operations and maintenance (operating).'[112]

How might other municipalities mobilise an electorate behind such measures, when rates may not be reduced, and pay-offs may not be evident within one or even two electoral cycles? Dresner suggests: 'When we visit Paris, the visions that we have for Paris – we bring the myth we carry with us. It's exactly the same for nature.' So what is achievable depends on the hopes and dreams of every citizen. If this becomes a collective ambition it can reset the nature–human relationship. From this, so much more community and civic endeavour can be leveraged. Education, fundamental at the start of Portland's awakening, remains paramount.

Bruce Clarkson, Deputy Vice-Chancellor Research at the University of Waikato, has led a programme of native vegetation restoration that exploits the corridor that is the Waikato River, running through his home town of Hamilton. In his teaching and his leisure time, he has embraced what has become a movement with momentum in mainland New Zealand. This is the restoration and maintenance of native biodiversity. Today there are 73 sanctuaries for native plants, birds and other wildlife on or near mainland Aotearoa, and hundreds of local biodiversity protection and enhancement programmes. Most of the sanctuaries deploy predator-proof fencing, and 42 of them are community rather than government-led.[113] Clarkson argues that over the next 30 years, as the world's population moves from 54 to 64 percent urban-dwelling, urban ecology is likely to become increasingly important. New Zealanders have always lived predominantly in towns and cities, mostly on the sensitive and vulnerable coastal fringe.

For New Zealand, at least, ecologists have calculated what it takes for restoration to add up to anything that supports the original life of an area; 'habitats that are between five and 10 percent of what was there are not helpful to the protection of species,' Clarkson says. 'As the remnants reduce to 10 percent or less, then species go into decline.' In the management of urban areas, it is Hamilton's 'neighbouring' city of New Plymouth that comes closest, with 8.9 percent, and even more planned. Western Australia's Indian Ocean capital, Perth, with 30 percent, does best in Australasia. Christchurch post the 2011 earthquake still has an opportunity, especially in its renewed wetland areas, for significant improvement and is now moving on creative ideas for its uninhabitable red zone.[114]

Currently, less than two percent of the entire basin of Hamilton's originally rich forest soils remain in their natural condition. Difficulties are further compounded with the city's 67 forest remnants averaging just 1.1 percent and, as elsewhere, 'habitat continues to be removed, piece by piece,' says Clarkson; 'cut-and-fill practices for housing development continue to remove its last vestiges'. Despite some legal constraints, this highly destructive practice continues unabated in most parts of Aotearoa. It makes for 'sitting muck' when runoff from the more frequent intense rainstorms flows into drains before discharging into the ocean. The resultant sediment build-up is impacting the ecological functioning of fisheries, in fresh and coastal waters. Integrative management is rarely practised and little recognised by authorities in Aotearoa.

'habitats that are between five and 10 percent of what was there are not helpful to the protection of species'

Clarkson can point to the riparian replanting that is now part of a national movement he calls 'biophilic cities'. Besides planting, dedicated restorators are trapping predators, 'daylighting culverts', restoring stream meanders, removing barriers for the progress of native fish, and recruiting those more plastic disciplines such as landscape architecture that can work creatively at the hard edges of engineering. And they are developing related policies such as the improvement of natural freshwater, which support natural ecosystems and restoration of wider riparian forest.

Portland city has close to 650,000 people and has been rapidly growing. Nevertheless, almost 30 percent of the city is in areas of natural biodiversity, with strict barriers to urban encroachment onto fertile land. This careful approach was abandoned by hell-bent deregulation in New Zealand from 1991 – remedying this is still a work in progress by the current government. In Hamilton, with quiet leadership and the application of the usual principles of community, collaboration, connectivity of people, place, ecosystems – so much more achievable along a river system – and catchment cohesion, progress in a short time has been remarkable. As Clarkson says, 'we have so many gullies, many of them off the Waikato River, and we can aim for 10 percent restoration, because in many cases they are also associated with mana whenua, for example in the Mangakōtukutuku gully'.

Any discussion about urban water management, especially where issues of runoff and disposal are to the fore, inevitably turns on the economics of infrastructure. At a seminar on modern water management in Hawai`i in 2008, Suez Water executive Didier Gaujous spoke of what is commonly referred to as 'the knee of the curve' – the effect by which costs escalate exponentially as a municipality lifts its wastewater management coverage from 80 to 90 percent or more of its catchment. A coverage of 85 percent is what the US EPA has been pushing for. It struck me that, instead of shouldering a two- or even threefold cost burden for inclusion of that last 10 percent of infrastructure, volunteering citizens' cooperation into management of their stormwater runoff might easily push a city over that barrier at no extra financial cost. Many of those actions, whether by incentive or by penalty, are already part of the mix of policies Portland has adopted. Yet such are the complexities of modern administration, Kurtz is not convinced even the most progressive, integrated policies necessarily yield notable savings: 'Robust asset management, system management, system planning and financial planning all have to go together.' Incentives and compliance programmes, too, must cohere. 'Even though you're trying to keep new infrastructure costs as low as possible, it's very seldom you're actually decreasing the size or operational cost of your system.' But surely companies and individuals educated to reducing pollution and mitigating runoff can contribute to better outcomes and at least help hold water rates? He did agree that all remedies need to be examined. To be effective, a stronger, place/community-based ethic will be essential. And – like Kiribati – for governments to set the focus of the single overarching issue of our epoch, climate change, to shape all other policy, from agriculture to health, population to transport.

Modern Portland is not yet reflecting its traditional First Nation history – just as there is not a predominant Polynesian culture in Hawai`i, Aotearoa or Rapanui. But if you take a social/political perspective, for all the above to be implemented, you still need those fundamentals of catchment, community and connection underpinned by indigenous knowledge, sound science and climate change mitigation. A change in consciousness is what's needed: politics as we know it is simply not enough.

On my last day in Portland, when I had completed my interviews in the university district, I planned to walk back to the Pearl District and down towards the Willamette to the 1890s heritage train station. But a swelling ruck of blackness overhead suddenly gave forth in a downpour. I jumped on a streetcar for shelter. Through its wrap-around front windows a rainbow sprang from the dark clouds over the river district spreading out below us. Water in its wondrous incarnations still inspires us to believe that – while timing is everything – anything is possible.

Previous page: The mighty sequoia tree, and this page the author beneath it, in his garden in Nelson, Aotearoa New Zealand. *Photos Aliscia Young*

Coda

It took place during the solstice of the 2018 southern winter and northern summer. Any measure of the international community's security seemed to be beset by the manifestations of our own frailties and fears. There were the raging fires in places with familiar names that were somewhat unlikely places for burning; wars that seemed never to end with their unprecedented refugee numbers; famine – sadly, predictable as well as unpredicted; and multiple outrages perpetrated by the vulgarity of what often passes for global leadership. All in all, much of the world seemed to teeter on profound incapacity and diminishing democracy, exacerbated by protracted and often shameless moral failure.

At the very moment when it seemed nothing could more distract us, an extraordinary allegory took place before the eyes of the billions of people worldwide who were wired into news. This event occurred at a far extremity of the Pacific, in a culture whose living traditions revere, exalt and ritualise water. For 23 days we all followed its playing out. With the plot structure of a whodunnit, this slowly unfolding event became something else. It took on all the elements of a modern 'morality play' conducted by 12 young Thai students and their teacher. Trapped in a cave in a labyrinth, they embodied in time-lapse experience the fate of all who currently occupy the planet: limited resources for survival, limited air and a certainty that, with an ever impending monsoon knocking at the door, rising waters would soon wash decisively through their perilous perch of habitation. Across the globe we watched, connected electronically as a single organism, shaken from its usual atomised distraction to occupy an edge-of-the-seat moment in a shared humanity. Everyone desperately wanted to see them live.

We all know that, miraculously, they survived. But what were the elements of their survival? First, a teacher and leader who, though quite young, was already instructed as a monk in the arts of discipline, restraint, kindness and care. He was Buddhist, but to Christians and perhaps even Muslims, he might just as easily have been a Christ-like figure. His constructive leadership under impossible pressure showed his training, extraordinary insight, compassion and foresight. The boys became his willing 12 disciples. By following his guidance on breath, meditation and discipline he took them from likely panic, disarray and almost certain death, to contemplation, composure, hope – and, finally, life. They were prepared with their tūrangawaewae: a way of seeing that was a way of being.

We know they were rescued, but only through the efforts of what lately have seemed too often to be discarded values: cooperation on an international level and a considerable degree of courage and sacrifice on the ground; for one rescuer, tragically, it meant supreme self-sacrifice. There was even a subplot about technology in this story. Yes, technology was essential and the rescue would have been impossible without pumps, pipes, neoprene and other special equipment. But Elon Musk, with all his ambition, wealth, impulsiveness and inventiveness, played an instructive role: he showed that extreme wealth and extreme technology are not going to grant us all the answers – most certainly not on their own – to deliver humanity from the critical issues we face at this moment.

To briefly recapitulate: this was a human story in which many of the traditional values invoked in this book were activated. To the fore were self-discipline and cooperation, community – all at work within a relational context defined by the exigencies of their 'sovereignty' – and, again, 'the catchment'. The other telling words, also drawn from big themes in the preceding chapters, are again alliterative. While this was clearly not a situation of the group's choosing, the operative words are 'connection' and 'consequences'. In the Tham Luang Nang Non Caves, with their ancient and Buddhist traditions, the importance of place was the other theme, and in ways that are obvious not just to the Thai people.

It is in water-defined places that, I believe, the potential lies for a dual restoration. One is of plants, animals – of ecosystems. The other is a reframing of the way in which we reorient our society, not tribal but definitely attentive to the earth and its nuances; not necessarily communal but certainly community-oriented; not Luddite but certainly inside a closed rather than a consumptive linear system that, on current trends, is not only wasteful and destructive but also, as if we do not already know, ultimately deadly.

Notes

PROLOGUE: PULSING WISDOM

1. While the story remains fresh in my mind, I am not at liberty to recount it.
2. Elsdon Best, *Maori Religion and Mythology* (Government Printer, Wellington, 1982), Part II, pp. 100, 114, 222, 195; Ngārangi Rangiuia, pers. comm., Ōtaki, 12 Dec. 2017.
3. Dale Stokes & Doc White, *The Fish in the Forest: Salmon and the web of life* (Berkeley: University of California Press, 2014), p. 73; see also discussion on the science of isotopes and mist in chapter (Alan Mark).
4. D.K. Rowe, J.P. Smith & S. Grayling, 'Status of koaro (*Galaxias brevipinnis*) populations in the Te Arawa lakes and options for their restoration', NIWZ client report, Aug. 2008: https://niwa.co.nz/sites/niwa.co.nz/files/koaro_status_report_0.pdf
5. Richard Walsh, 'Traditional games on North Pentecost, Vanuatu', in John Morrison, Paul Geraghty & Linda Crowl (eds), *Education, Language, Patterns and Policy*, Science of Pacific Peoples series (Suva: Institute of Pacific Studies, University of the South Pacific, 1994), pp. 218, 229.
6. David V. Williams, *'Te Kooti Tango Whenua': The Native Land Court 1864–1999* (Wellington: Huia Publishers, 1999), p. 114: 'Cultures pattern perceptions of reality into conceptualisations of what they perceive reality to be; of what is to be regarded as actual, probable, possible or impossible. These conceptions form what is termed "the world-view" of a culture. The world-view is the central systematisation of conception of reality to which members of the culture [subscribe and] from which stems their value system.'
7. 'Noam Chomsky: Indigenous people are our only hope for survival', *Ecologise.In*, 4 Nov. 2017: www.ecologise.in/2017/11/04/noam-chomsky-indigenous-people-are-the-only-hope-for-human-survival
8. Seamus Heaney, *Electric Light* (New York: Farrar, Straus & Giroux, 2001), p. 3.
9. Maualaivao Albert Wendt, 'Mauli', in T.M. Suaalii-Sauni (ed.), *Whispers and Vanities: Samoan indigenous knowledge and religion* (Wellington: Huia Publishers, 2014), p. 43.
10. Herman Melville, *Moby-Dick*, ch. 1, 'Loomings' (London: Constable, 1922).
11. Theodore Roethke, 'The Waking', in *The Collected Poems of Theodore Roethke*, 1953 (New York: Doubleday, 1966), p. 104: www.poetryfoundation.

org/poems/43333/the-waking-56d2220f25315/

12. David Young, *Woven by Water: Histories from the Whanganui River* (Wellington: Huia Publishers, 1998).
13. Marcus Thomas & Neil Silverwood, *Caves: Exploring New Zealand's subterranean wilderness* (Wellington: Whio Publishing, 2017).
14. David Abram, *Becoming Animal: An earthly cosmology* (New York: Vintage, 2011), focuses with considerable insight on becoming aware of this kind of phenomenon.
15. Anne Salmond, *Tears of Rangi: Experiments across worlds* (Auckland: Auckland University Press, 2017), p. 256.
16. Kanalu G. Terry Young, *Rethinking the Native Hawaiian Past* (New York/London: Routledge, 2012), ch. 2.
17. Ross Gibson, Centenary Professor of Creative & Cultural Research, University of Canberra, used the phrase at a seminar in Wellington in the 1990s.

1 SEQUOIA: LOCATION IN MEDITATION

1. David Henry Thoreau, *Walden: or Life in the Woods and On the Duty of Civil Disobedience* (New York: Simon & Schuster, 2004), p. 217.
2. Sequoia were planted from early on in rural parts of districts as diverse as the Waikato, Rotorua, Marlborough, Canterbury and Otago. A study in 1978–79 by the NZ Forest Service found that the quality and strength of timber from *Sequoia sempervirens* established in New Zealand were inferior to that from North America (Archives NZ, C395 675, 6506 Box 2/52).
3. Robert Macfarlane, 'Where the wild things are', *Guardian*, 30 Oct. 2004: www.theguardian.com/books/2004/oct/30/featuresreviews.guardianreview35; www.walden.org/what-we-do/library/thoreau/mis-quotations/
4. Callum Coates, Viktor Schauberger seminar, Taupō, 11–12 Nov. 2007.
5. Colin Tudge, *The Secret Life of Trees: How they live and why they matter* (London: Penguin Books, 2006), p. 255.
6. Richard Preston, *The Wild Trees* (New York: Random House, 2008), pp. 210–11.
7. Ibid., p. 212.
8. Ephrat Livni, 'The Japanese practice of "forest bathing" is scientifically proved to improve your health', 13 Oct. 2016: www.qz.com/804022/health-benefits-japanese-forest-bathing
9. Dave Butler, Tony Lindsay & Janet Hunt, *Paradise Saved* (Auckland: Random House, 2014), p. 204.
10. Other native myrtles, notably the ramarama, are also being affected.
11. en.wikipedia.org/wiki/2018_California_wildfires
12. Callum Coates, Viktor Schauberger seminar, Taupō, 11–12 Nov. 2007.
13. 'Summary for policymakers of IPCC Special Report on Global Warming of 15°C approved by governments', 8 Oct. 2018: www.ipcc.ch/news_and_events/pr_181008_P48_spm.shtml
14. Fog precipitates onto the forest greenery and then drips to the forest floor, providing a small amount of moisture during summer dry periods. Fog accounts for about 40 percent of the redwoods' moisture intake: www.nps.gov/redw/learn/nature/about-the-trees
15. Preston, *The Wild Trees*, p. 211.
16. J.T. Salmon, *The Native Trees of New Zealand* (Wellington: Reed, 1980), pp. 40–41.
17. Dale Smith, pers. comm., 21 Jan. 2016.
18. Dale Smith, pers. comm., 18 Jan. 2015.
19. Ibid.
20. Ibid.
21. Andrea Wulf, *The Invention of Nature: The adventures of Alexander von Humboldt, the lost hero of science* (London: John Murray, 2015), pp. 58–59.

2 BOTH SIDES NOW: FOG CAPTURE IN THE PACIFIC

1. Harry Broad (ed.), *Ngā Whenua Rāhui: Our nature, our people, our future* (Wellington: Ngā Whenua Rāhui, July 2015), p. 66.
2. Alan F. Mark & Katharine J.M. Dickinson, 'Maximising water yield with indigenous non-forest vegetation: A New Zealand perspective', *Frontiers in Ecology and the Environment*, vol. 6, no. 1, Feb. 2008, pp. 25–34: https://esajournals.onlinelibrary.wiley.com/doi/abs/10.1890/060130. Mark and Dickinson describe fog as 'an early-stage condensate from saturated air relatively close to the ground'. Fog is literally a cloud on the ground. It can cover vast areas and comes in a variety of types: fogs seen over

the sea are different to those found in valley lowland areas and over mountains: www.metlink.org/other-weather/science-in-the-sky/fog
3. Ngārangi Rangiuia, pers. comm., 12 Dec. 2017.
4. Broad (ed.), *Ngā Whenua Rāhui*, p. 66.
5. Rachel Kaufman, 'Giant redwoods may dry out: Warming to blame?', *National Geographic News*, 15 Feb. 2010: www.nationalgeographic.com/news/2010/2/100215-redwoods-california-global-warming/
6. Joel K. Bourne Jr, 'The super trees', *National Geographic Magazine*: www.nationalgeographic.com/magazine/2009/10/redwoods-earths-tallest-trees/
7. Di Lucas, 'Tussock planting time again', *Terra Nova*, Sept. 1991, pp. 53–54. The author is a highly respected advocate for the protection of the high country.
8. Javier Núñez, 'The Honolulu Seminar Pt 3, concluding on water management in islands, coastal and isolated areas'; 'Guidelines on water management in islands, coastal and isolated areas', Pacific Economic Cooperation Council, 5 Dec. 2008: www.pecc.org/resources/infrastructure-1/849-guidelines-on-water-management-in-islands-coastal-and-isolated-areas/file
9. Mark & Dickinson, 'Maximising water yield'.
10. George Gibbs, *Ghosts of Gondwana: The history of life in New Zealand* (Nelson: Potton & Burton, 2016), pp. 121–23, 126.
11. Alan F. Mark, 'Grasslands', Te Ara – the Encyclopedia of New Zealand: www.TeAra.govt.nz/en/grasslands
12. Mark & Dickinson, 'Maximising water yield'.
13. Eric Pawson & Tom Brooking (eds), *Environmental Histories of New Zealand* (Melbourne: Oxford University Press, 2002), pp. 25–27, 30–32. Māori agency in firing is determined by the sudden and otherwise uncharacteristic increase in the interval between firings following Māori settlement.
14. Mark, 'Grasslands'.
15. Ibid.; Kenneth B. Cumberland, 'A century's change: Natural to cultural vegetation in New Zealand', *Geographical Review*, vol. 31, no. 4, Oct. 1941; James Beattie, *Empire and Environmental Anxiety: Health, science, arts and conservation in South Asia and Australasia* (New York: Cambridge Imperial and Post-Colonial Studies Series, Palgrave Macmillan, 2011), p. 36.
16. Mark, 'Grasslands'; Robert Peden, 'Pastoralism and the transformation of the open grasslands', in Eric Pawson & Tom Brooking (eds), *Seeds of Empire: Environmental histories of New Zealand* (London/New York: I.B. Tauris, 2011), pp. 73–82.
17. Peden, 'Pastoralism', pp. 79–82, in which burning is seen as 'a management tool'.
18. Peter Holland, Paul Star & Vaughan Wood, 'Pioneer grassland farming: Pragmatism, innovation and experimentation', in Pawson & Brooking, *Seeds of Empire*, pp. 50, 51, 73–82.
19. Kelvin Lloyd, 'Ecological processes in the South Island high country', in *Change in the High Country: Ecological stewardship and tenure review* (Wellington: Parliamentary Commission for the Environment, 1 April 2009), p. 6: www.pce.parliament.nz/publications/change-in-the-high-country-environmental-stewardship-and-tenure-review
20. Peter Holland, *The Howling Wilderness: Settlers and the environment in southern New Zealand* (Auckland: Auckland University Press, 2013), p. 108.
21. Peden, 'Pastoralism', p. 75 – a term for leasehold high country from 1852.
22. Roberta McIntyre, *Whose High Country? A history of the high country of New Zealand* (Auckland: Penguin, 2008), p. 187.
23. Mark, *Above the Treeline* p. 342.
24. Mark & Dickinson, 'Maximising water yield'.
25. Alan F. Mark, *Standing My Ground: A voice for nature conservation* (Dunedin: Otago University Press, 2015), pp. 21, 26.
26. Alan Mark, oral history, 3 Feb. 2015, Alexander Turnbull Library, Wellington.
27. Ibid.
28. Ibid.
29. Mark, *Standing My Ground*, p. 46.
30. Ibid., pp. 40–45; Mark, oral history.
31. Mark, *Standing My Ground*, pp. 34–35.
32. Ibid., p. 47–50; Mark, oral history.
33. Mark, *Standing My Ground*, p. 49.
34. Mark, oral history.
35. Mark, *Standing My Ground*, p. 51.
36. Mark, oral history.
37. D.K. Holdsworth & A.F. Mark, 'Water and nutrient input:output: Effects of plants cover at seven sites in upland snow tussock grassland in Eastern and Central Otago, New Zealand', *Journal of the Royal*

Society of New Zealand, vol. 20, no. 1, 1990, pp. 1–24.
38. Mark & Dickinson, 'Maximising water yield', p. 40.
39. Lucas, 'Tussock planting time again'.
40. Neil L. Ingraham, Alan F. Mark & Russell D. Frew, 'Fog deposition by snow tussock grassland on the Otago uplands: Response to recent review of evidence', *NZ Journal of Hydrology*, vol. 47, no. 2, 2008, pp. 107–22.
41. Neil L. Ingraham, Alan F. Mark & Russell D. Frew, 'Short note: Assessing the contribution of fog to high water yields on snow tussock uplands of Otago: Another response', *NZ Journal of Hydrology*, vol. 47, no. 2, 2008, pp. 107–22.
42. Núñez, 'The Honolulu Seminar (Pt 3)'.
43. Aleszu Bajak, 'Fog catchers pull water from air in Chile's dry fields', *New Scientist*, 25 June 2014: www.newscientist.com/article/mg22229754.400-fog-catchers-pull-water-from-air-in-chiles-dry-fields/
44. 'These fog catcher devices trap humidity and collect drinkable water in Chile', *Wonderful Engineering*, 2 Sept. 2015: www.wonderfulengineering.com/these-fog-catcher-devices-trap-humidity-and-collect-drinkable-water-in-chile/
45. Amit Katwala, 'The devastating environmental impact of technological progress', 27 Oct. 2019: www.wired.co.uk/article/lithium-copper-mining-atacama-desert; Dave Sherwood, 'In Chilean desert, global thirst for lithium is fueling a "water war"', 29 Aug. 2018: www.reuters.com/article/us-chile-lithium-water/in-chilean-desert-global-thirst-for-lithium-is-fueling-a-water-war-idUSKCN1LE16T; Nick Wadhams, 'Better mining of energy minerals sought in US-backed effort', 27 Sept. 2019: www.bloomberg.com/news/articles/2019-09-26/better-mining-of-energy-minerals-sought-in-u-s-backed-effort
46. Ingraham, Mark & Frew, 'Fog deposition', pp. 114–15.
47. Mark & Dickinson, 'Maximising water yield'.
48. Ibid.
49. Ibid.
50. Jan Wright, *Change in the High Country: Ecological stewardship and tenure review* (Wellington: Parliamentary Commission for the Environment, 2009), p. 5: www.pce.parliament.nz/publications/change-in-the-high-country-environmental-stewardship-and-tenure-review
51. Mark & Dickinson, 'Maximising water yield'.
52. Ann Brower, 'For the sake of our high country, stop tenure review', 28 Apr. 2017: www.stuff.co.nz/national/politics/91935035/ann-brower-for-the-sake-of-our-high-country-stop-tenure-review; Alan F. Mark, 'Grasslands: Land ownership', Te Ara – the Encyclopedia of New Zealand: www.TeAra.govt.nz/en/grasslands/page-4
53. Wright, *Change in the High Country*, pp. 5–6; Ann Brower, *Who Owns the High Country? The controversial story of tenure review in New Zealand* (Nelson: Potton & Burton, 2008).
54. McIntyre, *Whose High Country?* pp. 336–47.
55. Harry Broad, 'Farming in a fragile landscape', *Terra Nova*, July 1992, pp. 42–43.
56. Eugenie Sage, 'Government to end tenure review', 14 Feb. 2019: www.beehive.govt.nz/release/government-end-tenure-review

3 FIRE AND WATER, ROCK AND FISH: ABORIGINAL MANAGEMENT OF AUSTRALIA

1. Keith Vincent Smith, *Mari Nawi: Aboriginal odysseys* (Australia: Rosenberg, 2010), p. 27.
2. Eleanor Dark, *The Timeless Land*, a trilogy (Australia: Macmillan, 1941); Tom Griffiths, *The Art of Time Travel: Historians and their craft* (Melbourne: Black Inc., 2016), ch. 1. Griffiths gives the name as 'Bennilong': http://adb.anu.edu.au/biography/bennelong-1769; the *Australian Dictionary of Biography* provides no first name; other sources, including *National Geographic*, give his first name as 'Woollaraware': www.nationalgeographic.com.au/history/woollarawarre-bennelong.aspx
3. Smith, *Mari Nawi*, pp. 28–33.
4. Ibid., p. 27.
5. Stephen Muecke, 'The tank stream': https://dictionaryofsydney.org/entry/the_tank_stream
6. David Collins, *An Account of the English Colony in New South Wales, vol. 2* (Sydney: University of Sydney Library), digital text, pp. 142–43: http://adc.library.usyd.edu.au/data-2/colacc1.pdf
7. Bill Gammage, *The Biggest Estate on Earth: How Aborigines made Australia* (Sydney: Allen & Unwin, 2011), pp. 239–42.

8. 'Aboriginal Water Rights', Mitchell Library 87170-AGG, quotes Dr Dunmore Lang, 15 Nov. 1840; Gammage, *The Biggest Estate on Earth*, is entirely about this relationship.
9. Michael Cathcart, *The Water Dreamers: The remarkable history of our dry continent* (Melbourne: Text, 2009), p. 16.
10. Collins, *An Account of the English Colony*, p. 152.
11. Ibid., p. 358.
12. Ibid., p. 370.
13. Smith, *Mari Nawi*, p. 34.
14. J. Arago, *Promenade autour du monde pendant les années 1817, 1818, 1819 et 1820 sur les corvettes* L'Uranie *et* La Physicienne ... (Paris: Leblanc, 1822), pp. 278–79; 'Extraction of one upper incisor, a sketch', Jacques Arago, Louis de Freycinet 1817–20, Bibliothèque cantonale et Universitaire, Lausanne.
15. 'Spencer & Gillen: A journey through Aboriginal Australia': http://spencerandgillen.net/spencerandgillen; 'Walter Baldwin Spencer': https://en.wikipedia.org/wiki/Walter_Baldwin_Spencer
16. Baldwin Spencer & F.J. Gillen, *The Native Tribes of Central Australia*, 1899 (e-text), pp. 212–14: www.sacred-texts.com/aus/ntca/index.htm
17. Ibid., pp. 189–93, 452–56.
18. Ibid., pp. 450–52.
19. Alan N. Williams, Sean Lem, Ian D. Goodwin & Mike Smith, 'Hunter-gatherer response to late Holocene climatic variability in Northern and Central Australia', *Journal of Quaternary Science*, vol. 5, 2010, pp. 831–38.
20. B. Brennan, 'Map of Indigenous massacres grows to include more sites of violence across Australia', 27 July 2018: www.abc.net.au/news/2018-07-27/map-of-indigenous-massacres-grows-to-include-more-sites/10040206; Judith Wright, *The Cry for the Dead* (Melbourne: Oxford University Press, 1982); Griffiths, *The Art of Time Travel*, Ch 5.
21. I.A.E. Bayly, 'Review of how indigenous people managed for water in desert regions of Australia', *Journal of the Royal Society of Western Australia*, vol. 82, 1999, p. 17.
22. Ibid., pp. 17–18.
23. Ibid., pp. 18, 23; quoting David Carnegie, *Spinifex and Sand: A narrative of five years' pioneering and exploration in Western Australia* (London: C. Arthur Pearson, 1898).
24. Bayly, 'Review of how indigenous people managed for water', p. 20.
25. Ibid., pp. 18, 21, 22.
26. Ibid., p. 22.
27. Cathcart, *The Water Dreamers*, pp. 16–18.
28. Gammage, *The Biggest Estate on Earth*, pp. 103–11, details evidence of soil management.
29. Tyson Yunkaporta, *Sand Talk: How Indigenous thinking can save the world* (Melbourne: Text Publishing, 2019), pp. 97–100.
30. In 1992 the Australian High Court decision in *Mabo vs Queensland* (175 CLR1, 1992) recognised the existence of Native title under law.
31. Robert Poirier & Doris Schartmueller, 'Indigenous water rights in Australia', *Social Science Journal*, vol. 49, 2012, pp. 317–24. Nevertheless, the significance of concepts such as 'cultural flows' have been recognised in law and practice for some years now.
32. Aboriginal Art and Culture: aboriginalart.com.au
33. 'Indigenous knowledge in the built environment: A guide for tertiary educators', Department of Education and Training, Australian Government: https://ltr.edu.au/resources/ID12-2418 Deakin_Jones_2018_Guide.pdf
34. The Anthropocene is a proposed geological epoch dating from the commencement of significant human impact on Earth's geology and ecosystems, including, but not limited to, anthropogenic climate change.
35. Thanks to Rosalie Hastwell, Team Leader Arts & Cultural Development, Maroondah City Council (see Acknowledgements).
36. Jack Banister, 'Beneath modern Melbourne, a window opens into its ancient history', *Guardian*, 25 Dec. 2019: www.theguardian.com/australia-news/2019/dec/26/beneath-modern-melbourne-a-window-opens-into-its-ancient-history?CMP=Share_iOSApp_Other
37. 'Brewarrina, NSW': www.aussietowns.com.au/town/brewarrina-nsw
38. Marcia Langton, 'Earth, wind, fire and water: The social and spiritual construction of water in Aboriginal societies', in B. David, B. Barker & J.J. Niven (eds), 'The social anthropology of Australian Indigenous societies', *American Anthropologist*, 2006, pp. 148–49.
39. L. Hughes, W. Steffen & A. Dean, 'Lethal consequences: Climate change impacts on the Great Barrier Reef', 5 July 2018: www.

climatecouncil.org.au/resources/climate-change-great-barrier-reef; Cathcart, *The Water Dreamers*, p. 247.

40. R.H. Mathews, 'The Aboriginal fisheries at Brewarrina', *Journal of the Royal Society of New South Wales*, vol. 37, 1903, pp. 146–54.
41. Monica Tan, 'The fish traps at Brewarrina are extraordinary and ancient structures', *Guardian*, 10 July 2015: www.theguardian.com/australia-news/2015/jul/10/fish-traps-brewarrina-extraordinary-ancient-structures-protection
42. Bruce Pascoe, *Dark Emu: Aboriginal Australia and the birth of agriculture* (Broome: Magabala Books, 2018), p. 75.
43. Mathews, 'The Aboriginal fisheries at Brewarrina'.
44. Pascoe, *Dark Emu*, p. 75.
45. Mathews, 'The Aboriginal fisheries at Brewarrina'; Tan, 'The fish traps at Brewarrina'.
46. Pascoe, *Dark Emu*, p. 153.
47. Ian James McNiven, Gunditj Mirring Traditional Owners Aboriginal Corporation, 'Dating Aboriginal stone-walled fish traps at Lake Condah, southeast Australia', *Journal of Archaeological Science*, vol. 39, no. 2, 2012, pp. 268–86.
48. Cathcart, *The Water Dreamers*, p. 17.
49. Peter Dargin, *Aboriginal Fisheries of the Darling-Barwon River* (Brewarrina: Brewarrina Historical Society, 1976).
50. Michael McGowan, 'Australia's drought crisis and farmers' stories of anxiety, fear and resilience', *Guardian*, 4 Aug. 2018: www.theguardian.com/australia-news/2018/aug/04/australias-drought-crisis-and-farmers-stories-of-anxiety-fear-and-resilience; *Last Call at the Oasis*, documentary film, dir. Jessica Yu, co-writer Alex Prud'homme, 2011.
51. Tim Flannery, *The Future Eaters* (Adelaide: Reed New Holland, 1999, first printed 1994), p. 372.
52. Joshua Busby, 'Climate change brought down another prime minister in Australia', *Washington Post*, 27 Aug. 2018: www.washingtonpost.com/news/monkey-cage/wp/2018/08/27/climate-change-brought-down-another-government-in-australia-heres-what-happened/
53. Flannery, *The Future Eaters*, ch. 29.
54. Ibid.
55. *Gurrumul*, dir. Paul Williams, prod. Shannon Swan, Madmen Entertainment, 2018.

4 HAWAI`I — AOTEAROA: THE SHAPE OF WATER

1. Martha Warren Beckwith, *Hawaiian Mythology* (Honolulu: University of Hawai`i Press, 1982), pp. 91–92.
2. Elsdon Best, *Spiritual and Mental Concepts of the Maori* (Wellington: Government Printer, 1954), pp. 14–15: 'The *aria* of Karukaru, an *atua* of the Whanganui district, is an owl – the common morepork. The *atua* achieved fame as a protector of human welfare, as, for instance, in warning persons of attempts being made to bewitch them.'
3. Dennis Kawaharada, *Storied Landscapes: Hawaiian literature and place* (Honolulu: Kalamakū Press, 1999), p. 33.
4. https://en.wikipedia.org/wiki/Ua_Mau_ke_Ea_o_ka_`Āina_i_ka_Pono; this phrase is ascribed to Kamehameha III in 1843, when Britain returned Hawaiian sovereignty to its people.
5. Waitangi Tribunal, *Ko Aotearoa Tēnei: A report into claims concerning New Zealand law and policy affecting Māori culture and identity: Te Taumata Tuatahi* (Wellington: Legislation Direct, 2011), p. 23.
6. Jacinta Ruru, Janet Stephenson & Mick Abbott (eds), 'Tension lines', in *Making Our Place: Exploring land-use tensions in Aotearoa New Zealand* (Dunedin: Otago University Press, 2011), p. 16.
7. Martin Holgate, *From Care to Action: Making a sustainable world* (London: Earthscan Publications, 1996), p. 123: 'People with righteous mind must be peaceful within themselves as individuals and they must be peaceful with the world outside. But how can they remain peaceful when the world is exploited to the point of threating humanity's own survival?'
8. Dennis Kawaharada, 'The discovery and settlement of Polynesia', *Malama/The Light of Knowledge: The magazine of the University of Hawai`i*, 21 Apr. 2011: http://archive.hokulea.com/ike/moolelo/discovery_and_settlement.html
9. See K.R. Howe (ed.), *Vaka Moana: Voyages of the ancestors: The discovery and settlement of the Pacific* (Auckland: David Bateman with Auckland Museum, 2006), pp. 94–95 (plant distribution), 74 (animals); Trevor Worthy & Richard N. Holdaway, *The Lost World of the Moa: Prehistoric life of New*

Zealand (Christchurch: Canterbury University Press, 2002), p. 562.

10. Ben Finney & Sam Low, 'Navigation', in Howe (ed.), *Vaka Moana*, Ch V.
11. Ministry for Culture and Heritage, 'Tuia – Encounters 250: Birds and navigation', video: www.thevoyage.co.nz/en/video/15_Birds-and-navigation
12. https://johnrakestraw.net/2017/06/04/bar-tailed-godwits/; www.nzbirdsonline.org.nz/species/sooty-shearwater
13. Rakato Te Rangiita, pers. comm., 27 June 2015; see also Paul Hamer, 'One in six? The rapid growth of Maori population in Australia', *New Zealand Population Review*, 2008, pp. 153–76: www.population.org.nz/app/uploads/2010/01/nzpr-vol-33-and-34_paul-hamer.pdf
14. E.S. Craighill Handy, *The Hawaiian Planter* (Honolulu: The Museum, 1940), p. 5; E.S. Craighill Handy & Mary Kawena Pukui, 'The Polynesian family system in Ka-`u, Hawai`i': www.ulukau.org/elib/cgi-bin/library?e=d-0pfs01-000Sec--11haw-50-20-frameset-book--1-010escapewin&a=d&d=D0&toc=0
15. Some Hawai`i islands do share the volcanic abruptness of Polynesian departure points in Tāhiti and the Marquesas.
16. Bob Brockie, 'Native plants and animals – overview: Species unique to New Zealand', Te Ara – the Encyclopedia of New Zealand: www.TeAra.govt.nz/en/native-plants-and-animals-overview/page-1
17. Samuel M. Gon et al, '*Aina momona, honua au loli* – Productive lands, changing world: Using the Hawaiian footprint to inform biocultural restoration and future sustainability in Hawai`i', *Sustainability*, vol. 10, 2018, p. 2: www.mdpi.com/2071-1050/10/10/3420
18. Howe (ed.) *Vaka Moana*, p. 80; Warren Judd, 'Polynesians in the southern ocean', *New Zealand Geographic*, iss. 069, Sept.–Oct. 2004: www.nzgeo.com/stories/polynesians-in-the-southern-ocean. This article describes how Enderby Island, in the Southern Ocean, had been at least temporarily colonised by Polynesian seafarers between AD 1300 and 1500. A much earlier voyage is recounted later in this publication.
19. Jonathan Kay Kamakawiwo`ole Osorio, *Dismembering Lāhui: A history of the Hawaiian Nation to 1887* (Honolulu: University of Hawai`i Press, 2002): chs 3, 4, 5 deal with this from different perspectives following the Great Māhele.
20. K. Martinko, 'Honolulu cracks down on single-use plastics': www.treehugger.com/environmental-policy/hawaiis-plastic-bag-ban-goes-into-effect-but.html. Hawai`i's ban was voted in, in 2012, prohibiting grocery stores from handing out certain types of plastic bags. However, the new law allows stores to provide reusable biodegradable plastic bags to customers. New Zealand enacted a single-use plastic bag ban in 2019.
21. David F. Stannard, *Before the Horror: The population of Hawai`i on the eve of Western contact* (Honolulu: University of Hawai`i, 1989), p. 57; William Ellis, *A Tour Around Hawaii* (Honolulu: Hawaiian Gazette, 1917), p. 79 refers to the population as 'extraordinarily healthy'; Natalie Kurashima, 'The potential of indigenous agricultural food production under climate change in Hawai`i', *Nature Sustainability*, 25 Feb. 2019, refers to a population of greater than 800,000.
22. James Macrae, 'With Lord Byron at the Sandwich Islands in 1825: Being extracts from the ms. diary of James Macrae, Scottish botanist' (Honolulu: J.F. Wilson, 1922). p. 30: https://archive.org/stream/withlordbyronat01macrgoog/withlordbyronat01macrgoog_djvu.txt
23. Beckwith, *Hawaiian Mythology*, p. 310.
24. Martha Warren Beckwith (ed. & trans.), *Kumulipo: A Hawaiian creation chant* (Chicago: University of Chicago Press, 1951): www.kauainenehcp.com/uploads/8/1/8/0/81802884/kumulipo-text.pdf
25. Beckwith, *Hawaiian Mythology*. It is important to note Beckwith's observation, pp. 309–10, that by the time the *Kumulipo* chant made its way into print in the late 1890s, its first division was 'a reworking from old material. The conception of the birth from one form of matter to another, and from one form of life to another, corresponds with the text of whare wānanga or schools of learning belonging to the East Coast of New Zealand [and to other chants collected in Tahiti, the Marquesas, and the Tuamotus]'.
26. Richard Taylor, *Te Ika a Maui, or New Zealand and its inhabitants, 1855* (Wellington: A.H. & A.W. Reed, 1974), pp. 13–15.
27. Carlos Andrade, pers. comm., Dec. 2008; Andrade is an author and former UHP professor.
28. Kanalu G. Terry Young, *Re-thinking the Native*

Hawaiian Past (Oxfordshire: Routledge, 1998), p. 44.

29. Osorio, *Dismembering Lāhui*, ch. 1.
30. Patrick Vinton Kirch, *A Shark Going Inland Is My Chief: The island civilization of ancient Hawai`i* (Berkeley: University of California Press, 2014), pp. 210–11, 289–90; Te Rangi Hīroa, *The Coming of the Maori* (Wellington: Maori Purposes Fund Board, 1949): http://nzetc.victoria.ac.nz/tm/scholarly/tei-BucTheC-t1-g1-t4-body1-d4-d3.html
31. Beckwith, *Hawaiian Mythology*, pp. 310–12 discusses the recitation in symmetry of land and sea organisms through the first part of the *Kumulipo*, followed by 'a thousand lines of genealogical pairs, husband and wife'.
32. Beckwith, *Kumulipo*, p. 60; Beckwith, *Hawaiian Mythology*, pp. 310–11. After its explanation of the spirit world, the *Kumilipo* moves in the second part to the world of humans, beginning with lower life forms, moving to birds and fish and 'upwards' until, after 1000 lines of genealogical pairs, it reaches husband and wife. However, Beckwith's PhD thesis at Columbia suggests, in a brief discussion, that this symmetry is at times also about word play. S.N. Hale`ole, *The Hawaiian Romance of Lā`ieikawai*, Martha Beckwith, trans., reprinted from the *Thirty-third Annual Report of the Bureau of American Ethnology* (Washington: Government Printing Office, 1918), p. 33: https://archive.org/stream/hawaiianromanceo00hale#page/n0/mode/1u; Handy, *The Hawaiian Planter*, p. 13 also discusses the sharing of names on land and sea.
33. Beckwith, *Hawaiian Mythology*, pp. 310–12.
34. Robert Louis Stevenson, *Travels in Hawaii*, A. Grove Day (ed.) (Honolulu: University of Hawai`i Press, 1991), pp. 6–7.
35. E.S. Craighill Handy, *The Hawaiian Planter*, Bernice P. Bishop Museum, Bulletin 161 (Honolulu: Bishop Museum, 1940), p. 34.
36. E.S.C. Handy & E.G. Handy, *Native Planters in Old Hawaii: Their life, lore and environment*, Bishop Museum, Bulletin 233 (Honolulu: Bishop Museum, 1972), pp. 57–58.
37. David Malo, *Hawaiian Antiquities* (Honolulu: Bishop Museum, 1951), p. 17 sets out names for the shape and purpose of a descending landscape.
38. Kirch, *A Shark Going Inland*, p. 85.
39. Handy, *The Hawaiian Planter*, pp. 36–37.
40. Kawaharada, *Storied Landscape*, p. 33. Lady Maria Calcott, Baron George Anson Byron & Richard Rowland Bloxham (contributors), *Voyage of the HMS* Blonde *to the Sandwich Islands, in the years 1824–1825* (London: John Murray, 1926, ATL, Pq 910 voy 1826), p. 120: several other techniques for making fishponds were used, some inland.
41. Fran Dieudonne (ed.), *The Pacific Islands and the Sea* (California: Neptune House, 2002), pp. 95, 97–98.
42. 'Auckland's lava caves': www.urbex.co.nz/lava-caves.html
43. Richard Smith, David J. Lowe & Ian Wright, 'Volcanoes', Te Ara – the Encyclopedia of New Zealand: www.teara.govt.nz/en/volcanoes; Taylor, *Te Ika a Maui*, p. 122.
44. 'Wailua River': https://en.wikipedia.org/wiki/Wailua_River
45. David Young, *Faces of the River* (Wellington: NZ Listener, 1984), appendix p. 211.
46. Ngārangi Rangiuia, pers. comm., 23 Oct. 2018 and 30 Apr. 2019. Hui Te Rangiora's son was Irawhaki, father of Māui. Hui Te Rangiora discovered Hawai`i and reached as far as South America and to Asia. After much searching for his lost father, Māui found him and met Hui Te Rangiora, who instructed him in celestial navigation, waka building and sailing, and ocean-reading. This is how Māui, having built Nuku Tai Memeha, became such a widely celebrated sailor in the Pacific and how he found Tuahiwi o Aotea (the broad shorelines of Aotea), his name for New Zealand.
47. Ngārangi Rangiuia, pers. comm., 23 Oct. 2018.
48. David Young, *Matahina: Power in the land* (Wellington: Steele Roberts, 1998), p. 50; Linda Te Aho, 'Waikato: River of life', in Jacinta Ruru, Janet Stephenson & Mick Abbott (eds), *Making Our Place: Exploring land-use tensions in Aotearoa New Zealand* (Dunedin: Otago University Press, 2011), pp. 147–49.
49. Kirch, *A Shark Going Inland*, pp. 219–23.
50. Kawika Winter, pers. comm., 22 Oct. 2017; Ngārangi Rangiuia, pers. comm., October 2018, argues that the idea of all Māori having a whakapapa status and 'no one is slave', is a new idea, even in his lifetime.
51. http://maoridictionary.co.nz/search?&keywords=whenua; strangely, Elsdon Best in his *Spiritual and Mental Concepts of the Maori*

(Wellington: Government Printer, 1954) does not raise this.
52. Kirch, *A Shark Going Inland*, pp. 219–23.
53. Ibid., pp. 225–30.
54. Paul D'Arcy, *Hawai`i: Balancing coercion and consent in eighteenth-century Kānaka Maoli statecraft* (Acton, ACT: Australian National University Press, 2018), p. 84.
55. Gavan Daws, *Shoal of Time: A history of the Hawaiian Islands* (Honolulu: University of Hawai`i Press, 1974), pp. 213–34, 312–13.
56. Osorio, *Dismembering Lāhui*, chs 7, 8.
57. R.M. McDowall, *Ikawai: Freshwater fishes in Māori culture and economy* (Christchurch: Canterbury University Press, 2011), pp. 144–93; Elsdon Best, *Fishing Methods and Devices of the Maori*, Dominion Museum Bulletin no. 12, 1929 (Wellington: Government Printer, 1977), pp. 131–52, 157–86; David Young, *Woven by Water: Histories from the Whanganui River* (Wellington: Huia Publishing, 1998), pp. 184–88, 186–88 (building an utu piharau or lamprey weir).
58. Ngārangi Rangiuia, pers. comm. 9 Dec. 2017: 'Rāhui is a safety mechanism, also to stop us from being spiritually interfered with. We talk about Mate Māori … where spiritually something might happen that might impinge.'
59. Alan M. Friedlander, Janna M. Shackeroff & John N. Kittinger, 'Traditional marine reserves and their use in contemporary Hawai`i,' in Tamatoa Bambridge (ed.), *The Rahui: Legal pluralism in Polynesian traditional management of resources and territories* (Canberra: ANU Press, 2016, p. 180): www.researchgate.net/publication/318318709_Traditional_marine_resources_and_their_use_in_contemporary_Hawai.%27i
60. Young, *Woven by Water*, pp. 25–26, 47–48, 59.
61. Ngārangi Rangiuia, pers. comm., 9 Dec. 2017.
62. Sally Murphy, 'Rāhui on Waitakere Ranges to protect kauri', 2 Dec. 2017: www.radionz.co.nz/news/national/345250/rahui-on-waitakere-ranges-to-protect-kauri; Te Ahukaramū Charles Royal (ed.), *The Woven Universe: Selected writings of Maori Marsden* (Te Wānanga-o-Raukawa, 2003), pp. 49–51, discusses rāhui as it was inside a traditionally holistic way of thinking.
63. Friedlander, Shackeroff & Kittinger, 'Traditional marine resources and their use', pp. 187, 189, 192–93.
64. Lorin Thorax, 'From traditional to modern management in Fakarava', in Bambridge (ed.), *The Rahui*, p. 161: http://press-files.anu.edu.au/downloads/press/p337293/pdf/book.pdf?referer=1708
65. Ngārangi Rangiuia, pers. comm., 9 Dec. 2017.
66. W.H. Skinner, 'Ancient Maori canals, Marlborough, N.Z.', *Journal of the Polynesian Society*, vol. 21, no. 3, 1912, pp. 105–07.
67. McDowall, *Ikawai*, p. 39, quoting Tipene O'Regan; Ngārangi Rangiuia, who has chiefly ancestral links to this whenua, suggests construction began earlier than this.
68. Skinner, 'Ancient Maori canals', p. 106.
69. Ibid., pp. 105–06.
70. Ibid., p. 107.
71. Bruce McFadgen, pers. comm., 14 Feb. 2020.
72. Bruce G. McFadgen & Peter Adds, 'Tectonic activity and the history of Wairau Bar, New Zealand's iconic site of earlier settlement', *Journal of the Royal Society New Zealand*, vol. 49, no. 4, 2019, pp. 4–6: www.tandfonline.com/doi/full/10.1080/03036758.2018.1431293
73. Bruce McFadgen, pers. comm., 14 Feb. 2020; McFadgen & Adds, 'Tectonic activity', pp. 4–6.
74. Elsdon Best, *Fishing Methods of the Maori* (Wellington: Government Printer, 1977), pp. 164–65; McDowall, quoting Marr, *Ikawai*, p. 195.
75. McDowall, quoting Adkin, *Ikawai*, pp.194–95, map PA1-q002-18, ATL; Paul Hamer, '"A tangled skein": Lake Horowhenua, Muaūpoko and the Crown, 1898–2000', Waitangi Tribunal report WAI 2200, no. A150, 2015: https://forms.justice.govt.nz/search/Documents/WT/wt_DOC_91916304/Wai%202200%2C%20A150.pdf
76. McDowall, *Ikawai*, pp. 192–95; Hamer, '"A tangled skein"', pp. 8–11.
77. Hamer, '"A tangled skein"'; 'Opening Te Waihora/Lake Ellesmere', Environment Canterbury: https://ecan.govt.nz/your-region/your-environment/river-and-drain-management/opening-te-waihoralake-ellesmere; David Young, *Rivers: New Zealand's shared legacy* (Auckland: Random House, 2013), p. 60.
78. I.G. Barber, 'Of boundaries, drains and crops: A classification system for traditional Maori horticultural ditches', *New Zealand Journal of Archaeology*, vol. 11, 1989, pp. 23–50: https://nzarchaeology.org/download/of-boundaries-

drains-and-crops-a-classification-system-for-traditional-maori-horticultural-ditches

79. Ibid., p. 40.
80. Michael K. Burtenshaw, 'Maori gourds: An American connection?', *Journal of the Polynesian Society*, vol. 108, Dec. 1999, p. 427.
81. Te Rangi Hiroa, *The Material Culture of the Cook Islands (Aitutaki)* (New Plymouth: Thomas Avery and Sons, 1927): http://nzetc.victoria.ac.nz/tm/scholarly/tei-BucMate-t1-body-d2-d7-d8.html
82. Ibid.
83. Ngārangi Rangiuia, pers. comm., 29 Apr. 2019; Hiroa, *The Material Culture of the Cook Islands*.
84. Lisa Matisoo Smith, in Alice A. Storey, Jose Miguel Ramirez, Daniel Quiroz, David V. Burley, David J. Addison, Richard Walter, Atholl J. Anderson, Terry L .Hunt, J. Stephen Athens, Leon Huynen & Elizabeth A. Matisoo-Smith, 'Radiocarbon and DNA evidence for a pre-Columbian introduction of Polynesian chickens to Chile', *Proceedings of the National Academy of Sciences*, vol. 104, no. 25, June 2007, pp. 10335–39.
85. Irving Jenkins, *The Hawaiian Calabash* (London: Kegan Paul, 1989), p. 179.
86. S. Percy Smith, 'Lore of the Whare-Wananga, or teachings of the Maori college on religion, cosmology and history, written down by H.T. Whatahoro from the teachings of Te Matohorohanga and Nepia Pohuhu, priests of the wharewananga of the East Coast, New Zealand', *Journal of the Polynesian Society*, Memoir 4, 1915, pp. 39–40, 61, 223: Smith's informant speaks of return voyages Taihiti–Hawai`i ending about 1150; Toi, a great navigator of Aotearoa, was involved.
87. Michael Stevens, 'Pōhā: A clever way of storing food', Te Kete Ipurangi, Ministry of Education: https://instructionalseries.tki.org.nz/Instructional-Series/School-Journal/School-Journal-Level-2-September-2014/Poha-A-Clever-Way-of-Storing-Food; 'Traditional Māori food gathering': www.tepapa.govt.nz/discover- collections/read-watch-play/maori/traditional-maori-food-gathering
88. Jenkins, *The Hawaiian Calabash*, p. 179.
89. Robert McNab, *Historical Records of New Zealand* (Wellington: Government Printer, 1914), pp. 361–63.
90. Anne Salmond, *Two Worlds: First meetings between Maori and Europeans 1643–1771* (Auckland: Viking, 1991), pp. 164, 179, 181.
91. Ngārangi Rangiuia, pers. comm., 12 Dec. 2017.
92. Helen Leach, *1,000 Years of Gardening in New Zealand* (Wellington: A.H. & A.W. Reed, 1984), p. 68. Leach sums up the alternatives: drainage, or trench boundary markers, widely used in Polynesia.
93. Salmond, *Two Worlds*, identified in her index under 'taro'.
94. Mark Horrocks, Ian W.G. Smith, Scott L. Nicol & Rod Wallace, 'Sediment, soil and plant microfossil analysis of Maori gardens at Anaura Bay, eastern North Island, New Zealand: Comparison with descriptions made in 1769 by Captain Cook's expedition', *Journal of Archaeological Science*, vol. 35, Sep. 2008.
95. William Colenso, 'On the vegetable food of the ancient New Zealanders before Cook's visit', *Transactions of the New Zealand Institute*, vol. 13, 1880, pp. 14–15, 86–87; Colenso refers to taro varieties *Colocasia antiquorum* and *Caladium esculenta*.
96. Louise Furey, *Maori Gardening: An archaeological perspective* (Wellington: Department of Conservation Science and Technical Publishing, 2006), p. 13.
97. Nancy Awhitu, evidence, Te Rohe Potae Inquiry District, *Environmental and Wahi Tapu Report*, Michael Belgrave et al., 2011, p. 57; Rāwiri Taonui, 'Ngā waewae tapu – Māori exploration – Western North Island', Te Ara – the Encyclopedia of New Zealand: www.TeAra.govt.nz/en/nga-waewae-tapu-maori-exploration/page-6
98. Nancy Swarbrick, 'Logging native forests', Te Ara – the Encyclopedia of New Zealand: www.TeAra.govt.nz/en/logging-native-forests
99. 'Value wetlands': www.wetlandtrust.org.nz/get-involved/value-wetlands/; *A Directory of Wetlands in New Zealand* (Wellington: Department of Conservation, 1996), p. ix; David Young, *Our Islands, Our Selves: A history of conservation in New Zealand* (Dunedin: Otago University Press, 2004), pp. 111–12, 160–62, 234.
100. Beckwith, *Hawaiian Mythology*, p. 16.
101. Kanalu G. Terry Young, *Rethinking the Native Hawaiian Past* (New York/London: Routledge, 2012), p. 49.
102. Beckwith, *Hawaiian Mythology*, p. 15.
103. Ibid., pp. 12–13.
104. Ibid., pp. 14–16, 54–55.
105. Ibid., p. 16.

106. John Dawson & Rob Lucas, *New Zealand's Native Trees* (Nelson: Craig Potton Publishing, 2013), p. 338; George Gibbs, *Ghosts of Gondwana: The history of life in New Zealand* (Nelson: Potton & Burton, 2016), pp. 187–89. Gibbs explains that there are three lineages, one from New Caledonia; Phillip G. Simpson, 'Pohutukawa and biodiversity', *Conservation Advisory Science Notes*, no. 100, Department of Conservation, Wellington, 1994: the ancestor probably occupied the mobile islands of the southwest Pacific in the early Tertiary period (the Palaeocene, 65–55mya), perhaps colonising the volcanic or sedimentary rocks as they emerged from the sea. This ability is still common to Pacific species, including Hawai`i's `ōhi`a lehua, New Zealand's pōhutukawa and the Kermadec Island pōhutukawa. Malo, *Hawaiian Antiquities*, comments that New Zealand 'has been important, therefore, in both the origin and diversification of *Metrosideros*, pōhutukawa and rātā' (p. 84).
107. Hector Busby, 'Thor Heyerdahl', *Living Treasures*, TV One, 1990; Ngārangi Rangiuia, pers. comm., 23 Oct. 2018.
108. Dilys Johns, Geoffrey J. Irwin & Yun K. Sung, 'An early sophisticated East Polynesian voyaging canoe discovered on New Zealand's coast', *Proceedings of the National Academy of Sciences*, vol. 111, no. 41, 2014: www.pnas.org/content/111/41/14728.full; Young, *Our Islands, Our Selves*, p. 51.
109. John Patterson, *Pacific Parables: Learning from Māori tradition* (Wellington: Steele Roberts, 2014), p. 23; Katharine Luomala, *Voices on the Wind* (Honolulu: Bishop Museum Press, 1986), pp. 161, 166–68, 170, 173–76, 178, 191. Beckwith, *Hawaiian Mythology*, pp. 266–76 gives a Tahitian version.
110. Luomala, *Voices on the Wind*, pp. 161, 165–70, carries a detailed account of Rātā's lessons in the need for reciprocal and respectful relationship with higher powers.
111. Howe (ed.), *Vaka Moana*, p. 146.
112. Luomala, *Voices on the Wind*, pp. 171–78.
113. Ibid., pp. 19, 151–52, 161.
114. Ibid., pp. 170, 178.
115. Smith, 'Lore of the Whare-Wananga', pp. 206–07, 210–11.

5 WAIKĪKĪ: GHOSTS, GLITZ AND GOBIES

1. Gary Y. Okihiro, *Island World: A history of Hawaii and the United States* (Berkeley: University of California, 2008), chs 1, 2, 6, 7.
2. Gaye Chan & Andrea Feeser, *Waikiki: A history of forgetting and remembering* (Honolulu: University of Hawai`i Press, 2006), p. 11; Lord George Anson Byron with Richard Rowland Bloxham, *Voyage of the HMS* Blonde *to the Sandwich Islands, in the years 1824–1825* (London: John Murray, 1926), pp. 106–07; www.diamondheadhike.org/militaryhistory.html
3. Robert Louis Stevenson, *Travels in Hawaii*, ed. Arthur Grove Day (Honolulu: University of Hawai`i Press, 1991), p. 41.
4. www.hawaii.edu/news/2013/08/23/planned-learning-center-will-enhance-kanewai-cultural-garden/; https://en.wikipedia.org/wiki/Diamond_Head,_Hawaii
5. Jonathan Kamakawiwo`ole Osorio, *Dismembering Lāhui: A history of the Hawaiian nation to 1887* (Honolulu: University of Hawai`i Press, 2002), ch. 7.
6. Kamehameha Schools: www.ksbe.edu/assets/ks_official_facts_and_statistics.pdf
7. Kamehameha Schools, 'Our land kuleana', `Āina: www.ksbe.edu/land/
8. Sacred Texts, Chant One: www.sacred-texts.com/pac/ku/ku12.htm#chant01; see also Martha Warren Beckwith, *The Kumulipo: A Hawaiian creation chant* (Honolulu: University of Hawai`i Press, 1951).
9. www.nationalgeographic.org/encyclopedia/wetland
10. Beckwith, *The Kumulipo*, pp. 59–60; https://archive.org/stream/pelehiiakamythfr00emeriala/pelehiiakamythfr00emeriala_djvu.txt
11. Haunani-Kay Trask, *Light in the Crevice Never Seen* (Corvallis, OR: Calyx Books, 1994), p. 76.
12. Fran Dieudonne (ed.), *The Pacific Islands and the Sea* (Honolulu: Neptune House, 2002), p. 103.
13. Benjamin Pang Jeng Lo, Martin Inn, Susan Foe, Robert Amaker, *The Essence of T'ai Chi Chu'an: The literary tradition* (North Atlantic Books, 1981), pp. 31–38; John Patterson (trans.) *Lao Tzu: The art of peace* (Wellington: Steele Roberts, 2007).

14. R.M. McDowall, *Ikawai: Freshwater fishes in Maori culture and economy* (Christchurch: Canterbury University Press, 2011), pp. 264–65.
15. Martha Warren Beckwith, *Hawaiian Mythology* (Honolulu: University of Hawai`i Press, 1976), p. 63. Hawaiian native freshwater fish seem to have been even less studied than those in Aotearoa. Of the five species (all anadromous), one is stated to be 'near threatened' and two others are 'data deficient'; the status of the other three is not given: https://en.wikipedia.org/wiki/Endemism_in_the_Hawaiian_Islands
16. Kava Society New Zealand: https://kavasociety.nz/blog/2018/2/5/kava-vs-kawakawa-and-the-maori-memories-of-kava
17. Ngārangi Ranguia, pers. comm., 10 Dec. 2018.
18. Houston Wood, *Displacing Natives: The rhetorical production of Hawaii* (Lanham, US: Rowman & Littlefield, 1999).
19. Chan & Feeser, *Waikīkī*, pp. 25–35; Dennis Kawaharada, *Storied Landscapes: Hawaiian literature and place* (Kalamaku Press, 1999).
20. Jonathan Osorio, 'The value of Hawaii': www.civilbeat.org/2010/07/2902-the-value-of-hawaii-hawaiian-issues-by-jon-osorio/
21. Haunani-Kay Trask, pers. comm., Dec. 2008.
22. Rev. P. Walsh, 'The effect of deer on the New Zealand bush: A plea for the protection of our forest reserves', *Transactions of the New Zealand Institute*, vol. 25, 1892, p. 394.
23. *Star-Bulletin*, 31 March 2006 & 30 April 2006: http://archives.starbulletin.com/2006/03/31/news/story01.html; 'Line break turns Honolulu canal into an open sewer', *Environment News Service*: www.ens-newswire.com/ens/mar2006/2006-03-28-09.html; Wikipedia.org/wiki/Ala-Wai-Canal
24. https://chroniclingamerica.loc.gov/lccn/sn83025121/1903-04-17/ed-1/seq-6/
25. Chan & Feeser, *Waikiki*, p. 44.
26. Ibid.
27. Trask, *Light in the Crevice Never Seen*, pp. 60–61.
28. J. Houston, *Hawaiian Son*, p. 223 carries an account of a pinao and the relationship Eddie Kamo felt with this animal.
29. Margaret Titcomb & Mary Kawena Pukui, 'Native Use of Fish in Hawaii', Memoir No. 29, *Journal of the Polynesian Society*, vol. 60, no. 195, p. 105.
30. Mary Kawena Pukui, *`Olelo No`eau: Hawaiian proverbs and poetical sayings* (Honolulu: Bernice P. Bishop Museum Press, Special Publication, 1983), p. 146.
31. Lawrence Miike, *Water and the Law in Hawaii* (Honolulu: University of Hawai`i Press, 2004), p. 15; Pukui, *`Olelo No`eau,* p. 146.
32. G. Mair, 'Notes on fishes in Upper Whanganui River', *TNZI*, 12 May 1879, pp. 315–16; McDowall, *Ikawai*, p. 167.
33. Theodore Schwenk, *Sensitive Chaos* (East Wessex: Rudolph Steiner Press, 2008).
34. Elsdon Best, *Spiritual and Mental Concepts of the Maori*, Monograph No. 2 (Dominion Museum: Government Printer, 1954), pp. 29–34.
35. M.I. Bonnett, R.M. McDowall & J.R.E. Sykes, 'Critical habitats for conservation of giant kokopu. *Galaxaxias argenteus* (Gmelin, 1789)', *Science for Conservation* 206, Department of Conservation, Wellington, 2002: www.doc.govt.nz/globalassets/documents/science-and-technical/sfc206.pdf
36. http://downwindproductions.com/tours/streams/kaneloa.html; http://manoa.hawaii.edu/coe/kulia/resources/curriculum_resources/Aquatic%20Project%20Wild%20.pdf
37. https://nativeplants.Hawaii.edu
38. https://waimaori.maori.nz; https://teohu.maori.nz
39. www.doc.govt.nz/get-involved/funding/nga-whenua-rahui; Harry Broad, pers. comm., 12 May 2009.
40. Catherine Wilson, 'The new battle for Bougainville's Panguna Mine', *The Interpreter*, 21 August 2018: www.lowyinstitute.org/the-interpreter/new-battle-bougainville-s-panguna-mine
41. www.nzherald.co.nz/nz/news/article.cfm?c_id=1&objectid=12029780
42. https://waimaori.maori.nz; https://teohu.maori.nz/who-we-work-with/wai_maori/
43. https://whc.unesco.org/en/events/812/
44. Huhana Smith, 'Hei Wai Ora: A photo essay', in Gary Williams & Betsan Martin (eds), 'Watersheds: Responsible governance of water and the waterways of life, global issues in local contexts', Horowhenua NZ Symposium, Nov. 2009, pp. 103–09; Huhana Smith, pers. comm., 26 January 2019; www.stuff.co.nz/business/farming/74364241/wetland-lies-close-to-the-heart. Geoff Park, in *Ngā Uruora: Ecology and history in a New Zealand landscape* (Wellington: Victoria University Press, 1995), ch. 4, describes in detail the scope and

variety of this landscape of water, trees and wildlife largely lost to farming.

45. www.stuff.co.nz/business/farming/74364241/wetland-lies-close-to-the-heart
46. www.odt.co.nz/news/national/maori-knowledge-brought-bear-problem
47. www.stuff.co.nz/business/farming/74364241/null
48. Smith, 'Hei Wai Ora'.
49. www.mtm.ac.nz/oranga-taiao-oranga-tangata/outputs/
50. Ann Graeme, 'Linking land to sea', *Forest & Bird*, Dec. 2019, pp. 46–48.
51. Ministry for the Environment, 'Status of our freshwater species', 2017: www.mfe.govt.nz/publications/fresh-water/fresh-water-report-2017-ecosystems-habitats-and-species/status-of-our; Mike Joy, 'Paradise squandered: New Zealand's environmental asset stripping', 2014 Bruce Jesson Memorial Lecture, pp. 15–17.
52. www.stats.govt.nz/indicators/conservation-status-of-indigenous-freshwater-species; www.mfe.govt.nz/publications/fresh-water/fresh-water-report-2017-ecosystems-habitats-and-species/status-of-our
53. Lawrence Miike, pers. comm., Honolulu, Dec. 2008; Carlos Anrade, pers. comm., Mānoa, Honolulu, Dec. 2008.
54. Miike & Andrade, ibid.
55. Andrade, pers. comm., Dec. 2008, Mānoa.
56. Beckwith, *Hawaiian Mythology*, p. 51 points out that this legend has similarities to the story of Kana, who must journey to the underworld 'through which the sun passes each night from west to east'; he encounters Pele's brother, Kahoali`i the shark, to obtain, with blackened hands, the sun, stars and cockerel of the dawn.
57. Mary Kawena Pukui, *The Water of Kane* (Honolulu: Kamehameha Publishing, 1994), pp. 3–22; Beckwith, *Hawaiian Mythology*, p. 492–95 describes the legend as being 'a Jason type' and offers several versions: for her, the story offers an explanation as to the way things are, but, being a construct of a holistic culture, Aukele's detailed story is also associated with the significance of the dawn.
58. Gary Snyder, *The Practice of the Wild* (San Francisco: North Point Press, 1990), p. 21.

6 HAWAI`I ISLAND: GODS, GARDENS, FIRE AND WATER

1. Patricia Grace & Waiariki Grace, *Earth, Sea, Sky: Images and Māori proverbs from the natural world of Aotearoa New Zealand* (Nelson: Craig Potton, 2006), p. 22.
2. J.C. Beaglehole, *The Life of Captain James Cook* (London: A. & C. Black, 1974), p. 649.
3. David Malo, *Hawaiian Antiquities (Moolelo Hawaii)*, trans. N.B. Emerson (Honolulu: Bernice P. Bishop Museum, Special Publication No. 2, 1898), pp. 144–45; www.onlyinyourstate.com/hawaii/incredible-rivers-hi/; Martha Warren Beckwith, *Hawaiian Mythology* (Honolulu: University of Hawai`i Press, 1976), pp. 31–33.
4. 'The Hawaiian Moon Calendar and the Ways of the Practitioners (Kahuna)', rewritten from articles by Lyons Kai`ioho Naone III, published in *Na Po`e Hawai`i Magazine*, Spring 1995–Winter 1996: www.fmpug.com/images/upload_files/Hawaiian%20Moon%20Calendar%20and%20Ways%20of%20the%20Practitioners.pdf
5. Beaglehole, *The Life of Captain James Cook*, pp. 660–63.
6. Anne Salmond, *The Trial of the Cannibal Dog: Captain Cook in the South Seas* (Auckland: Penguin, 2004), pp. 389–416; Beaglehole, *The Life of Captain Cook*, pp. 662–72.
7. Thomas Farber, *Face of the Deep* (San Francisco: Mercury House, 1998), p. 153.
8. Malo, *Hawaiian Antiquities*, pp. 53, 160, 165, 189; Anne Salmond, *Aphrodite's Island: The European discovery of Tahiti* (Auckland: Viking, 2009), pp. 227, 230–32, 295; Dennis Kawaharada suggests Hawai`i's cannibalism may be allegorical: 'The traditions of O`ahu: Waialua, The Killing of the Cannibal King', Asia–Pacific Digital Library: https://guides.library.kapiolani.hawaii.edu/apdl/oahu/stories/waialua
9. Ngārangi Rangiuia, pers. comm., Ōtaki, 9 Dec. 2017.
10. Beckwith, *Hawaiian Mythology*, p. 335: Tūrehu is also a Hawaiian term for their fairy people.
11. Ibid., pp. 521, 528; Ngārangi Rangiuia, pers. comm., Ōtaki, 9 Dec. 2017.
12. Mary Kawena Pukui, retold by Caroline Curtis, *The Water of Kane and other legends of the Hawaiian Islands* (Honolulu: Kamehameha Publishing, 1994), pp. 85–90.

13. Ngārangi Rangiuia, pers. comm., Ōtaki, 9. Dec. 2017.
14. Peter Buck, *Vikings of the Sunrise* (Christchurch: Whitcombe & Tombs, 1975), p. 279–80; Ngārangi Rangiuia, pers. comm., Ōtaki, 9 Dec. 2017; Elsdon Best, *Maori Religion and Mythology*, Pt II (Wellington: Government Printer, 1982), pp. 294, 306, 418.
15. Best, *Maori Religion and Mythology*, Pt II, pp. 294, 306, 418.
16. Nancy Swarbrick, 'Waikato region – Māori settlement', Te Ara – the Encyclopedia of New Zealand: www.TeAra.govt.nz/en/object/27074/uenuku
17. Ngārangi Rangiuia, pers. comm., Ōtaki, 9 Dec. 2017.
18. David Young, *Rivers: New Zealand's shared legacy* (Auckland: Random House, 2013), pp. 280, 287, 299.
19. Ngārangi Rangiuia, pers. comm., Ōtaki, 9 Dec. 2017.
20. Best, *Maori Religion and Mythology*, Pt II, pp. 194–95, 416.
21. Ibid., pp. 414–15; http://nzetc.victoria.ac.nz/tm/scholarly/tei-Bes02Reli-t1-body-d4-d5-d10.html
22. Ngārangi Rangiuia, pers. comm., Ōtaki, 9 Dec. 2017: the relationship between the two brothers had a different emphasis in Hawai`i.
23. Beckwith, *Hawaiian Mythology*, pp. 53–59, 64.
24. James Mollison, *Permaculture: A designer's manual* (NNSW, Aus.: Tagari Publications, 1988) describes permaculture in definitive detail; its definitions are rooted in sustainability and working with nature, both of which the ancient Hawaiians did successfully. The great loss of forests to dryland and wetland taro growing would tend to limit the extent to which both criteria might stand in a modern evaluation.
25. Patrick Vinton Kirch, *A Shark Going Inland Is My Chief: The island civilization of ancient Hawaii* (Berkeley: University of California Press, 2012), pp. 140, 146, 203.
26. https://evols.library.manoa.hawaii.edu/bitstream/10524/47256/3-GonschorBeamer.pdf
27. Kawika Winter et al. argue in favour of the more inclusive word 'moku', or district: 'The *Moku* system: Managing biocultural resources for abundance within social-ecological regions in Hawai`i': https://ideas.repec.org/a/gam/jsusta/v10y2018i10p3554-d173697.html
28. Lorenz Gonschor & Kamanamaikalani Beamer, 'Toward an inventory of ahupua`a in the Hawaiian kingdom', *Hawaiian Journal of History*, vol. 48, 2014: https://evols.library.manoa.hawaii.edu/bitstream/10524/47256/1/3-GonschorBeamer.pdf
29. Ibid. (emphasis in the original).
30. Kirch, *A Shark Going Inland*, pp. 154–55.
31. E.S. Craighill Handy, *The Hawaiian Planter*, Bernice P. Bishop Museum Bulletin No. 161 (Honolulu: Bishop Museum, 1940), p. 35.
32. Kirch, *A Shark Going Inland*, p. 131.
33. Gavan Daws, *Shoal of Time: A history of the Hawaiian Islands* (Honolulu: University of Hawai`i Press, 1968), pp. 29–44, 264–80; Jonathan Kay Kamakawiwo`ole Osorio, *Dismembering Lāhui: A history of the Hawaiian nation to 1887* (Honolulu: University of Hawai`i Press, 2002), ch. 8.
34. Alan M. Friedlander, Janna M. Shackeroff & John N. Kittinger, 'Traditional marine resources and their use in contemporary Hawai`i', in Tamatoa Bambridge (ed.), *Rahui: Legal pluralism in Polynesian traditional management of resources and territories* (Canberra: ANU Press, 2012), p. 181.
35. Natalie Kurashima et al.: https://www.ksbe.edu/land/about-us/
36. Ibid.
37. David F. Stannard, *Before the Horror: The population of Hawaii on the eve of Western contact* (Honolulu: University of Hawai`i Press, 1989); Kirch, *A Shark Going Inland*, pp. 158–59, 168.
38. Ian Pool & Tahu Kukutai, 'Taupori Māori – Māori population change – Population changes, 1769–1840', Te Ara – the Encyclopedia of New Zealand: www.TeAra.govt.nz/en/taupori-maori-maori-population-change/page-1; https://en.wikipedia.org/wiki/List_of_islands_of_New_Zealand; https://en.wikipedia.org/wiki/Hawaiian_Islands
39. www.sciencedaily.com/releases/2019/02/190226144342.htm
40. Kawika Winter, 'Kalo varieties': www.researchgate.net/publication/265294634_Kalo_Hawaiian_Taro_Colocasia_esculenta_L_Schott_Varieties_An_assessment_of_nomenclatural_synonymy_and_biodiversity. For a much earlier view, see Handy, *The Hawaiian Planter*, p. 32.
41. Lord George Anson Byron & R.R. Bloxham, *Voyage of the HMS* Blonde *to the Sandwich Islands, in the years 1824–1825* (London: John Murray, 1926), p. 16.

42. Daws, *Shoal of Time*, pp. 73–74.
43. Byron, *Voyage of the HMS* Blonde, pp. 13–14.
44. James D. Houston with Eddie Kamae, *Hawaiian Son: The life and music of Eddie Kamae* (Honolulu: `Ai Pohaku Press, 2004), pp. 191–92.
45. Richard Taylor, *Te Ika a Maui*: *New Zealand and its inhabitants* (Wellington: A.H. & A.W. Reed, 1974), pp. 196–97; Kawika Winter, pers. comm., 18 March 2019.
46. Malo, *Hawaiian Antiquities*, pp. 242–43, 44.
47. Best, *Maori Religion and Mythology*, Pt II, pp. 39, 299: 'It was also Watea who separated land from water'; Malo, *Hawaiian Antiquities*, p. 16: 'the separation of Wakea from his wife Papa, which according to Southern Polynesian myth was the lifting up of the vault of heaven from the plain of the earth'; Ngārangi Rangiuia, pers. comm., Oct. 2018.
48. Ngārangi Rangiuia, pers. comm., Oct. 2018.
49. Taylor, *Te Ika a Maui*, p. 18; Elsdon Best, *Maori Myth and Religion*, Pt 1 (Wellington: Government Printer, 1982), p. 69.
50. Kirch, *A Shark Going Inland*, p. 143.
51. Byron, *Voyage of the HMS* Blonde, p. 160. Sowans is an old Scottish meal – a sort of porridge, made from the grains or husks of oats: www.scotslanguage.com/articles/view/id/2022. The preparation of poi is described in Mary Kawena Pukui, 'Poi and its preparation', 'Hawaiian Ethnological Notes', Bishop Museum, Honolulu: www.hawaiialive.org/resources/manuscript/503.pdf
52. Malo, *Hawaiian Antiquities*, p. 244.
53. Bradford Haami, 'Kiore – Pacific rats', Te Ara – the Encyclopedia of New Zealand: www.TeAra.govt.nz/en/kiore-pacific-rats; Best, *Maori Religion and Mythology*, Pt II, pp. 319, 399. Best, p. 335, draws a comparison with Māuitikitiki of Polynesia and Rongomāui's roles in bringing the kūmara and taro respectively to Earth.
54. 'The Maori as a digger. Using the *ko* or digging stick', photo by A. Hamilton, in Best, *Maori Agriculture* (Wellington: Government Printer, 1925), p. 77.
55. T. Rigg & J. Bruce, 'The Maori gravel soil of Waimea West, Nelson, New Zealand', *Journal of the Polynesian Society*, vol. 32, 1923, pp. 85–90; Richard Cassels, 'Human ecology in the prehistoric Waikato', *Journal of the Polynesian Society*, vol. 81, no. 2, June 1972, pp. 196–247 describes similar work conducted in the Waikato.
56. Rigg & Bruce, 'The Maori gravel soil', pp. 85–90; www.researchgate.net/profile/Don_Graves; Cassels, 'Human ecology', pp. 196–247.
57. Best, *Maori Agriculture*, p. 42.
58. Ngārangi Rangiuia, pers. comm., Oct. 2018.
59. *Carbon News*, 7 February 2019.
60. 'Turning bio-waste into char', 'Report dishes the dirt on world soil health': www.massey.ac.nz/massey/aboutmassey/news/article.cfm?mnarticle_uuid=22CDB6D1-FBDB-E435-F6F6-F8AA4A7295AA
61. Don Graves, pers. comm., Motueka, 5 Nov. 2018.
62. www.agmrc.org/renewable-energy/biofuelsbiorefining-general/biochar-a-multitude-of-benefits
63. *Biowastes Project*, 'News from the Programme Manager', iss. 3, 2011: www.cibr.org.nz/assets/Uploads/Issue-No.-3-DEC-2011-WEB.pdf. Ngāti Tūwharetoa's Tuaropaki Trust has been prominent in the exploration of char as a soil conditioner, and for other carbon-neutralising practices.
64. Martin Heine, pers. comm., 7 Feb. 2019.
65. 'Pyrolysis is the thermal decomposition of materials at elevated temperatures in an inert atmosphere': https://en.wikipedia.org/wiki/Pyrolysis
66. Don Graves, pers. comm., 9 Nov. 2018.
67. *International Journal of Agronomy*, 2018, Article ID 6837404: https://doi.org/10.1155/2018/6837404; Don Graves, 'A comparison of methods to apply biochar to temperate soils', in Natalia Ladygina & Francois Rineau (eds), *Biochar and Soil Biota* (Florida: CRC Press, 2013), pp. 202–61: www.researchgate.net/profile/Don_Graves
68. Te Ahukaramū Charles Royal (ed.), *The Woven Universe: Selected writings of Maori Marsden* (Te Wānanga o Raukawa, 2003), p. 56: 'Cultures pattern perceptions of reality into conceptualisations of what they perceive reality to be; of what is to be regarded as actual, probable, possible or impossible. These conceptions form what is termed "the world-view" of a culture. The world-view is the central systematisation of conception of reality to which members of the culture [subscribe and] from which stems their value system.'
69. Ngārangi Rangiuia, pers, comm., 23 Oct. 2018.

70. Basil Keane, 'Riri – traditional Māori warfare – Atua – origins of wars', Te Ara – the Encyclopedia of New Zealand: www.TeAra.govt.nz/en/riri-traditional-maori-warfare/page-1
71. Beckwith, *Hawaiian Mythology*, pp. 423, 518, 519, 520.
72. Malo, *Hawaiian Antiquities*, p. 182; Katherine Luomala, *Voices on the Wind* (Hawai`i: Bishop Museum Press, 1986), p. 67.
73. Malo, *Hawaiian Antiquities*, p. 184.
74. Te Ahukaramū Charles Royal, 'Te Waonui a Tāne – forest mythology – Tāne and his forests', Te Ara – the Encyclopedia of New Zealand: www.TeAra.govt.nz/en/te-waonui-a-tane-forest-mythology/page-1. Tāne is sometimes given different names to reflect his different roles: he is Tānemahuta as god of the forest, Tānetewānanga as the bringer of knowledge, and Tānenuiarangi as bringer of higher consciousness.
75. Taylor, *Te Ika a Maui*, p. 22; Ron Matthews, 'Māori spirituality', lecture series, Ministry for the Environment, Wellington, 1987; Best, *Maori Religion and Mythology*, Pt II, pp. 221–23.
76. Best, *Maori Religion and Mythology*, Pt II, p. 222; Malo, *Hawaiian Antiquities*, p. 96.
77. Matthews, 'Māori spirituality'.
78. Ngārangi Rangiuia, pers. comm., Ōtaki, 12 Dec. 2017.
79. Best, *Maori Religion and Mythology*, Pt II, p. 222.
80. Ibid., pp. 294, 306, 418.
81. R.M. McDowall, *Ikawai* (Christchurch: Canterbury University Press, 2011), pp. 181–95; Matiu Marekura, 'Te Awa Tupua: Our sacred cord of unity', in Āneta Hinemihi Rāwiri (ed.), *Mouri Tū, Mouri Ora: Water for wisdom and life* (Whanganui: Te Atawhai o Te Ao, 2009), p. 37.
82. Geoff Park, *Theatre Country: Essays on landscape and whenua* (Wellington: Victoria University Press, 2006), ch 4.
83. Turuhia (Jim) Edmonds, 'Learning from the river, learning from the people of the river', in Rāwiri (ed.), *Mouri Tū, Mouri Ora*, p. 21; Colin Richards, 'Water for wisdom and life', in Rāwiri (ed), *Mouri Tū, Mouri Ora*, p. 17.
84. McDowall, *Ikawai*, chs 7–17.
85. Ngārangi Rangiuia, pers. comm., Tasman, 23 Oct. 2018, observed that, traditionally, by no means all in Aotearoa had ready access to their whakapapa; even as a child of rangatira in the 1960s he observed sharp social divisions within his own community.
86. Beckwith, *Hawaiian Mythology*, p. 64.
87. 'The Water of Kane', in Nathaniel B. Emerson (ed.), *Unwritten Literature of Hawaii: The sacred songs of the hula* (Washington: Government Printing Office, 1909, reprinted 1977): www.sacred-texts.com/pac/ulh/ulh45.htm
88. Dennis Kawaharada, *Storied Landscapes: Hawaiian literature and place* (Honolulu: Kalamaku Press,1999), pp. 75–78.
89. Emerson, *Unwritten Literature of Hawaii*, pp. 257–59; Kawaharada, *Storied Landscapes*, pp. 75–78.
90. Beckwith, *Hawaiian Mythology*, pp. 65–66.
91. Byron, *Voyage of HMS* Blonde, p. 31; Patrick Kirch, *A Shark Going Inland*, pp. 268–86.
92. Beckwith, *Hawaiian Mythology*, p. 64. There are two genera of this tree endemic to Hawai`i.
93. Ibid., p. 65.
94. Sereno Edwards Bishop, 'Reminiscences of Old Hawaii', Historical Series (Honolulu: Hawaiian Gazette, 1936): https://archive.org/details/reminiscencesofo00bish, p. 12; map.
95. Ibid., pp. 13–14.
96. Ibid., p. 25.
97. Anna Johnston,' The strange career of William Ellis', *Victorian Studies*, vol. 49, no. 3, 2007 (Indiana University Press), p. 491, William Ellis, ms-papers-2862, ATL.
98. William Ellis, *Polynesian Researches: Society Islands, Tubuai Islands and New Zealand* (Rutland, VT: Charles E. Tuttle, 1969; reprint of original, 1929), pp. 74–75.
99. Ibid., pp. 99–111.
100. Ibid., pp. 104–05.
101. Ibid.
102. William Ellis, *Journal of a Tour around Hawaii, the largest of the Sandwich Islands* (Boston: Crocker & Brewster, 1825), p. 160. (Both the township of Kaimū and its beach were destroyed in a major eruption in 1990.)
103. Ibid., pp. 166–67.
104. Ibid.
105. www.britannica.com/science/kipuka
106. www.express.co.uk/news/world/970921/hawaii-volcano-update-kapoho-bay-before-and-after-aerial-pictures; https://en.wikipedia.org/wiki/Kapoho,_Hawaii

107. Ellis, *Journal of a Tour around Hawaii*, p. 337.
108. Ibid., pp. 186, 337; Mary Kawena Pukui, *`Olelo No`eau: Hawaiian proverbs & poetical sayings* (Honolulu: Bernice Bishop Museum Press, 1983), Special Publication No. 71, para 1245.
109. Ellis, *Journal of a Tour around Hawaii*, pp. 186, 337; Pukui, *`Olelo No`eau*, para 1245.
110. Ellis, *Journal of a Tour around Hawaii*, pp. 185–87.
111. Ibid., p. 187.
112. Ibid., pp. 199–201.
113. Ibid., p. 199.
114. E.S. Craighill Handy & Mary Kawena Pukui, 'The Polynesian family system in Ka-`u, Hawai`i', *Journal of the Polynesian Society*, vol. 59, no. 3, 1950, p. 234.
115. Ibid., pp. 234–35.
116. Ibid. p. 235.
117. Ibid., p. 236.
118. Evelyn Stokes, 'The legacy of Ngatoroirangi: Maori customary use of geothermal resources', (Hamilton: University of Waikato, 2000), pp. 24–25: www.boprc.govt.nz/media/509127/the-legacy-of-ngatoroirangi-maori-customary-use-of-geothermal-resources.pdf; Martin Wikaira, 'Ngāti Tūwharetoa – The journeys of Ngātoroirangi and Tia', Te Ara – the Encyclopedia of New Zealand: www.TeAra.govt.nz/en/ngati-tuwharetoa/page-2
119. Stokes, 'The legacy of Ngatoroirangi'.
120. *Kahui Maunga Report* (Wellington: Waitangi Tribunal, Legislation Direct, 2013), vol. 1, pp. 36–38; details of slightly differing accounts are given in Stokes, 'The legacy of Ngatorirangi', pp. 23–38; Wikaira, 'Ngāti Tūwharetoa', names Kuiwai and Haungaroa.
121. Keith Wood, 'Tangata whenua water resource management: Ngāti Rangi and the Tongariro Power Scheme', in Rāwiri, *Mouri Tū, Mouri Ora*, p. 43.
122. Ellis, *Journal of a Tour around Hawaii*, p. 138; Peter Buck, 'Lectures on traditional Hawaiian history and culture', MS SC Buck 1301, 1932 contains 'Orion chant for Raiatea', which appears to be a navigational guide from Tahiti to the active volcano of Big Island.
123. Beckwith, *Hawaiian Mythology*, pp. 31–32.
124. Malo, *Hawaiian Antiquities*, p. 141; Kirch, *A Shark Going Inland*, pp. 140, 155, 314.
125. Byron, Voyage of the HMS *Blonde*, p. 198.

7 KEEPERS OF THE LONG VIEW: WHANGANUI RIVER

1. T.W. Downes, 'Maori mentality regarding the lizard and *taniwha* in the Whanganui River area', *Journal of the Polynesian Society*, vol. 46, no. 4 (184), 1937, pp. 207–08. Downes noted that, on the Whanganui, 'There is considerable difficulty in separating the *taniwha* and lizard as the two seem to be connected not only by shape but by reputed habits.'
2. Titi Tihu, pers. comm., Cherry Grove, Feb. 1984; see David Young, *Faces of the River* (Wellington: TVNZ/NZ Listener, 1986), pp. 11–14. This is also the name given by Downes; and by Elsdon Best in *Maori Religion and Mythology*, Pt II, (Wellington: Government Printer, 1982), p. 506.
3. T.W. Downes, *Tapu Rock at Petipetiaaurangi*, watercolour, ?1938: https://natlib.govt.nz/records/22764141
4. Downes, 'Maori mentality', p. 209.
5. David Young, *Woven by Water: Histories from the Whanganui River* (Wellington: Huia, 1998), Ch. 7, especially pp. 136–39.
6. Downes, 'Maori mentality', p. 211. It is clear that Downes' informant for several of his papers in the *Journal of the Polynesian Society* was Titi Tihu, then a young man. Motu Kākaho has not been identified, but is said to have been from Northland.
7. Young, *Woven by Water*, pp. 231–45.
8. *The Whanganui River Report*, Waitangi Tribunal (Wellington: GP Publications, 1999).
9. Peter Buck, *Vikings of the Sunrise* (Christchurch: Whitcombe & Tombs, 1975), pp. 282–83 details *Aotea*'s voyage, including its departure 'in an off-season', and its being blown on to the Kermadecs, where the crew first discovered that the karaka seeds they carried would be redundant in their new home. Downes recorded that, on the river, Tūtangatakino was accompanied by a second taniwha (Downes, 'Maori mentality', p. 219). Moko-hikuwaru/Marakihau is the name Petuere gives to the taniwha's co-navigator from Hawaiki.
10. Master carver Paki Harrison. For a detailed description of Ra`iātea and the Taputapuātea marae and its associations with the wider Pacific, including Aotearoa, see Anne Salmond, *Aphrodite's Island: The European discovery of Tahiti* (Auckland: Penguin Viking, 2009), pp. 25–27. The name

Tūtangatakino does not seem to be known today in Tahiti.

11. Rāwiri Taonui, 'Canoe traditions – Canoes of the west coast and lower North Island', Te Ara – the Encyclopedia of New Zealand: www. TeAra.govt. nz/en/canoe-traditions/page 7
12. Best, *Maori Religion and Mythology*, Pt II, p. 508. While guardians such as whales accompanied some voyaging waka to New Zealand, the *Takitimu* was another that was guided by taniwha, in this case by two: Araiteuru and Ruamano.
13. David Young, *Whio: Saving New Zealand's endangered blue duck* (Nelson: Craig Potton Publishing, 2003), pp. 60–73, 79–81.
14. www.legislation.govt.nz/act/ public/2017/0007/ latest/whole. html#DLM6831453; Klaus Bosselmann; *The Principle of Sustainability: Transforming law and governance*, 2nd edn (Oxford: Taylor & Francis, 2016), pp. 164–65; https://elgar. blog/2017/06/22/rivers-as-people/; www.theguardian.com/australia-news/2018/ apr/01/its-only-natural-the-push-to-give-rivers-mountains-and-forests-legal-rights
15. Petuere Kiwara, pers. comm., 30 May 2015.
16. A similar entity, Sedna, is described by the Inuit, with what appear to be uncannily similar qualities of retribution and tenderness towards her charges. James Houston, 'Goddess of the sea: The story of Sedna', *Canadian Encyclopedia*: www. thecanadianencyclopedia.ca/en/article/the-goddess-of-the-sea-the-story-of-sedna
17. Judith Binney, *Redemption Songs: A life of Te Kooti Arikirangi Te Turuki* (Auckland & Wellington: Auckland University Press/Bridget Williams Books, 1995), pp. 442– 43, 460–62. The houses were Takitimu, Wairarapa; Tokanganui-a-noho, a Tainui whare at Te Kuiti; and the third house, Te-Whai-a-te-motu, built for Te Kooti closer to home at Ruatahuna, has the marakihau Te Putaanga and Te Tahi o te Rangi.
18. Petuere Kiwara, pers. comm., 30 May 2015.
19. T.W. Downes, 'A Tuahu on the Whanganui River', *Journal of the Polynesian Society*, vol. 37, 1928, pp. 165–68.
20. Ibid., p. 167; Downes, 'Maori mentality', pp. 206–24.
21. *The Whanganui River Report*, p. 77.
22. T.W. Downes, *Old Whanganui* (Hawera: W.A. Parkinson and Co., 1915), pp. 102–04.
23. David Young, *NZ Listener*, 31 March 1984, pp. 19–21.
24. Young, *Woven by Water*, p. 17.
25. Downes, 'Maori mentality', pp. 206–24.
26. W.J. Phillipps, *Carved Meeting Houses of Western and Northern Areas of New Zealand* (Wellington: Goverment Printer, 1954), pp. 84–86.
27. Ibid., pp. 185–86; R.M. McDowall, *Ikawai: Freshwater fishes in Maori culture and economy* (Christchurch: Canterbury University Press, 2011), pp. 160–62, 165.
28. Young, *Woven by Water*, pp. 153–54.
29. 'The Hawaiian Moon Calendar and the Ways of the Practitioners (*Kahuna*)', rewritten from articles by Lyons Kai`ioho Naone III in *Na Po`e Hawai`i*, Spring 1995–Winter 1996.
30. Young, *Woven by Water*, pp. 153–54.
31. www.nzgeo.com/stories/masterpiece-theatre/; www.stuff.co.nz/taranaki-daily-news/ news/67462225/motunui-panels-returned-to-taranaki; Sydney M. Mead and Athol McCredie (eds), *Te Maori: Maori art from New Zealand collections* (Auckland: Heinemann, with the American Federation of Arts, 1984), p. 226.

8 KĪPUKA OF INFLUENCE: LIMAHULI LOOKS BACK TO SEE FORWARD

1. Mary Kawena Pukui, *`Olelo No`eau: Hawaiian proverbs and poetical sayings* (Honolulu: Bernice P. Bishop Museum, 1983), no. 2390, pp. 78–79.
2. Thanks to New Zealanders Robert Sullivan (Ngāpuhi) and Anne Kennedy, who were teaching at the University of Hawai`i, Mānoa, at the time, who told me about this place.
3. www.worldatlas.com/articles/the-ten-wettest-places-in-the-world.html
4. 'Limahuli Garden', pamphlet [n.d.], p. 51; https:// ntbg.org/gardens/limahuli. Under the National Tropical Botanical Garden legislation are two others, another on Kaua`i, one on Maui and just two more on the US mainland; James D. Houston, *Hawaiian Son: The life and music of Eddie Kame* (Honolulu: `Ai Pohaku Press, 2004).
5. Kawika Winter, pers. comm., October 2018.
6. Ralph S. Kuykendal, *The Hawaiian Kingdom* (Honolulu: University of Hawai`i, 1938), pp. 88–92; Rhys Richards, pers. comm., 2008: in a big year,

e.g. 1825, more than 11,000 tonnes of sandalwood was exported; Diane Lee Rhodes, 'Pu`uhonua o Honaunau National Historical Park: A cultural history of three traditional Hawaiian sites on the West Coast of Hawai`i Island': www.nps.gov/parkhistory/online_books/kona/history5e.htm
7. The place has been chartered by the US Congress since 1964. The mission of the National Tropical Botanical Garden is to enrich life through discovery, scientific research, conservation and education, by perpetuating the survival of plants, ecosystems and cultural knowledge of tropical regions: http://en.wikipedia.org/wiki/National_Tropical_Botanical_Garden; https://nbtg.org.news/saving-kaua's-seabirds/
8. J.C. Beaglehole, *The Life of Captain James Cook* (London: Adam & Charles Black, 1974), p. 580.
9. Katharine Luomala, *Voices on the Wind: Polynesian myths and chants* (Hawai`i: Bishop Museum Press, 1986) pp. 5, 154–55.
10. Mary Kawena Pukui with Laura C.S. Green, *Folktales of Hawai`i/He Mau Ka`ao Hawai`i* (Honolulu: Bishop Museum Press, 2003), p. 1; www.kahunaresearchgroup.org/case-log/legend-of-the-mu-people-hawaiis-original-inhabitants
11. Tina Ngata, 'Wai Māori', in Mike Joy (ed.), *Mountains to Sea: Solving New Zealand's freshwater crisis* (Wellington: Bridget Williams Books, 2018), pp. 18–29.
12. http://dlnr.hawaii.gov/dar/files/2015/08/Haena_Mgmt_Plan_6.1.11.pdf
13. Kenneth R. Wood, 'Summary of the Limahuli Flora and Vascular Plant Species Checklist, Kaua`i, Hawai`i. Prepared for the U.S. Fish and Wildlife Service, Honolulu, Hawai`i & the Kaua`i Endangered Seabird Recovery Project', report no. 298, Affiliation: National Tropical Botanical Garden, Oct. 2012: https://researchgate.net/figure/National-Tropical-Botanical-Garden-Limahuli-Valley-Kauai-Hawaiian-Islands_fig1_266138686; www.researchgate.net/publication/266138686_Summary_of_the_Limahuli_Flora_and_Vascular_Plant_Species_Checklist_Kauai_Hawaii_Prepared_for_the_US_Fish_and_Wildlife_Service_Honolulu_Hawaii_the_Kauai_Endangered_Seabird_Recovery_Project_43_pp
14. Zach Royer, 'Legends of the Mu people: Hawaii's original inhabitants', Kahuna Research Group, 3 September 2014: www.kahunaresearchgroup.org/case-log/legend-of-the-mu-people-hawaiis-original-inhabitants. Koko Willis & Pali Jae Lee, *Tales from the Night Rainbow: Mo`olelo a Na Po Makole: The story of a woman, a people, and an island: An oral history as told by Kaili`ohe Kame`ekua of Kamalo, Moloka`i, 1816–1931* (Honolulu: Paia-Kapela-Willis, Ohana, 1984).
15. Patrick Kirch, 'When did the Polynesians settle Hawaii?', Departments of Anthropology and Integrative Biology, University of California, Berkeley, 2000: www.academia.edu/1034367/When_Did_the_Polynesians_Settle_Hawaii
16. Jonathan Kamakawiwo`ole Osorio, *Dismembering Lāhui: A history of the Hawaiian nation to 1887* (Honolulu: University of Hawai`i Press, 2002), ch. 3; see also https://himonarchy.weebly.com/the-great-mahele.html
17. Lawrence Miike, *Water and the Law in Hawai`i* (Honolulu: University of Hawai`i Press, 2004), pp. 74–77, 82–96; www.environment-hawaii.org/?p=4008; Summer Sylva, 'Indigenizing water law in the 21st century: Na Moku Aupuni o Ko`olau Hui: A Native Hawaiian case study', Cornell Journal of Law and Public Policy, vol. 16, issue 3, 2007, p. 579 on public trust doctrine.
18. Carlos Andrade, *Hā`ena: Through the eyes of the ancestors* (Honolulu: University of Hawai`i Press, 2008), p. 79.
19. www.hawaiihistory.org/index.cfm?fuseaction=ig.page&PageID=288: 'Under the reign of Kamehameha III, Hawaii's traditional system of land use underwent drastic change. Guided by foreign advisors, the king divided lands that had formerly been held in common and administered by chiefs and their konohiki, or overseers. The Māhele allocated 23% of land in the islands to the king (called crown lands); 40% comprised konohiki lands to be divided among 245 chiefs; and 37% was declared government lands, to be awarded to commoners who worked the land as active tenants. An appointed Land Commission and Court of Claims administered the land division. The Māhele was followed in 1850 by the Kuleana Act which established fee simple ownership of land. Historical land tenants were required to document their claims to specific parcels in order to gain permanent title. Once granted, a kuleana plot was entirely independent of the traditional ahupua`a in which it was situated and it could also

be sold to parties with no historical ties to the area.'

20. Andrade, *Hā`ena*, pp. 113, 114–15.
21. Ibid., p. 79.
22. Carlos Andrade, pers. comm., Dec. 2008.
23. The rule does still hold in New Zealand law, however. The late John Hanita Paki led a long legal battle for rights to a non-navigable section of the mid-Waikato, as part of a Māori entity and law case known as Pouakani, in which they made significant gains in 2006: www.courtsofnz.govt.nz/cases/paki-and-others-v-attorney-general-of-new-zealand. But further litigation did not advance their interests: www.courtsofnz.govt.nz/assets/cases/2010/sc-7-2010-john-hanita-paki-and-ors-v-attorney-general-leave-judgment.pdf; Miike, *Water and the Law in Hawaii*, p. 104.
24. Andrade, *Hā`ena*, pp. 111–12; Miike, *Water and the Law in Hawai`i*, pp. 102–05, discusses such rights in the *Reppun v Board of Water Supply* case.
25. Ha`ena Fisheries Committee, 'Proposed management plan for the Ha`ena community-based and fishing regulations for … Kaua`i', 1 June 2011: http://dlnr.hawaii.gov/dar/files/2015/08/Haena_Mgmt_Plan_6.1.11.pdf
26. Elisha Atkinson, 'Why has Hawaii banned suntan lotion?', 13 July 2018, *Telegraph*: www.telegraph.co.uk/travel/news/hawaii-sun-tan-lotion-ban/
27. Stacy Jupiter, Kostantinos A. Stamoulis, Robert Whittler, Alan Friedlander, Kawika Winter et al., 'Place-based management can reduce human impacts on coral reefs in a changing climate', *Bulletin of the Ecological Society of America*, vol. 100, no. 3, July 2019 (Honolulu: University of Hawai`i, Institute of Marine Biology).
28. Burdett A. Rich, *Lawyers' reports, Annotated, 1893* (London: Forgotten Books, Feb. 2019), vol. 21, p. 195.
29. The decision in *McBryde v Robinson*, 1973; reversed by the US District Court in 1977 because it violated 'the private property rights of water users', it was sent back for reconsideration and partially confirmed by the US Circuit Court of Appeals. Miike, *Water and the Law in Hawaii*, pp. 82–85, 101; www.environment-hawaii.org/?p=4008
30. Miike, *Water and Law in Hawai`i*, pp. 90–96.
31. Sylva, 'Indigenizing water law': https://scholarship.law.cornell.edu/cjlpp/vol16/iss3/5/
32. Miike, *Water and the Law in Hawai`i*, pp. 74, 78.
33. Ibid., p. 130
34. https://en.wikipedia.org/wiki/Public_trust_doctrine#Origins
35. Jim Lichatowich et al., 'Wild Pacific salmon: A threatened legacy', 2017: https://flyfishersinternational.org/Portals/0/Documents/Conservation/Pacific_Salmon_Legacy_Final_7-27-17.pdf?ver=2017-07-27-111242-000
36. Miike, *Water and the Law in Hawai`i*, p. 213.
37. www.nzherald.co.nz/nz/news/article.cfm?c_id=1&objectid=12109430
38. Miike, *Water and the Law in Hawai`i*, pp. 82–96, 130.
39. www.hawaiipublicradio.org/post/east-maui-sees-largest-stream-restoration-hawaii-history
40. Ibid.
41. David Young, 'Curated cross-roads heritage', *Phanzine Magazine*, Wellington, Aug. 2018; 'Taro growers, practitioners elated with E. Maui water rights decision', *Maui News*: www.mauinews.com/news/local-news/2018/06/taro-growers-practitioners-elated-with-e-maui-water-rights-decision/
42. Ku`uwehi Hiraishi, 'East Maui sees largest stream restoration in Hawai`i history', Hawai`i Public Radio: www.hawaiipublicradio.org/post/east-maui-sees-largest-stream-restoration-hawaii-history
43. 'Taro growers, practitioners elated with E. Maui water rights decision', *Maui News*, 21 June 2018: www.mauinews.com/news/local-news/2018/06/taro-growers-practitioners-elated-with-e-maui-water-rights-decision/
44. Ibid.; Ku`uwehi Hiraishi, 'East Maui sees largest stream restoration in Hawai`i history', Hawai`i Public Radio, 25 June 2018: www.hawaiipublicradio.org/post/east-maui-sees-largest-stream-restoration-hawaii-history
45. 'Taro growers, practictioners elated': www.mauinews.com/news/local-news/2018/06/taro-growers-practitioners-elated-with-e-maui-water-rights-decision/; www.khon2.com/news/local-news/landmark-decision-restores-flow-in-10-maui-streams-after-more-than-a-century/125290345; www.mauinews.com/news/local-news/2016/01/a-b-water-diversion-permits-ruled-invalid/
46. www.huffingtonpost.com/entry/massive-ab-corporate-water-theft-faces-public-comment_us_58aa48b6e4b026a89a7a2d55
47. Ngāti Apa v Attorney-General, *New Zealand Law Reports*, vol. 3, para 661.

48. Tamatoa Bambridge (ed.), 'Conclusion: What are the lessons to be learnt from the *rahui* and legal pluralism? The political and environmental efficacy of legal pluralism', in T. Bambridge (ed.), *The Rahui: Legal pluralism in Polynesian traditional management of resources and territories* (Canberra: ANU Press, Pacific Series, 2016), pp. 227–30.
49. While the 1840 Treaty recognised Māori rights over their lands, fisheries, forests and estates, it was not until the 1990s, when Treaty principles were agreed that required the Crown to, among other things, act in good faith towards Māori in an act of 'partnership', that real progress began to be made.
50. *Paki v Crown* challenged this ruling in respect of the Waikato River; the case was won in the Supreme Court, 2012.
51. www.legislation.govt.nz/act/public/2017/0007/latest/whole.html#DLM6831453; Klaus Bosselmann, *The Principle of Sustainability: Transforming law and governance* (2nd edn) (Oxford: Taylor & Francis, 2016), pp. 164–65; www.tpk.govt.nz/en/mo-te-puni-kokiri/kokiri-magazine/kokiri-33-2016/te-awa-tupua
52. Catherine Iorns Magallanes, 'A world where the rivers are people too': https://elgar.blog/2017/06/22/rivers-as-people/
53. Jane Gleeson-White, 'It's only natural: The push to give rivers, mountains and forests legal rights', *Guardian*, 1 Apr. 2018: www.theguardian.com/australia-news/2018/apr/01/its-only-natural-the-push-to-give-rivers-mountains-and-forests-legal-rights
54. Robert Macfarlane, 'Should this valley have rights?', *Guardian Weekly*, 8 Nov. 2019, pp. 41–44.
55. Catherine Iorns Magallanes, 'From rights to responsibilities using legal personhood and guardianship for rivers', in B. Martin, L. Te Aho & M. Humphries-Kil (eds), *ResponsAbility: Law and governance for living well with the earth* (London: Routledge, 2019), pp. 216–39.
56. 'Introduction: River as ethnographic subject', in John R. Wagner & Jerry K. Jacka (eds), *Island Rivers: Fresh water and place in Oceania* (Canberra: ANU Press, 2018), p. 5.
57. Rachel Kennard, 'The potential for Maori customary claims to freshwater', LLB dissertation, University of Otago, Dunedin, p. 46: www.otago.ac.nz/law/research/journals/otago036259.pdf
58. 'Te Mana o te Wai: The health of our wai, the health of our nation', Kāhui Wai Māori report to Hon. Min. David Parker, 2019: www.mfe.govt.nz/sites/default/files/media/Fresh%20water/kahui-wai-maori-report.pdf
59. 'Best practice iwi/hapū engagement for collaborative freshwater management': https://landcareresearch.co.nz/__data/assets/pdf_file/0012/127020/Best-practice-iwi-and-hapu-engagement-for-collective-freshwater-management.pdf. The word 'kaitiakitanga' was first given legal recognition in the 1991 Resource Management Act.
60. Kennard, 'The potential for Maori Customary Claims', p. 52.
61. Guardians of Fiordland's Fisheries & Marine Environment Inc., Laurel Teirney (principal author), 'Fiordland Marine Conservation Strategy', 2003: www.mfe.govt.nz/sites/default/files/media/Marine/Fiordlandmarineconservationstrategy2003final.pdf; Michael Begon, Colin R. Townsend & John L. Harper, *Ecology: From individuals to ecosystems* (Oxford: Wiley-Blackwell, 2006), ch. 20; David Young, *Keeper of the Long View: Sustainability and the PCE* (Wellington: Parliamentary Commissioner for the Environment, 2007), pp. 109–10.
62. https://thespinoff.co.nz/atea/atea-otago/27-11-2017/if-the-hills-could-sue-jacinta-ruru-on-legal-personality-and-a-maori-worldview/; Colin W. Townsend, *Ecological Applications: Towards a sustainable world* (US: Wiley & Blackwell, 2019), Ch. 20.
63. Robert J. Hommon, 'Social evolution in ancient Hawai'i', in Patrick V. Kirch (ed)., *Island Societies: Archaeological approaches to evolution and transformation* (Cambridge: Cambridge University Press, 1986), pp. 60–61; Patrick Kirch, in *A Shark Going Inland* (Berkeley: University of California Press, 2014), pp. 294–96, is still describing the construction of significant irrigation systems in windward Moloka`i and O`ahu 'between the 13th and 15th century'.
64. NIWA Taihoro Nukurangi, 'Past climate variations over New Zealand': https://niwa.co.nz/our-science/climate/information-and-resources/clivar/pastclimate
65. Dennis Kawaharada, *Storied Landscapes: Hawaiian literature and place* (Honolulu: Kalamaku Press, 1999), pp. 79–80.

66. Van James, *Ancient Sites of Hawaii: Archaeological places of interest on the Big Island* (Honolulu: Mutual Publishing, 1995).
67. Kawaharada, *Storied Landscapes*, pp. 79–80.
68. Ibid., p. 81.
69. Kawika Winter, pers. comm., 31 Oct. 2018.
70. In 2015 New Zealand's then prime minister took out the 'Fossil of the Day' award, given by the Climate Change Action Group at the UN Paris Climate Talks, for the nation's spending on fossil fuel subsidies. 'NZ wins first 'Fossil' award at Paris talks', *New Zealand Herald*, 1 Dec. 2015: www.nzherald.co.nz/elementmagazine/news/article.cfm?c_id=1503340&objectid=11553913

9 SĀMOA, NIUE: THE MEMORY OF WATER

1. Robert Louis Stevenson, *Vailima Letters*, ch. 1, 2 Nov. 1890: http://pinkmonkey.com/dl/library1/digi256.pdf
2. R.L. Stevenson, *A Footnote to History: Eight years of trouble in Samoa* (Honolulu: University of Hawai`i Press, 1996; original 1892). Stevenson arrived in Sāmoa in December 1889, nine months after the cyclone.
3. Stevenson, *Vailima Letters*, ch. 1, 2 Nov. 1890.
4. H.J. Moors, *With Stevenson in Samoa* (London: T. Fisher Unwin, 1911), p. 12.
5. Stevenson, *Vailima Letters*, ch. 18: https://en.wikisource.org/wiki/Vailima_Letters/Chapter_XVIII
6. Hummingbirds do not exist in the southern hemisphere, but some large moths exhibiting similar movements do: http://cookislands.bishopmuseum.org/showarticle.asp?id=8
7. D. Kear & B.L. Wood, 'The hydrology of Upolu: The geology and hydrology of Western Samoa', *New Zealand Geological Survey Bull. 63*, Wellington, 1959, p. 68.
8. Stevenson to Colvin: https://en.wikisource.org/wiki/Vailima_Letters/Chapter_I, 2 Nov. 1890 suggests this battle had occurred perhaps 50–60 years earlier; Gavin Bell, *In Search of Tusitala: Travels in the Pacific after Robert Louis Stevenson* (London: Picador, 1994), pp. 288–89.
9. Stevenson to Colvin, *Vailima Letters*, ch. 1, 2 Nov. 1890: http://pinkmonkey.com/dl/library1/digi256.pdf; ; further references to the recurring invasion of their garden arise through the letters.
10. See David Young, *Rivers: New Zealand's shared legacy* (Auckland: Random House, 2012), p. 236.
11. Stevenson, *Vailima Letters*, ch. 1, 2 Nov. 1890.
12. M.D. Olson, 'Re-constructing landscapes: The social forest, nature and spirit-world in Samoa', *Journal of the Polynesian Society*, vol. 106, 1997, pp. 8, 12, 16; K.R. Howe (ed.), *Vaka Moana: Voyages of the ancestors* (Auckland: David Bateman & Auckland Museum, 2006), p. 38.
13. Martha Beckwith, *Hawaiian Mythology* (Honolulu: University of Hawai`i Press, 1976), p. 271.
14. Stevenson, *Vailima Letters*, ch. 1, 5 Nov. 1890: http://pinkmonkey.com/dl/library1/digi256.pdf
15. Ibid., ch. 3, 2[?] Dec. 1890 to Sidney Colvin: https://en.wikisource.org/wiki/Vailima_Letters/Chapter_III
16. Ibid., ch. 3, 27 Dec. 1890.
17. D. Kear & B.L. Wood, 'The geology and hydrology of Western Samoa'; Manuto in the story is given as Manatu (Bay), fig. 40, endmaps.
18. Ibid., pp. 21–23.
19. Brother Herman, *Tala o le Vavau: Samoan legends* (Pago Pago: Association of the Marist Brothers' Old Boys, 1955), p. 55; Paul Theroux, 'Some misconceptions about RLS', *Journal of Pacific History*, vol. 16, no. 3, July 1981, pp. 164–66.
20. Stevenson, *Vailima Letters*, ch. 3, 27 Dec. 1890.
21. M.D. Olson, 'Reconstructing landscapes: The social forest, nature and spirit-world in Samoa', *Journal of the Polynesian Society*, vol. 106, 1997, pp. 18–19.
22. Ibid., p. 9.
23. https://rainforests.mongabay.com/deforestation/archive/Samoa.htm; Cedric Schuster, Arthur Whistler & Tapulolou Siuli Tuailemafua, *The Conservation of Biological Diversity in Upland Ecosystems of Samoa* (Sāmoa: Division of Environment and Conservation, 1999), pp. 37–42.
24. Ibid., pp. 27, 28; Paul Alan Cox, 'Forests, flying foxes and Samoans: An ecological triangle', *Plant Talks*, vol. 31, p. 21.
25. Schuster et al., *The Conservation of Biological Diversity*, pp. 37–39, 41–43, 46, 48, 50–52.
26. Fred Pearce, 'Rivers in the sky: How deforestation is affecting global water cycles', Yale Environment 360: https://e360.yale.edu/features/how-deforestation-affecting-global-water-cycles-climate-change

27. Wren Green, pers. comm., 20 Nov. 2015. Wren returned to Sāmoa in February 2018, when the photographs were taken, only to confirm his impressions of a playground forever lost.
28. Gavin Bell, *In Search of Tusitala: Travels in the Pacific after Robert Louis Stevenson* (London: Picador, 1994), p. 289.
29. Kear & Wood, 'The geology and hydrology of Western Samoa', p. 68 (250–750cm p.a.); Brian Kouvelis, pers. comm., 16 Oct. 2018 makes the important point that statistical data on rainfall and streamflows should be reset at 1990, in order to better comprehend the swiftness of change now upon us – not fully reflected in today's data averages that are still attached to the averages of old cycles.
30. www.worldometers.info/world-population/samoa-population/
31. Samoa's 5th National Report 2014 to the Convention on Biological Biodiversity, Ministry of Natural Resources and Environment, Apia, 2014, introduction.
32. https://en.wikipedia.org/wiki/1889_Apia_cyclone; Stevenson, *A Footnote to History*, Ch. X; https://en.wikipedia.org/wiki/Samoan_crisis
33. Stevenson, *Footnote to History*, pp. 120–29; https://en.wikipedia.org/wiki/1889_Apia_cyclone; casualty figures between the two sources differ markedly. I have been guided by Wikipedia's more conservative ones.
34. https://en.wikipedia.org/wiki/Cyclone_Val
35. https://en.wikipedia.org/wiki/Category:Tropical_cyclones_in_Samoa; Paul Alan Cox, *Nafanua: Saving the Samoan rainforest* (US: W.H. Freeman, 1999).
36. Cox, *Nafanua*, pp. 12, 67, 70, 80, 82, 140–42.
37. Erick Brenstrum, pers. comm., 18 Sept. 2018.
38. Ibid. Brenstrum writes of a major unnamed cyclone that devastated much of the North Island in 1936, in 'The ghost of cyclones past', *New Zealand Geographic*, Jan. 1997: www.nzgeo.com/stories/the-ghost-of-cyclones-past
39. Eileen McSaveney, 'Floods – Southland floods and Cyclone Bola', Te Ara – the Encyclopedia of New Zealand: www.TeAra.govt.nz/en/floods/page-4; 'Cyclone Bola strikes', Ministry for Culture and Heritage: https://nzhistory.govt.nz/page/cyclone-bola-strikes
40. Young, *Rivers*, p. 214.
41. Te Ahukaramū Charles Royal (ed.), *The Woven Universe: Selected writings of Maori Marsden* (Te Wānanga-o-Raukawa, 2003), p. 46.
42. Jonathan White, *Tides: The science and spirit of the ocean* (San Antonio: Trinity University Press, 2017), pp. 152, 264–65.
43. https://en.wikipedia.org/wiki/1936_New_Zealand_cyclone; https://iwonderweather.co.nz/story/Cyclone-meets-wildfire-March-1918/6376/topic/extreme-weather; www.niwa.co.nz/news/climate-change-challenge-for-small-islands-in-pacific
44. 'Our Atmosphere and Climate 2017: Data to 2016', New Zealand's Environmental Reporting Series, New Zealand Government: www.mfe.govt.nz/sites/default/files/media/Environmental%20reporting/our-atmosphere-and-climate-2017.pdf, pp. 17, 29 ('the drier soils' is in an addendum, 17 April 2018).
45. NIWA, 'Climate change scenarios for New Zealand': www.niwa.co.nz/our-science/climate/information-and-resources/clivar/scenarios
46. Geoff Park, Rod Hay et al., 'The National Ecological Survey of Western Samoa: The conservation of biological diversity in the coastal lowlands of Western Samoa', Dept of Conservation, Wellington, 1992, p. ix.
47. Ibid., pp. ix, x.
48. Ibid.; Rod Hay, pers. comm., 24 Sept. 2018: 'Recovery following the Apia Cyclone was uninhibited by the choking effect of mile-a-minute (Mikania micrantha) and other vine weeds, which were introduced into Samoa much more recently.'
49. Olson, 'Re-constructing landscapes', p. 13.
50. 'Samoa's 5th National Report 2014 to the Convention on Biological Diversity', para 3.5: www.cbd.int/doc/world/ws/ws-nr-05-en.pdf
51. http://nzjf.org.nz/free_issues/NZJF28_3_1983/AF61A68A-F9EB-43A1-BB2E-28CD03AA46D1.pdf
52. Nu`ufou Petaia, 'Incorporating traditional knowledge in development activities: Western Samoa', in John Morrison, Paul Geraghty & Linda Crowl (eds), *Science of Pacific Island Peoples*, vol. 4, *Education, Language, Patterns & Policy* (Institute of Pacific Studies, University of the South Pacific, 1994), pp. 173–84.
53. 'Samoa's 5th National Report 2014', para 2.6.
54. 'Stronger roots: Growing resilient forests in Samoa', UN Development Programme: www.undp.org/content/undp/en/home/ourwork/ourstories/

stronger-roots--growing-resilient-forests-in-samoa-.html; 'Samoa's 5th National Report 2014'.

55. 'Mangrove Ecosystems for Climate Change Adaptation and Livelihood (MESCAL) Samoa Project', Le Asaga Bay Mangroves Biodiversity Report, May 2013: www.iucn.org/sites/dev/files/mangroves_biodiveristy_assessment_report_samoa.pdf
56. Tui Atua Tupua Tamasese Ta`isi Efi, 'An ethic of responsibility in Samoan customary law', keynote address, Law, Ethics & Responsibility Symposium, Hopuhopu, Ngāruawāhia, Waikato–Tainui College for Research and Development, 25 Nov. 2014, pp. 4–5. This is one of several deeply informed essays on these themes by the same author, including 'Whispers and Vanities in Samoan Indigenous Religious Culture', Tui Atua Tupua Tamasese Ta`isi Efi, Head of State of the Independent State of Samoa, World Parliament of Religions, Melbourne, Australia, 3 Dec. 2009. See also T.D. Salesa, 'When the waters met: Some shared histories of Christianity and ancestral Samoan spirituality', in Tamasailau Suaalii-Sauni et al. (eds), *Whispers and Vanities: Samoan indigenous knowledge and religion* (Wellington: Huia Publishers, 2009), pp. 143–58.
57. Karen Lupe, 'Le Ala i le Mafa e, The Path Towards Enlightenment', in Suaalii-Sauni, *Whispers and Vanities*, p. 231.
58. Salesa, 'When the waters met', p. 157.
59. Meredith, pers. comm., Sāmoa, June 2002; Katherine Luomala, *Voices of the Wind: Polynesian myths and chants* (Honolulu: Bishop Museum Press, 1986), despite her considerable experience of Sāmoa and her attention to Kāne/Tāne through much of the Pacific Islands, gives few details of his influence in Sāmoa; similarly with Salesa, 'When the waters met', pp. 150, 157; Olson, 'Re-constructing landscapes', pp. 7–32; Ngārangi Rangiuia, pers. comm., 23 Nov. 2018: 'Tangaloa is their main deity.'
60. Olson, 'Re-constructing landscapes', p. 9.
61. Sylvia Earle, 'Oceans: The well of life', *Time*, 28 Oct. 1996, pp. 52–53, quoted in Epeli Hau`ofa, 'The ocean in us', *Contemporary Pacific*, vol. 10, no. 2, 1998 (Honolulu: University of Hawai`i Press), pp. 391, 410, 402–03: https://scholarspace.manoa.hawaii.edu/bitstream/10125/13234/v10n2-392-410-dialogue.pdf
62. https://en.wikipedia.org/wiki/Niue#cite_note-28
63. J.C. Beaglehole, *A Life of Captain James Cook* (London: Adam & Charles Black, 1974), pp. 389–90; Richard Sidey, 'Fruit Bat and Coconut Crabs', vimeo, Sept. 2018; Aliscia Young, pers. comm., Sept. 2018.
64. https://en.wikipedia.org/wiki/Niue; Beaglehole, *The Life of Captain James Cook*, pp. 389–90; https://mfat.govt.nz/en/countries-and-regions/pacific/niue/; Aliscia Young, pers. comm., Sept. 2018.
65. *Atlas of the South Pacific* (Wellington: New Zealand Government Printing, 1982), p. 32; Oliver Sutherland, *Paikea: The life of I.L.G. Sutherland* (Christchurch: Canterbury University Press, 2013), p. 264.
66. Michael Hoare, Royal Society talk, Nelson, 3 September 2018; Aliscia Young & Richard Sidey, pers. comm., Sept. 2018.
67. https://forestry.gov.vu/projects/project-1-forestry-protected-area-management/
68. https://intechopen.com/books/applied-bioremediation-active-and-passive-approaches/persistence-and-bioaccumulation-of-persistent-organic-pollutants-pops. Paraquat was officially annexed as a POP in 2019 in the Rotterdam Convention Prior Informed Consent List; Phillipa Stevenson, 'The scandal of our PCP dumps', 'PCPs: Crunch time for the timber industry', *Terra Nova*, June, July, Aug. 2002; John O'Grady, pers. comm., 13 Jan. 2020.
69. Bev Doole, pers. comm., 27 Sep. 2018; www.cookislandsnews.com/item/64480-action-on-pesticide-problem/64480-action-on-pesticide-problem

10 RAPANUI/EASTER ISLAND: RATS, FORESTS AND GOOD WATER

1. Aliscia Young, pers. comm., spring 2014: https://alisciayoung.com/videos.html; Alfred Métraux, *Ethnology of Easter Island*, Bishop Museum Bulletin no. 160, Honolulu, 1940, p. 16. These crickets seem to be an introduced species; *Discover* claims that 10 insects remain of the island's original fauna: Nathaniel Scarping, 'Extinction looms for Easter Island's only remaining native species', *Discover*, 20 July 2016: https://blogs.discovermagazine.com/planet-earth/extinction-looms-for-easter-islands-only-remaining-native-species

2. Terry Hunt & Carl Lipo, *The Statues That Walked: Unravelling the mystery of Easter Island* (New York: Free Press, 2011), pp. 29–30; C.M. Stevenson, J.M. Ramirez, F.J. Morin & N. Barbacci (eds), *The Reñaca Papers: VI International Conference on Rapa Nui and the Pacific* (Los Osos, CA: Easter Island Foundation and University of Valparaíso, 2004), p. 62. The tree survives from seed at Kew Gardens, *Jubaea disperta* was the former name of the Easter Island palm.
3. Hunt & Lipo, *The Statues That Walked*, p. 31; www.discovermagazine.com/planet-earth/extinction-looms-for-easter-islands-only-remaining-native-species
4. Steven Roger Fischer, *Island at the End of the World: The turbulent history of Easter Island* (London: Reaktion Books, 2005), p. 81.
5. J.C. Beaglehole, *The Life of Captain James Cook* (London: Adam & Charles Black, 1974), p. 374. Of course the Polynesian route was not from New Zealand, but from eastern Tahiti; K.R. Howe (ed.), *Vaka Moana: Voyages of the ancestors* (Auckland: David Bateman & Auckland Museum, 2006), pp. 252–53, records Roggeveen making a similar observation when he first reached Easter Island.
6. Métraux, *Ethnology of Easter Island*, p. 23.
7. Ibid., p. 11.
8. Beaglehole, *The Life of Captain James Cook*, pp. 372, 374–75.
9. Métraux, *Ethnology of Easter Island*, pp. 13, 38.
10. Rhys Richards, *Easter Island 1793–1861: Observations by early visitors before the slave raids* (Los Osos, CA: Easter Island Foundation, 2008), p. 17.
11. John Linton Palmer, 'A visit to Easter Island, or Rapa Nui, in 1868', *Journal of the Royal Geographic Society of London*, vol. 40, 1870, pp. 167–81: www.jstor.org/stable/pdf/1798641.pdf
12. Métraux, *Ethnology of Easter Island*, p. 163.
13. Richards, *Easter Island 1793 to 1861*, p. 54.
14. Beaglehole, *The Life of Captain James Cook*, p. 373.
15. Fischer, *Island at the End of the World*, p. 44.
16. Paul Bahn & John Flenley, *The Enigmas of Easter Island: Island on the edge* (Oxford: Oxford University Press, 2003), p. 167.
17. Palmer, 'A visit to Easter Island', pp. 168–67: www.jstor.org/stable/pdf/1798641.pdf
18. Ibid., p. 168.
19. Métraux, *Ethnology of Easter Island*, pp. 12–14 (quoting Roggeveen).
20. Bahn & Flenley, *The Enigmas of Easter Island*, pp. 149, 150.
21. Beaglehole, *The Life of Captain James Cook*, pp. 372–73.
22. Bahn & Flenley, *The Enigmas of Easter Island*, pp. 87, 149.
23. Hunt & Lipo, *The Statues That Walked*, p. 32.
24. Fischer, *Island at the End of the World*, p. 57.
25. Stephen Fischer, *Rongorongo, the Easter Island Script: History, traditions, texts* (Oxford, England: Clarendon Press, 1997).
26. Bahn & Flenley, *The Enigmas of Easter Island*, pp. 12, 13.
27. Ibid., p. 16.
28. Hunt & Lipo, *The Statues That Walked*, p. 182.
29. Ibid., p. 182.
30. Francis Mazière, *Mysteries of Easter Island* (London: Collins, 1969), p. 97.
31. Palmer, 'A visit to Easter Island', p. 171.
32. Fischer, *Island at the End of the World*, p. 42.
33. Bahn & Flenley, *The Enigmas of Easter Island*, p. 16.
34. Palmer, 'A visit to Easter Island', p. 171; Mazière, *Mysteries of Easter Island*, p. 97.
35. Hunt & Lipo, *The Statues That Walked*, pp. 43–44; Hilary Mitchell & John Mitchell, *Te Tau Ihu o Te Waka: A history of Māori of Nelson and Marlborough* (Wellington/Nelson: Huia Publishers/Wakatu Incorporation, 2004): vol. 1, pp. 52–53, describes rock mulching together with extensive use of wood ash, gravel and sand on the Waimea Plains, Nelson, by Waitaha and Te Rapuwai in gardens from about 1460–1650.
36. Fischer, *Island at the End of the World*, p. 42.
37. Paul Bahn & John Flenley, *Easter Island, Earth Island* (London: Thames & Hudson, 1992), p. 30.
38. Mazière, *Mysteries of Easter Island*, p. 57.
39. Stevenson et al., *The Reñaca Papers*, p. 62.
40. Aliscia Young, pers. comm., spring 2014.
41. Fischer, *Island at the End of the World*, p. 42.
42. Bahn & Flenley, *Enigmas of Easter Island*, p. 30.
43. Métraux, *Ethnology of Easter Island*, p. 11.
44. Ibid.
45. Palmer, 'A visit to Easter Island', p. 168.
46. Ibid.
47. Bahn & Flenley, *The Enigmas of Easter Island*, pp. 137–38.

48. *Atlas of the South Pacific* (Wellington: New Zealand Government Printing, 1982), pl. 20; Bahn & Flenley, *The Enigmas of Easter Island*, facing p. 10. Tuff is a form of vented, hardened volcanic ash.
49. Métraux, *Ethnology of Easter Island*, p. 11.
50. Fischer, *Island at the End of the World*, p. 42.
51. Edmundo Edwards & Alexandra Edwards, *When the Universe Was an Island: Exploring the cultural and spiritual cosmos of ancient Rapa Nui* (Santiago: Hangaroa Press, 2010), p. 119. Hiro the god on Rapanui seems somewhat different from Hiro (Aotearoa) and Hilo, (Hawai`i).
52. Ibid., p. 119.
53. Métraux, *Ethnology of Easter Island*, pp. 13, 20, 37, 38, 86.
54. Edwards, *When the Universe Was an Island*, p. 119: www.smithsonianmag.com/smart-news/easter-island-statues-may-have-marked-sources-fresh-water 180971244/
55. Ibid., p. 115.
56. Ibid.
57. Métraux, *Ethnology of Easter Island*, pp. 12–17.
58. Stevenson et al., *The Reñaca Papers*, p. 62.
59. Ibid., p. 62.
60. Ibid.
61. www.windfinder.com/windstatistics/easter_island; Howe, *Vaka Moana*, pp. 70–71; Métraux, *Ethnology of Easter Island*, pp. 52–54.
62. Stevenson et al., *The Reñaca Papers*, p. 62.
63. Hunt & Lipo, *The Statues That Walked*, p. 31.
64. Bahn & Flenley, *Enigmas of Easter Island*, p. 192.
65. Valentí Rull, Núria Cañellas-Boltá, Olga Margalef, Sergi Pla-Rabes, Alberto Sáez & Santiago Giralt, 'Three millennia of climatic, ecological and cultural change on Easter Island: An integrative overview', *Frontiers in Ecology and Evolution*, 2016, p. 4: www.frontiersin.org/articles/10.3389/fevo.2016.00029/full
66. Hunt & Lipo, *The Statues That Walked*, p. 31; www.landcareresearch.co.nz/science/plants-animals-fungi/ecosystems/prehistoric-settlement/human-arrival
67. Trevor Worthy & Richard N. Holdaway, *The Lost World of the Moa: Prehistoric life in New Zealand* (Christchurch: Canterbury University Press, 2002), pp. 262–64; Bahn & Flenley, *Enigmas of Easter Island*, p. 160. Rhys Richards suggests that sometimes kiore were packed for voyaging simply as cultural items; they became items of food largely only in colder climates.
68. J. Stephen Athens, Timothy Rieth & Thomas S. Dye, 'A paleoenvironmental and archaeological model-based age estimate for the colonization of Hawai`i', *American Antiquity*, vol. 79, no. 1, pp. 144–55.
69. Hunt & Lipo, *The Statues That Walked*, pp. 26–27.
70. www.aphis.usda.gov/wildlife_damage/nwrc/publications/14pubs/14-138%20shiels.pdf
71. Bahn & Flenley, *Enigmas of Easter Island*, pp. 159–61.
72. Hunt & Lipo, *The Statues That Walked*, pp. 28–29; Worthy & Holdaway, *The Lost World of the Moa*, pp. 479–80, 562; J. Diamond, 'Rats as agents of extermination', *Nature*, vol. 318, 1985, pp. 602–03: this report was based largely on the groundbreaking research of New Zealand scientist Ian Atkinson.
73. Rhys Richards, pers. comm., July 2018; http://tpo.tepapa.govt.nz/viewtopicexhibitdetail.asp?exhibitid=0x000a4e26
74. The extermination of kiore by New Zealand's Department of Conservation has often been at odds with Māori who sometimes describe this animal as a taonga (treasure). Thus far Māori interests have been recognised by the setting aside of some tiny islands for the exclusive benefit of the kiore and Ngātiwai.
75. Hunt & Lipo, *The Statues That Walked*, p. 167: www.researchgate.net/publication/256545599_The_Human_Transformation_of_Rapa_Nui_Easter_Island_Pacific_Ocean
76. www.doc.govt.nz/nature/pests-and-threats/predator-free-2050/; Hawai`i as a territory lags behind New Zealand, but, with its considerably smaller land area, potentially could achieve predator-free status more quickly: https://dlnr.hawaii.gov/dofaw/featured/nr15-139f/
77. Marcelo Flores, Pedro Lazo, Graciela Campbell & Alejandro Simeone, 'Breeding status of the red-tailed tropicbird and threats to its conservation', *Pacific Science*, April 2017: www.bioone.org/doi/abs/10.2984/71.2.4
78. Kevin Hackwell, Forest and Bird, pers. comm., 13 April 2018; Predator Free New Zealand was established by the government in 2016 to oversee this goal by coordinating seed funding and private investment: https://pf2050.co.nz/predator-free-2050-limited/
79. https://newzealandecology.org/nzje/3258.pdf

80. www.doc.govt.nz/news/media-releases/2017/doc-plans-for-auckland-islands-predator-eradication/
81. http://conservationvolunteers.co.nz
82. www.visitzealandia.com
83. David Butler, Tony Lindsay & Janet Hunt, *Paradise Saved: The remarkable story of New Zealand's wildlife sanctuaries and how they are stemming the tide of extinction* (Auckland: Random House, 2014).
84. www.doc.govt.nz/our-work/eradicating-mice-from-antipodes-island/

11 KIRIBATI: LIFE ON AN OCEAN WAVE

1. Nakibae Teuatabo, in *Climate Change in the Pacific: Scientific assessment and new research*, ch. 6, 'Kiribati' (Australian Government): www.pacificclimatechangescience.org/wp-content/uploads/2013/09/Kiribati.pdf
2. Matthieu Rytz (dir.), *Anote's Ark*, EyeSteelFilm, Toronto, 2018.
3. https://link.springer.com/article/1007/540572-018-0199-7
4. Katherine Luomala, *Voices on the Wind: Polynesian myths and chants* (Honolulu: Bishop Museum Press, 1986), ch. 5.
5. Robert Louis Stevenson, *Island Landfalls*, Jenni Calder (ed.) (Edinburgh: Canongate Classic, 1987), p. 47.
6. Ibid.
7. Ibid.
8. *Atlas of the South Pacific* (Wellington: New Zealand Government Printing, 1982), p. 16.
9. https://en.wikipedia.org/wiki/Geography_of_Kiribati
10. Barrie Macdonald, *Cinderellas of the Empire: Towards a history of Kiribati and Tuvalu* (Canberra: ANU Press, 1982), p. 2.
11. 'Anote Tong: It's not just the climate that's at stake', interviewed by Tommaso Perrone, 22 Dec. 2017: www.lifegate.com/people/news/anote-tong-kiribati-interview; Grant Wyeth, 'For Pacific Island states, climate change is an existential threat', 5 June 2017: https://thediplomat.com/2017/06/for-pacific-island-states-climate-change-is-an-existential-threat/
12. Katerina Martina Teaiwa, *Consuming Ocean Island: Stories of people and phosphate from Banaba* (Bloomington, Indiana: Indiana University Press, 2015).
13. https://en.wikipedia.org/wiki/International_Date_Line#Description
14. *Atlas of the South Pacific*, [n.p.].
15. www.lifegate.com/people/news/anote-tong-kiribati-interview
16. Kinaai Kairo, pers. comm., Dec. 2018.
17. Katherine Luomala, 'Porpoise and taro in Gilbert Islands: Myths and customs', *Fabula* (Berlin), vol. 18, nos 3–4, 1977, p. 205.
18. Ibid., pp. 201, 205, 211. Jon Hather and Marshall Weisner report some confusion over distinguishing different species of Cyrtosperma, in 'Prehistoric giant swamp taro (*Cyrtosperma chamissonis*) from Henderson Island, Southeast Polynesia', *Pacific Science*, vol. 54, no. 2, 2000, pp. 150–51: https://core.ac.uk/download/pdf/5094228.pdf
19. S.J. Whitmee, *A Missionary Cruise in the South Pacific*, London, 1871 (ATL PAM 174), pp. 10–11; Roger Lawrence, *Tamana*, *Atoll Economy: Social change in Kiribati and Tuvalu*, Islands on the Line report, no. 4 (Canberra: Australian National University, Development Studies Centre, 1977), pp. 15–16.
20. Macdonald, *Cinderellas of the Empire*, p. 8; 'Banapa, Kiribati, Pacific Waste Project', EU/SPREP, Aug. 2014, John O'Grady, pers. comm., 27 July 2018.
21. Lawrence, pers. comm., 6 June 2018.
22. Lawrence, *Tamana*, p. 7.
23. Macdonald, *Cinderellas of the Empire*, pp. 3, 10–12, 24. Today it is around 1.8m: http://en.climatedata.org/asia/japan/kumamoto/tamana-5451/
24. Arthur Grimble, *Return to the Islands* (London: John Murray, 1957). This seems to have been a pattern in Tuvalu: 'every coconut palm over ten years old [on the island] carried the record of at least one drought upon it' (p. 37).
25. Lawrence, *Tamana*, p. 13.
26. Roger Lawrence, pers. comm., 12 July 2018.
27. Lawrence, *Tamana*, pp. 7–9; Teaiwa, *Consuming Ocean Island*, pp. 12–13, sets out other characteristations.
28. Lawrence, *Tamana*, pp. 8–9; Lawrence now queries the appropriateness of the word 'soil', pers. comm., May 2018.
29. Lawrence, pers. comm., May 2018.
30. Lawrence, *Tamana*, pp. 9, 11.

31. Lawrence, pers. comm., May 2018.
32. *Atlas of the South Pacific*: Banaba is the correct usage; https://en.wikipedia.org.wiki/Banaba.
33. H.C. Maude & H.E. Maude, *The Book of Banaba: From the Maude and Grimble Papers* (Suva: Institute of Pacific Studies, University of the South Pacific, 1994), p. 84; Arthur Grimble, 'The migrations of a Pandanus people as traced from a preliminary study of food, food-traditions, and food-rituals in the Gilbert Islands', *Journal of the Polynesian Society*, vol. 42, Memoir No. 12, 1933, pp. 67–70.
34. https://en.wikipedia.org/wiki/Banaba_Island
35. https://en.wikipedia.org/wiki/List_of_islands_of_Tuvalu; Maude, *The Book of Banaba*, introduction.
36. K.R. Howe, *Vaka Moana: Voyages of the ancestors, the discovery and settlement of the Pacific* (Auckland: David Bateman & Auckland Museum, 2006), pp. 168 69, 172.
37. https://en.wikipedia.org/wiki/Frigatebird
38. https://en.wikipedia.org/wiki/Tuvalu; Arthur Grimble, *Return to the Islands* (London: John Murray, 1957), p. 36; Teaiwa, *Consuming Ocean Island*, p. 11; Barrie Macdonald, *Cinderella of Empire*, p. 3: in 1942 the Fiji island of Rabi was purchased for their deployment.
39. Maude, *The Book of Banaba*, pp. 8, 17, 39.
40. Luomala, *Voices on the Wind*, p. 106. Luomala gives his Samoan original name as Sinilau – his name in the Gilbert Islands, now Kiribati – as Tinirau; mentioned in David Young, *Woven by Water: Histories from the Whanganui River* (Wellington: Huia Publishers, 1998), p. 62.
41. Damon Salesa, quoted in Tom Brooking & Eric Pawson, *Seeds of Empire: The environmental transformation of New Zealand* (London: I.B. Tauris, 2011), pp. 186–87.
42. http://archive.stats.govt.nz/browse_for_stats/environment/environmental-reporting-series/environmental-indicators/Home/Atmosphere-and-climate/nz-greenhouse-gas-emissions.aspx
43. Mike Joy, *Polluted inheritance: New Zealand's freshwater crisis* (Wellington: Bridget Williams Books, 2013), p. 24; http://archive.stats.govt.nz/browse_for_stats/environment/environmental-reporting-series/environmental-indicators/Home/Fresh%20water/river-water-quality-phosphorus.asp
44. Teaiwa, *Consuming Ocean Island*, pp. 133–34.
45. www.newscientist.com/article/2183901-a-freak-1870s-climate-event-caused-drought-across-three-continents/
46. Grimble, *Return to the Islands*, p. 34.
47. Macdonald, *Cinderellas of the Empire*, p. 3.
48. Grimble, *Return to the Islands*, p. 37.
49. Stevenson, *Island Landfalls*, p. 47.
50. Kathryn Edgerton-Tarpley, *Tears from Iron: Cultural responses to famine in 19th-century China* (Berkeley: University of California Press, 2008), p. 40.
51. Grimble, *Return to the Islands*, pp. 34–35; the calophyllum is a large evergreen, native to Asia and East Africa.
52. Ibid., p. 35.
53. Some of Grimble's understanding and terminology of the features and functioning of the coral and caves has been overtaken by modern science; Roger Lawrence has parsed this section, 16 July 2018.
54. Teaiwa, *Consuming Ocean Island*, p. 247.
55. Grimble, *Return to the Islands*, pp. 36–37, 39.
56. Ibid., pp. 41–43.
57. Ibid., p. 43.
58. Harry Maude, 'History of Banaba', in Alaima Talu & Max Quanchi (eds), *Messy Entanglements: The papers of the 10th Pacific History Association Conference*, Tarawa, Kiribati Pacific History Association, Brisbane, 1995, p. 87.
59. Paul S. Kench, Murray R. Ford & Susan D. Owen, 'Patterns of island change and persistence offer alternate adaptation pathways for atoll nations', *Nature Communications* 9, Article 605, 2018: www.nature.com/articles/s41467-018-02954-1
60. Ibid.
61. Curt D. Storlazzi et al., 'Most atolls will be uninhabitable by the mid-21st century …', *Science Advances*, 25 Apr. 2018: http://advances.sciencemag.org/content/4/4/eaap9741.full
62. Ibid.
63. Lawrence, *Tamana*, p. 13.
64. http://en.climatedata.org/asia/japan/kumamoto/tamana-5451/; https://theconversation.com.drought-and-flooding-rains-already-more-likely-as-climate change
65. Whitmee, *A Missionary Cruise*, p. 9.
66. Ibid., pp. 12–13; Luomala, 'Porpoise and taro in Gilbert Islands', p. 205.

12 AND IN THE NORTHWEST: THE PACIFIC AS ECOSYSTEM

1. James Lichatowich, *Salmon, People, and Place: A biologist's search for salmon recovery* (Corvallis, OR: Oregon State University Press, 2013); Allen S. Gottesfeld & Ken A. Rabnett, *Skeena River Fish and Their Habitat* (Portland, OR: Ecotrust & OSU Press, 2008), introduction.
2. Paul Alaback & Jim Pojar, 'Vegetation from ridgetop to seashore', in Peter K. Schoonmaker, Bettina von Hagen & Edward C. Wolf (eds), *The Rainforests of Home: Profile of a North American bioregion* (Washington, DC: Island Press, 1997), pp. 72–76.
3. Alaback & Pojar, 'Vegetation from ridgetop to seashore', pp. 73–76.
4. John Dawson & Rob Lucas, *New Zealand's Native Trees* (Nelson: Craig Potton Publishing, 2011), p. 20.
5. George Gibbs, *Ghosts of Gwondana: The history of life in New Zealand* (Nelson: Craig Potton Publishing, 2006), pp. 38–41 makes a similar biogeographic argument about bears in the northern polar regions and penguins in the southern.
6. The Australian spotted eel (*Anguilla reinhardtii*) has been an occasional visitor to New Zealand: Department of Conservation, 'Freshwater eels in New Zealand': http://doc.govt.nz/nature/native-animals/freshwater-fish/eels/freshwater-eels-in-new-zealand/
7. Department of Conservation, 'Eels': http://doc.govt.nz/nature/native-animals/freshwater-fish/eels; National Institute of Water and Atmospheric Research (NIWA), 'Longfin eel': www.niwa.co.nz/freshwater-and-estuaries/nzffd/NIWA-fish-atlas/fish-species/longfin_eel; www.niwa.co.nz/aquaculture/aquaculture-species/eel
8. Elsdon Best, *Fishing Methods and Devices of the Maori* (Wellington: Government Printer, 1977), pp. 95–100; Downes acknowledges in this that his five-page list is 'incomplete'.
9. Bob Brockie, 'Native plants and animals – overview: Alpine, wetland and coastal animals', Te Ara – the Encyclopedia of New Zealand: www.TeAra.govt.nz/en/photograph/10605/galaxiid
10. R.M. McDowall, 'The galaxiid fishes of South America', *Zoological Journal of the Linnaean Society*, vol. 50, no. 1, June 2008: www.researchgate.net/publication/230122651_The_galaxiid_fishes_of_South_America
11. Jim Lichatowich, pers. comm., 22 Feb. 2019 states that this applies to most populations of salmon species, but not to spring and summer chinook.
12. David R. Montgomery, *King of Fish: The thousand-year run of salmon* (Boulder, CO: Westview Press, 2003), p. 28; www.nature.com/articles/srep41649
13. Montgomery, *King of Fish*, pp. 30–31, 35–38; Dale Stokes & Doc White, *The Fish in the Forest: Salmon and the web of life* (Berkeley: University of California Press, 2014), p. 25.
14. Montgomery, *King of Fish*, pp. 25–26.
15. Willa Nehlsen & James A. Lichatowich, 'Pacific salmon: Life histories, diversity, productivity', in Schoonmaker et al. (eds), *The Rain Forests of Home: Profile of a North American bioregion* (Washington, DC: Island Press, 1997), p. 215.
16. Stokes & White, *The Fish in the Forest*, p. 27.
17. Nehlsen & Lichatowich, 'Pacific salmon', p. 213.
18. www.doc.govt.nz/nature/native-animals/freshwater-fish/eels/
19. NIWA, 'Tuna: Life cycle and leptocephalii': www.niwa.co.nz/te-kūwaha/tuna-information-resource/biology-and-ecology/life-cycle-and-leptocephalii; DOC, 'Eels': www.doc.govt.nz/nature/native-animals/freshwater-fish/eels/
20. David Young, *Rivers: New Zealand's shared legacy* (Auckland: Random House, 2012), pp. 262–63.
21. Bob McDowall, 'Freshwater fish', Te Ara – the Encyclopedia of New Zealand: www.TeAra.govt.nz/en/freshwater-fish/print
22. NIWA, 'Lamprey': www.niwa.co.nz/our-science/freshwater/tools/fishatlas/species/fish-species/lamprey
23. TuckRussell, 'The lamprey, close to extinction, could bring down NW salmon too', *Crosscut*, 1 March 2012: https://crosscut.com/2012/03/the-lamprey-close-extinction-could-bring-down-nw-s
24. McDowall, *Ikawai*, pp. 118–40; David Young, *Woven by Water: Histories from the Whanganui River* (Wellington: Huia Publishing), pp. 186–90; Gilbert Mair, 'Notes on fishes and Upper Whanganui River', TNZI, 12 (May 1879), pp. 315–16; www.stuff.co.nz/environment/89537160/central-otago-lamprey-breeding-grounds-discovered-for-first-time; www.niwa.co.nz/our-science/freshwater/tools/fishatlas/species/fish-species/lamprey

25. R.M. McDowall, *Ikawai: Freshwater fishes in Māori culture and economy* (Christchurch: Canterbury University Press, 2007), p. 155.
26. Mae Noble et al., 'Culturally significant fisheries: Keystones for management of freshwater social-ecological systems', *Ecology and Society*, vol. 21, no. 2, art. 22: http://dx.doi.org/10.5751/ES-08353-210222. In ecological terms, 'keystone' refers to a species on which many other biological expressions of organisms – both floral and faunal – depend, and may fail without. A 'cultural keystone species' is one that is capable of influencing a culture 'via the species' role in subsistence, spirituality and/or indigenous economies'.
27. Nehlsen & Lichatowich, 'Pacific salmon', p. 214.
28. Tony Taylor, *Fishing the River of Time* (Melbourne: Text Publishing, 2012).
29. Stokes & White, *The Fish in the Forest*, pp. 126–130; Jim Lichatowich, *Salmon Without Rivers: A history of the Pacific salmon crisis* (Washington, DC: Island Press, 2001), p. 27.
30. Stokes & White, *The Fish in the Forest*, pp. 103–30.
31. Trevor H. Worthy & Richard N. Holdway, *The Lost World of the Moa: Prehistoric life in New Zealand* (Christchurch: Canterbury University Press, 2002), p. 441.
32. Mike Joy, *Polluted Inheritance: New Zealand's freshwater crisis* (Wellington: Bridget Williams Books, 2015) provides an excellent summary of this issue; Mike Joy (ed.), *Mountains to Sea: Solving New Zealand's freshwater crisis* (Wellington: Bridget Williams Books, 2018).
33. Lichatowich, *Salmon Without Rivers*, p. 25.
34. Richard Sidey, pers. comm., 1 Nov. 2017; Richard Sidey, 'Forest spirits: Encounters with ghosts in the Great Bear Rainforest': www.richardsidey.com/spirit-bears
35. Lichatowich, *Salmon Without Rivers*, p. 26.
36. Robin Kimmerer, 'Reclaiming the honorable harvest', TEDx Sitka: www.tecumsehproject.org/blog/robin-kimmerer-reclaiming-the-honorable-harvest; Robin Wall Kimmerer, 'Corn tastes better on the honor system', *Emergence*: www.emergence magazine/robin Kimmerer.
37. Lichatowich, *Salmon Without Rivers*, p. 37.
38. T.W. Downes, *A History and Guide to the Wanganui River* (Whanganui: Wanganui Herald, 1921).
39. Māori Marsden & T.A. Henare, *Kaitiakitanga: A definitive introduction to the holistic world view of the Māori* (Wellington: Ministry for the Environment, 1992), p. 20.
40. Lewis Hyde, *The Gift: Creativity and the artist in the modern world* (New York: Vintage, 2007), p. 33.
41. Lichatowich, *Salmon Without Rivers*, pp. 34–37.
42. Ibid., p. 37.
43. Hyde, *The Gift*, p. 33.
44. Ibid., pp. 33–34; Stokes & White, *The Fish in the Forest*, pp. 128–29; Lichatowich, *Salmon Without Rivers*, pp. 35–37.
45. Hyde, *The Gift*, pp. 34–35.
46. Tim Flannery, *The Future Eaters* (Sydney: Reed New Holland, 1994), p. 236.
47. Worthy & Holloway, *The Lost World of the Moa*, ch. 3.
48. Geoff Park, *Ngā Uruora – The groves of life: Ecology and history in a New Zealand landscape* (Wellington: Victoria University Press, 1995), pp. 178–79, 205–06.
49. Jared Diamond, *Guns, Germs and Steel: A short history of everybody for the last 13,000 years* (London: Vintage, 1998).
50. Alfred W. Crosby, *Ecological Imperialism: The biological expansion of Europe, 900–1900* (Cambridge: Cambridge University, 2000) uses this term.
51. Montgomery, *King of Fish*, ch. 7; R.M. McDowall, *Gamekeepers for the Nation: The story of the New Zealand's acclimatisation societies, 1861–1990* (Christchurch: Canterbury University Press, 1994), ch 9.
52. Montgomery, *King of Fish*, ch. 6.
53. McDowall, *Gamekeepers for the Nation*, pp. 117–27.
54. Hans D. Radtke, Shannon W. Davis, Rebecca L. Johnson & Kreg Lindberg, 'Economic and democratic transition on the Oregon coast', in Schoonmaker et al. (eds), *The Rain Forests of Home*, p. 329.
55. Ibid.; Lichatowich, *Salmon Without Rivers*, p. 48, Table 3.1. Lichatowich's table has been modified for the purposes of this book.
56. G. Nabhan, *Cross-Pollinations: The marriage of science and poetry* (Minneapolis: Credo, Milkweed Editions, 2004).
57. Lichatowich, *Salmon Without Rivers*, p. 48, Table 3.1.
58. Gary Williams, 'Natural economics: Being present living in our time and place, a transformative vision', (Gary Williams, Ōtaki, 2012), pp. 2–8.

59. James Lichatowich, 'Foreword', in Gottesfeld & Rabbnet, 'Skeena River fish and their habitat': http://osupress.oregonstate.edu/book/skeena-river-fish-and-their-habitat; this process is also detailed in Lichatowich, *Salmon Without Rivers*, especially chs 5 & 6.
60. Jim Lichatowich & Richard Williams, 'Faith in nature: The missing element in salmon management and mitigation programmes', *Osprey* (International Federation of Fly Fishers), May 2015, pp. 1, 4–7.
61. James Lichatowich et al., 'Wild Pacific Salmon: A threatened legacy': www.salmonhistory.com/Combined_cover_and_doc_2018.pdf
62. https://en.wikipedia.org/wiki/List_of_dams_in_the_Columbia_River_watershed; Marc Reisner, *Cadillac Desert: The American West and its disappearing water* (New York: Penguin Books, 1993).
63. Ted Simon, *The River Stops Here: How one man's battle to save his valley changed the fate of California* (New York: Random House, 1994).
64. Waitangi Tribunal, *Te Kāhui Maunga Report: The National Park District Inquiry* (Wellington: Legislation Direct, 2013), p. 1126; www.waitangitribunal.govt.nz/assets/Documents/Publications/WT-Vol-3-Kahui-Maunga-Report.pdf
65. NIWA, 'The status of longfin eels in New Zealand: An overview of stocks and harvest': www.pce.parliament.nz/media/1237/jellyman-report-final2.pdf, pp. 24–25.
66. Young, *Rivers*, p. 239; David Young, *Matahina: A power in the land* (Wellington: Steele Roberts, 1999), pp. 59–73.
67. McDowall, *Ikawai*, pp. 222–24; Young, *Rivers*, p. 120.
68. Montgomery, *King of Fish*, ch. 6 particularly.
69. Lichatowich, *Salmon Without Rivers*, p. 54.
70. Johnny Sundstrom: www.siuslaw.org/johnny-sundstrom/
71. Joy, *Polluted Inheritance*, p. 19.
72. Mike Joy referred me to the Allee Effect – sometimes described as 'undercrowding' – analogous (or even considered synonymous by some) to 'depensation' in the field of fishery sciences.
73. www.pce.parliament.nz/media/1237/jellyman-report-final2.pdf, p. 40 confirms this view.
74. https://niwa.co.nz/news/declining-native-eel-population-causes-concern
75. David Young, *Whio: New Zealand's endangered blue duck* (Nelson: Craig Potton, 2006).
76. Lichatowich, *Salmon Without Rivers*, p. 9.
77. https://digitalcommons.law.seattleu.edu/cgi/viewcontent.cgi?article=1130&context=sulr
78. 'Willamette Valley treaties', *Oregon Encyclopedia*: https://oregonencyclopedia.org/articles/willamette_valley_treaties/#XORBwC1h2qA; Eric Cain & John Rosman, 'Broken Treaties: An oral history tracing Oregon's Native population', OPB in the News, 2017: www.opb.org/artsandlife/series/brokentreaties/oregon-tribes-oral-history-broken-treaties/
79. Waitangi Tribunal, *Ko Aotearoa Tenei: Report on the Wai 262 claim* (Wellington: Waitangi Tribunal, Legislation Direct, 2011); https://law.jrank.org/pages/8750/Native-American-Rights-Hunting-Fishing-Rights.html
80. Lichatowich et al., 'Wild Pacific Salmon': www.researchgate.net/publication/319112496_WILD_PACIFIC_SALMON_A_THREATENED_LEGACY, fig. 1, pp. 6, 26 shows a sustained abundance of salmon in the Pacific Northwest for 5000 years through First Nation occupation, followed by a plummeting decline now almost at asymptote in the past 150 years.
81. www.klamathfalls.city/232/Geothermal
82. 'Klamath Falls, Oregon: Water rights controversy': https://en.wikipedia.org/wiki/Klamath_Falls,_Oregon#Water_rights_controversy
83. Ibid.
84. David Young, *Faces of the River* (Wellington: Listener/TVNZ, 1986), pp. 33–44; Young, *Rivers*, p. 119.
85. Young, *Rivers*, p. 119; www.sacredland.org/mccloud-river-watershed-united-states; www.friendsoftheriver.org/wp-content/uploads/2018/09/Shasta-Dam-raise-referenced-fact-sheet-9-12-2018.pdf
86. www.critfc.org/fish-and-watersheds/climate, p. 10.
87. https://audubonportland.org/issues/habitat/urban/ross-island
88. Lichatowich et al., 'Wild Pacific Salmon', prologue, p. 1.
89. www.beehive.govt.nz/release/taking-action-improve-water-quality

90. Joy, *Mountains to Sea*.
91. Lichatowich, 'Wild Pacific Salmon', p. 11.
92. Joy, *Mountains to Sea*, p. 164.
93. Catherine Knight, *New Zealand's Rivers: An environmental history* (Christchurch: Canterbury University Press, 2016), pp. 239–40.
94. Schoonmaker et al., *The Rain Forests of Home*, introduction, p. 1.
95. Maggy Wassilieff, 'Trees in the rural landscape – Macrocarpa and other conifers', Te Ara – the Encyclopedia of New Zealand: www.TeAra.govt.nz/en/trees-in-the-rural-landscape/page-3; Denis Hocking, 'Keeping faith with the old faithful', *NZ Tree Grower*, Feb. 2006: www.nzffa.org.nz/farm-forestry-model/resource-centre/tree-grower-articles/february-2006/macrocarpa-keeping-faith-with-the-old-faithful/
96. Alaback & Pojar, 'Vegetation from ridgetop to seashore', p. 73.
97. https://en.wikipedia.org/wiki/2018_Oregon_wildfires
98. http://beta.latimes.com/local/lanow/la-me-fire-winds-20171207-story.html; www.popsugar.com/news/California-Wildfires-Los-Angeles-Ventura-December-2017-44335740; www.oregonlive.com/wildfires/
99. Tim Kurtz, City of Portland Environmental Services, pers. comm. Oct. 2017; www.portlandoregon.gov/bes/66158
100. 'China to get vertical gardens in 2018 to help tackle pollution': www.iflscience.com/environment/china-to-get-vertical-gardens-in-2018-to-help-tackle-smog/
101. 'The Willamette: Oregon's big river', *River Journeys 3* (Brisbane: International River Foundation, 2012), pp. 32–33: https://riverfoundation.org.au/wp-content/uploads/2016/10/River-Journeys-4_FINAL-for-web.pdf
102. https://en.wikipedia.org/wiki/Robert_W._Straub#Governor_of_Oregon
103. https://en.wikipedia.org/wiki/Willamette_Greenway
104. http://willamette-riverkeeper.org/basicsfacts/
105. https://en.wikipedia.org/wiki/Water_supply_and_sanitation_in_New_Zealand
106. www.ncbi.nlm.nih.gov/pmc/articles/PMC3262001/
107. www.museumofthecity.org/project/portlands-green-streets/
108. www.portlandoregon.gov/bes/article/201795
109. www.portlandoregon.gov/bps/49989
110. Philippa Stevenson, 'The scandal of the PCP dumps', *Terra Nova*, June 1992, pp. 10–15.
111. www.historicalthreads.blogspot.co.nz/2011/05/dioxin-and-willamette-river-pollution.html
112. Tim Kurtz, pers. comm., Dec. 2017.
113. Bruce Clarkson, 'Bringing indigenous nature back into cities', Cawthron Institute Centennial lecture, Nelson, 27 Sep. 2017.
114. https://doi.org/10.1080/03036758.2019.1637352

Bibliography

BOOKS

Abram, David, *Becoming Animal: An earthly cosmology* (New York: Vintage, 2011)

Alaback, Paul and Jim Pojar, 'Vegetation from ridgetop to seashore', in Peter K. Schoonmaker, Bettina von Hagen and Edward C. Wolf (eds), *The Rainforests of Home: Profile of a North American bioregion* (Washington DC: Island Press, 1997)

Andrade, Carlos, *Ha`ena: Through the eyes of the ancestors* (Honolulu: University of Hawai`i Press, 2008)

Arago, Jacques E.V., *Promenade autour du monde pendant les années 1817, 1818, 1819 et 1820 sur les corvettes* L'Uranie *et* La Physicienne … (Paris: Leblanc, 1822)

Atlas of the South Pacific (Wellington: Government Printer, 1982)

Bahn, Paul and John Flenley, *Easter Island, Earth Island* (London: Thames & Hudson, 1992)

——, *The Enigmas of Easter Island: Island on the edge* (New York: Oxford University Press, 2003)

Bambridge, Tamatoa (ed.), *The Rāhui: Legal pluralism in Polynesian traditional management of resources and territories*, Pacific Series (Canberra: Australian National University Press, 2016)

Beaglehole, J.C., *The Life of Captain James Cook* (London: A. & C. Black, 1974)

Beattie, James, *Empire and Environmental Anxiety: Health, science, arts and conservation in South Asia and Australasia* (New York: Palgrave Macmillan, 2011)

Beckwith, Martha Warren, *Hawaiian Mythology* (Honolulu: University of Hawai`i Press, 1982)

—— (ed. & trans.), *Kumulipo: A Hawaiian creation chant* (Chicago: University of Chicago Press, 1951)

Begon, M., C. Townsend and J. Harper, *Ecology: From individuals to ecosystems* (Oxford: Wiley-Blackwell, 2006)

Bell, Gavin, *In Search of Tusitala: Travels in the Pacific after Robert Louis Stevenson* (London: Picador, 1994)

Best, Elsdon, *Fishing Methods and Devices of the Maori* (Wellington: Government Printer, 1977)

——, *Maori Religion and Mythology* (Wellington: Government Printer, 1982)

——, *Spiritual and Mental Concepts of the Maori* (Wellington: Government Printer, 1955)

Binney, Judith, *Redemption Songs: A life of Te Kooti Arikirangi Te Turuki* (Wellington: Bridget Williams Books, 1995)

Bishop, Sereno Edwards, *Reminiscences of Old Hawaii*, Historical Series (Honolulu: Hawaiian Gazette Company, 1916)

Bosselmann, Klaus, *The Principle of Sustainability: Transforming law and governance*, 2nd edn (Oxford: Taylor & Francis, 2016)

Broad, Harry (ed.), *Ngā Whenua Rāhui: Our nature our people our futur*e (Wellington: Ngā Whenua Rāhui, July 2015)

Brooking, Tom and Eric Pawson, *Seeds of Empire: The environmental transformation of New Zealand* (London: I.B. Tauris, 2011)

Buck, Peter, *Vikings of the Sunrise* (Christchurch: Whitcombe & Tombs, 1975)

Butler, David, Tony Lindsay and Janet Hunt, *Paradise Saved: The remarkable story of New Zealand's wildlife sanctuaries and how they are stemming the tide of extinction* (Auckland: Random House, 2014)

Calcott, Lady Maria, Baron George Anson Byron and Richard Rowland Bloxham, *Voyage of the HMS* Blonde *to the Sandwich Islands, in the years 1824–1825* (London: John Murray, 1926)

Cathcart, Michael, *The Water Dreamers: The remarkable history of our dry continent* (Melbourne: Text Publishing, 2010)

Chan, Gaye and Andrea Feeser, *Waikiki: A history of forgetting and remembering* (Honolulu: University of Hawai`i Press, 2006)

Collins, David, *An Account of the English Colony in New South Wales from its First Settlement, in January 1788, to August 1801* (Sydney: University of Sydney Library, digital text, 2003)

Cox, Paul Alan, *Nafanua: Saving the Samoan rain forest* (New York: W.H. Freeman, 1999)

Crosby, Alfred W., *Ecological Imperialism: The biological expansion of Europe, 900–1900* (Cambridge: Cambridge University, 2000)

Dargin, Peter, *Aboriginal Fisheries of the Darling-Barwon River* (Brewarrina: Brewarrina Historical Society, 1976)

Dark, Eleanor, *The Timeless Land* (Australia: Macmillan, 1941)

Daws, Gavan, *Shoal of Time: A history of the Hawaiian Islands* (Honolulu: University of Hawai`i Press, 1974)

Dawson, John and Rob Lucas, *New Zealand's Native Trees* (Nelson: Craig Potton Publishing, 2011)

Diamond, Jared, *Guns, Germs and Steel: A short history of everybody for the last 13,000 years* (London: Vintage, 1998)

Dieudonne, Fran (ed.), *The Pacific Islands and the Sea* (California: Neptune House, 2002)

Downes, T.W., *Old Whanganui* (Hawera: W.A. Parkinson, 1915)

Downes, Thomas W., *A History and Guide to the Wanganui River* (Whanganui: The Wanganui Herald Newspaper Co., 1921)

Drew, Robert, *The Drowner* (Australia: Penguin Macmillan, 1996)

Edgerton-Tarpley, Kathryn, *Tears from Iron: Cultural responses to famine in 19th-century China* (Berkeley: University of California Press, 2008)

Edwards, Edmundo and Alexandra Edwards, *When the Universe Was an Island* (Santiago: Hanga Roa Press, 2010)

Ellis, William, *Journal of a Tour Around Hawaii, the Largest of the Sandwich Islands* (Boston: Crocker and Brewster, 1825)

——, *A Tour Around Hawaii* (Honolulu: Hawaiian Gazette Co., 1917)

——, *Polynesian Researches: Society Islands, Tubuai Islands and New Zealand*, new edn (Rutland: C.E. Tuttle, 1969)

Emerson, Nathaniel B., *Unwritten Literature of Hawaii: The sacred songs of the hula* (Washington: Government Printing Office, 1909; reprinted 1977)

Farber, Thomas, *Face of the Deep* (San Francisco: Mercury House, 1998)

Finney, Ben and Sam Low, 'Navigation', in K.R. Howe, *Vaka Moana: Voyages of the ancestors: The discovery and settlement of the Pacific* (Auckland: David Bateman and Auckland Museum, 2006)

Fischer, Steven Roger, *Island at the End of the World: The turbulent history of Easter Island* (London: Reaktion Books, 2005)

Fischer, Steven Roger, *Rongorongo, the Easter Island Script: History, traditions, texts* (England: Clarendon Press, 1997)

Flannery, Tim, *The Future Eaters* (Adelaide: Reed New Holland, 1999; first printed 1994)

Friedlander, A., J. Shackeroff and J. Kittinger, 'Traditional marine resources and their use in contemporary Hawai`i', in Tamatoa Bambridge (ed.), *Rahui: Legal pluralism in Polynesian traditional management of resources and territories*, Pacific Series, (Canberra: Australian National University Press, 2012)

Furey, Louise, *Maori Gardening: An archaeological perspective* (Wellington: Department of Conservation, 2006)
Gammage, Bill, *The Biggest Estate on Earth: How Aborigines made Australia* (Sydney: Allen & Unwin, 2011)
Gibbs, George, *Ghosts of Gondwana: The history of life in New Zealand* (Nelson: Potton & Burton, 2016)
Gilmore, Mary, *Old Days, Old Ways* (Australia: Angus and Robertson, 1963)
Gottesfeld, A. and K. Rabnett, *Skeena River: Fish and their habitat* (Oregon: Ecotrust/Oregon State University Press, 2008)
Grace, Patricia and Waiariki Grace, *Earth, Sea, Sky: Images and Māori proverbs from the natural world of Aotearoa New Zealand* (Nelson: Craig Potton Publishers, 2006)
Griffiths, Tom, *The Art of Time Travel: Historians and their craft* (Melbourne: Black Inc., 2016)
Grimble, Arthur, *Return to the Islands* (London: John Murray, 1957)
Handy, E.S. Craighill, *The Hawaiian Planter* (Honolulu: The Museum, 1940)
Heaney, Seamus, *Electric Light* (New York: Farrar, Straus & Giroux, 2001)
Herman, Brother, *Tala o le Vavau: Samoan legends* (Pago Pago: Association of the Marist Brothers' Old Boys, 1955)
Holgate, Martin, *From Care to Action: Making a sustainable world* (London: Earthscan Publications, 1996)
Holland, Peter, Paul Star and Vaughan Wood, 'Pioneer grassland farming: Pragmatism, innovation and experimentation', in Tom Brooking and Eric Pawson (eds), *Seeds of Empire: The environmental transformation of New Zealand* (London/New York: I.B. Tauris, 2011)
Holland, Peter, *Home in the Howling Wilderness: Settlers and the environment in southern New Zealand* (Auckland: Auckland University Press, 2013)
Hommon, Robert J., 'Social evolution in ancient Hawai`i', in Patrick Vinton Kirch (ed.), *Island Societies: Archaeological approaches to evolution and transformation* (Cambridge: Cambridge University Press, 1986)
Hope Jeannette and Gary Vines, *Brewarrina Aboriginal Fisheries Conservation Plan* (Sydney: J. Hope, Brewarrina Aboriginal Cultural Museum, 1994)
Houston, James D. and Eddie Kamae, *Hawaiian Son: The life and music of Eddie Kamae* (Honolulu: `Ai Pohaku Press, 2004)
Howe, K.R., *Vaka Moana: Voyages of the ancestors: The discovery and settlement of the Pacific* (Auckland: David Bateman and Auckland Museum, 2006)
Hunt, Terry and Carl Lipo, *The Statues That Walked: Unravelling the mystery of Easter Island* (New York: Free Press, 2011)
Hyde, Lewis, *The Gift: Creativity and the artist in the modern world* (New York: Vintage, 2007)
Iorns Magallanes, Catherine J., 'From rights to responsibilities using legal personhood and guardianship for rivers', in B. Martin, L. Te Aho and M. Humphries-Kil (eds), *Responsibility: Law and governance for living well with the earth* (London: Routledge, 2019)
James, Van, *Ancient Sites of Hawaii: Archaeological places of interest on Big Island* (Honolulu: Mutual Publishing, 1995)
Jenkins, Irving, *The Hawaiian Calabash* (London: Kegan Paul International, 1989)
Joy, Mike, *Polluted Inheritance: New Zealand's freshwater crisis* (Wellington: Bridget Williams Books, 2015)
—— (ed.), *Mountains to Sea: Solving New Zealand's freshwater crisis* (Wellington: Bridget Williams Books, 2018)
Kawaharada, Dennis, *Storied Landscapes: Hawaiian literature and place* (Honolulu: Kalamakū Press, 1999)
Kemmerer, Robin, *Reclaiming the Honorable Harvest* (Cambridge, MA: Harvard University Press, 2014)
Kesey, Ken, *Sometimes a Great Notion* (New York: Penguin Books, 1977)
Kirch, Patrick Vinton, *A Shark Going Inland Is My Chief: The island civilization of ancient Hawai`i* (Berkeley: University of California Press, 2012)
Knight, Catherine, *New Zealand's Rivers: An environmental history* (Christchurch: Canterbury University Press, 2016)
Kuykendal, Ralph S., *The Hawaiian Kingdom* (Honolulu: University of Hawai`i Press, 1938)
Langton, Marcia, 'Earth, wind, fire and water: The social and spiritual construction of water in Aboriginal societies', in Bruno David, Bryce Barker and Ian J. McNiven (eds), *The Social Archaeology of Australian Indigenous Societies* (Canberra: Aboriginal Studies Press, 2008)

Lawrence, Roger, *Tamana – Atoll Economy: Social change in Kiribati and Tuvalu*, Islands on the Line report, no. 4 (Canberra: Australian National University, Development Studies Centre, 1977)
Leach, Helen, *1,000 Years of Gardening in New Zealand* (Wellington: A.H. and A.W. Reed, 1984)
Lichatowich, J., R. Williams, B. Bakke, J. Myron, D. Bella, B. McMillan, J. Stanford and D. Montgomery, 'Wild Pacific Salmon: A threatened legacy', funded by Fly Fishers International and Wild Fish Conservancy (St Helens, OR: Bemis Printing, 2017)
Lichatowich, James, *Salmon, People, and Place: A biologist's search for salmon recovery* (Corvallis, OR: Oregon State University Press, 2013)
Lichatowich, Jim, *Salmon Without Rivers: A history of the Pacific salmon crisis* (Washington, DC: Island Press, 2001)
Lloyd, Kelvin, 'Ecological processes in the South Island high country', in *Change in the High Country: Ecological stewardship and tenure review* (Wellington: Parliamentary Commission for the Environment, 2009)
Luomala, Katherine, *Voices of the Wind: Polynesian myths and chants* (Honolulu: Bishop Museum Press, 1986)
Lupe, Karen, 'Le ala i le mafa e: The path towards enlightenment', in T.M. Suaalii-Sauni, Maualaivao Albert Wendt and Naomi Fuamatu (eds), *Whispers and Vanities: Samoan indigenous knowledge and religion* (Wellington: Huia Publishers, 2009)
Macdonald, Barrie, *Cinderellas of the Empire: Towards a history of Kiribati and Tuvalu* (Canberra: ANU Press, 1982)
Macrae, James, 'With Lord Byron at the Sandwich Islands in 1825; Being extracts from the ms. diary of James Macrae, Scottish biologist' (Honolulu: J.F. Wilson, 1922)
Malo, David, *Hawaiian Antiquities* (Honolulu: Bishop Museum, 1951)
Marekura, Matiu, 'Te Awa Tupua: Our sacred cord of unity', in Āneta Hinemihi Rāwiri (ed.), *Mouri Tū, Mouri Ora: Water for Wisdom and Life* (Whanganui: Te Atawhai o Te Ao, 2009)
Mark, Alan F., *Above the Treeline: A nature guide to alpine New Zealand* (Nelson: Craig Potton Publishing, 2012)
——, *Standing My Ground: A voice for nature conservation* (Dunedin: Otago University Press, 2015)
Marsden, Maori and T.A. Henare, *Kaitiakitanga: A definitive introduction to the holistic world view of the Maori* (Wellington: Ministry for the Environment, 1992)
Maude, H.C. and H.E. Maude, *The Book of Banaba: From the Maude and Grimble Papers* (Suva: Institute of Pacific Studies, University of the South Pacific, 1994)
Mazière, Francis, *The Mysteries of Easter Island* (London: Collins, 1969)
McDowall, R.M., *Gamekeepers for the Nation: The story of New Zealand's acclimatisation societies, 1861–1990* (Christchurch: Canterbury University Press, 1994)
——, *Ikawai: Freshwater fishes in Maori culture and economy* (Christchurch: Canterbury University Press, 2011)
McIntyre, Roberta, *Whose High Country? A history of the high country of New Zealand* (Auckland: Penguin, 2008)
McNab, Robert, *Historical Records of New Zealand* (Wellington: Government Printer, 1914)
Melville, Herman, *Moby Dick* (London: Constable, 1922)
Miike, Lawrence H., *Water and the Law in Hawai`i* (Honolulu: University of Hawai`i Press, 2004)
Mitchell, Hilary and John Mitchell, *Te Tau Ihu o Te Waka: A history of Māori of Nelson and Marlborough* (Wellington/Nelson: Huia Publishers/ Wakatu Inc., 2004)
Mollison, James, *Permaculture: A designer's manual* (Tyalgum, NSW: Tagari Publications, 1988)
Montgomery, David R., *King of Fish: The thousand-year run of salmon* (Boulder, CO: Westview Press, 2003)
Moors, H.J., *With Stevenson in Samoa* (London: T. Fisher Unwin, 1911)
Nabhan, Gary P., *Cross-Pollinations: The marriage of science and poetry* (Minneapolis: Credo, Milkweed Editions, 2004)
Nehlsen, W. and J. Lichatowich, 'Pacific salmon: Life histories, diversity, productivity', in Schoonmaker et al. (eds), *The Rain Forests of Home: Profile of a North American bioregion* (Washington, DC: Island Press, 1997)
Ngata, Tina, 'Wai Māori: A Māori perspective on the freshwater debate', in Mike Joy (ed.), *Mountains to Sea: Solving New Zealand's freshwater crisis* (Wellington: Bridget Williams Books, 2018)
Okihiro, Gary Y., *Island World: A history of Hawaii*

and the United States (Berkeley: University of California, 2008)

Osorio, Jonathan Kay Kamakawiwo`ole, *Dismembering Lāhui: A history of the Hawaiian nation to 1887* (University of Hawai`i Press, 2002)

Park, Geoff, *Theatre Country: Essays on landscape and whenua* (Wellington: Victoria University Press, 2006)

——, *Ngā Uruora, The Groves of Life: Ecology and history in a New Zealand landscape* (Wellington: Victoria University Press, 1995)

Park, Geoff, Rod Hay, A. Whistler, T. Lovegrove and P. Ryan, 'The National Ecological Survey of Western Samoa: The conservation of biological diversity in the coastal lowlands of Western Samoa', report by NZ Department of Conservation for the Ministry of External Relations and Trade, 1992

Pascoe, Bruce, *Dark Emu: Aboriginal Australia and the birth of agriculture* (Broome: Magabala Books, 2018)

Patterson, John, *Pacific Parables: Learning from Māori tradition* (Wellington: Steele Roberts, 2014)

Pawson, Eric and Tom Brooking (eds), *Environmental Histories of New Zealand* (Melbourne: Oxford University Press, 2002)

Peden, Robert, 'Pastoralism and the transformation of the open grasslands', in Eric Pawson & Tom Brooking (eds), *Seeds of Empire: Environmental histories of New Zealand* (London/New York: I.B. Tauris, 2011)

Preston, Richard, *The Wild Trees* (New York: Random House, 2008)

Puku`i, Mary Kawena, *`Olelo No`eau: Hawaiian proverbs & poetical sayings*, Bishop Museum Special Publication, no. 71 (Honolulu: Bishop Museum Press, 1983)

Puku`i, Mary Kawena, 'Poi and its preparation', 'Hawaiian Ethnological Notes', Bishop Museum, Honolulu

Puku`i, Mary Kawena, retold by Caroline Curtis, *The Water of Kane and other legends of the Hawaiian Islands* (Honolulu: Kamehameha Press, 1994)

Puku`i, Mary Kawena and Laura C.S. Green, *Folktales of Hawai`i/He Mau Ka`ao Hawai`i* (Honolulu: Bishop Museum Press, 2003)

Rāwiri, Āneta Hinemihi (ed.), *Mouri Tū, Mouri Ora, Water for Wisdom and Life* (Whanganui: Te Atawhai o Te Ao, 2009)

Rhodes, Diane Lee, 'Pu`uhonua o Honaunau National Historical Park: A cultural history of three traditional Hawaiian sites on the West Coast of Hawai`i Island': www.nps.gov/parkhistory/online_books/kona/history5e.htm

Rich, Burdett A., *Lawyers' Reports, Annotated, 1893, vol. 21* (London: Forgotten Books, Feb. 2019)

Richards, Rhys, *Easter Island 1793–1861: Observations by early visitors before the slave raids* (Los Osos, CA: Easter Island Foundation, 2008)

Roethke, Theodore, *Collected Poems* (New York: Doubleday, 1941)

Royal, Te Ahukaramū Charles (ed.), *The Woven Universe: Selected writings of Maori Marsden* (Te Wānanga-o-Raukawa, 2003)

Ruru, Jacinta, Janet Stephenson and Mick Abbott (eds), 'Tension lines', in *Making Our Place: Exploring land-use tensions in Aotearoa New Zealand* (Dunedin: Otago University Press, 2011)

Salesa, T.D., 'When the waters met: Some shared histories of Christianity and ancestral Samoan spirituality', in Tamasailau Suaalii-Sauni et al. (eds), *Whispers and Vanities: Samoan indigenous knowledge and religion* (Wellington: Huia Publishers, 2009)

Salmond, Anne, *Two Worlds: First meetings between Maori and Europeans 1643–1771* (Auckland: Viking, 1991)

——, *The Trial of the Cannibal Dog: Captain Cook in the South Seas* (Auckland: Penguin, 2004)

——, *Aphrodite's Island: The European discovery of Tahiti* (Auckland: Viking, 2009)

——, *Tears of Rangi: Experiments across worlds* (Auckland: Auckland University Press, 2017)

Salmond, J.T., *The Native Trees of New Zealand* (Wellington: Reed, 1980)

Schooner, P., B. Von Hagen and E. Wolf (eds), *The Rain Forests of Home: Profile of a North American bio-region* (Washington DC: Island Press, 1997)

Schuster, C., A. Whistler, T. Tuailemafua et al., 'The conservation of biological diversity in upland ecosystems of Samoa' (Division of Environment and Conservation, Samoa, 1999)

Schwenk, Theodore, *Sensitive Chaos* (East Wessex: Rudolph Steiner Press, 2008)

Simon, Ted, *The River Stops Here: How one man's battle to save his valley changed the fate of California* (New York: Random House, 1994)

Smith, Keith Vincent, *Mari Nawi: Aboriginal odysseys* (NSW, Australia: Rosenberg, 2010)

Snyder, Gary, *The Practice of the Wild* (San Francisco: North Point Press, 1990)
Spencer, Baldwin and F.J. Gillen, *The Native Tribes of Central Australia* (London: Macmillan, 1899)
Stannard, David F., *Before the Horror: The population of Hawaii on the eve of Western contact* (Honolulu: University of Hawai`i Press, 1989)
Stevenson, Christopher M., 'The socio-political structure of the southern coastal areas of Easter Island: AD 1300–1864', in Patrick V. Kirch (ed.), *Island Societies: Archaeological approaches to evolution and transformation* (Cambridge: Cambridge University Press, 1986)
Stevenson, Robert Louis, *A Footnote to History: Eight years of trouble in Samoa* (Honolulu: University of Hawai`i Press, 1996, original 1892)
——, *Island Landfalls*, ed. Jenni Calder (Edinburgh: Canongate Classic, 1987)
——, *Travels in Hawaii* (Honolulu: University of Hawai`i Press, 1991)
——, 'Vailima Letters': http://pinkmonkey.com/dl/library1/digi256.pdf
Stokes, Dale and Doc White, *The Fish in the Forest: Salmon and the web of life* (Berkeley: University of California Press, 2014)
Stokes, Evelyn, *The Legacy of Ngatorirangi: Māori customary use of geothermal resources* (Hamilton: University of Waikato, 2000)
Sutherland, Oliver, *Paikea: The life of I.L.G. Sutherland* (Christchurch: Canterbury University Press, 2013)
Taylor, Richard, *Te Ika a Maui: New Zealand and its inhabitants* (Wellington: A.H. and A.W. Reed, 1974)
Taylor, Tony, *Fishing the River of Time* (Melbourne: Text Publishing, 2012)
Te Aho, Linda, 'Waikato: River of life', in Jacinta Ruru, Janet Stephenson and Mick Abbott (eds), *Making Our Place: Exploring land-use tensions in Aotearoa New Zealand* (Dunedin: Otago University Press, 2011)
Teaiwa, Katerina Martina *Consuming Ocean Island: Stories of people and phosphate from Banaba* (Bloomington: Indiana University Press, 2015)
Thomas, Marcus and Neil Silverwood, *Caves: Exploring New Zealand's subterranean wilderness* (Wellington: Whio Publishing, 2017)
Thoreau, David Henry, *Walden: or, Life in the woods and on the duty of civil disobedience* (New York: Simon & Schuster, 2004)
Townsend, Colin W., *Ecological Applications: Towards a sustainable world* (US: Wiley & Blackwell, 2019)
Trask, Haunani-Kay, *Light in the Crevice Never Seen* (Corvalis, OR: Calyx Books, 1994)
Tudge, Colin, *The Secret Life of Trees: How they live and why they matter* (London: Penguin Books, 2006)
Wagner, J., J. Jacka, E. Hviding, A. Mawyer and M. Muru-Lanning, 'Introduction: River as ethnographic subject', in John R. Wagner & Jerry K. Jacka (eds), *Island Rivers: Fresh water and place in Oceania* (Canberra: ANU Press, 2018)
Walsh, Richard, 'Traditional games on North Pentecost, Vanuatu', in John Morrison, Paul Gerragty and Linda Crowl (eds), *Science of Pacific Island Peoples: Education, language, patterns and policy* (Suva: Institute of Pacific Studies, University of the South Pacific, 1994)
Wendt, Maualaivao Albert, 'Mauli', in T.M. Suaalii-Sauni et al. (eds), *Whispers and Vanities: Samoan indigenous knowledge and religion* (Wellington: Huia Publishers, 2014)
White, Jonathan, *Tides: The science and spirit of the ocean* (San Antonio: Trinity University Press, 2017)
Whitmee, Rev. S.J., *A Missionary Cruise in the South Pacific* (Sydney: Joseph Cook, 1871)
Wood, Houston, *Displacing Natives: The rhetorical production of Hawai`i* (Oxford: Rowman and Littlefield, 1999.
Wood, Keith, 'Tangata Whenua Water Resource Management: Ngati Rangi and the Tongariro power scheme', in Āneta Rāwiri (ed.), *Mouri Tū, Mouri Ora: Water for wisdom and life* (Whanganui: Te Atawhai o te Ao, 2009)
Worthy, Trevor H. and Richard N. Holdaway, *The Lost World of the Moa: Prehistoric life in New Zealand* (Christchurch: Canterbury University Press, 2002)
Wright, Judith, *The Cry for the Dead* (Melbourne: Oxford University Press, 1982)
Wulf, Andrea, *The Invention of Nature: The adventures of Alexander von Humbolt, the lost hero of science* (London: John Murray, 2015)
Young, David, *Woven by Water: Histories from the Whanganui River* (Wellington: Huia Publishers, 1998)
——, *Matahina: Power in the land* (Wellington: Steele Roberts, 1998)
——, *Whio: New Zealand's endangered blue duck* (Nelson: Craig Potton Publishing, 2003)
——, *Our Islands, Our Selves: A history of conservation*

in *New Zealand* (Dunedin: Otago University Press, 2004)
——, *Keeper of the Long View: Sustainability and the PCE* (Wellington: Parliamentary Commissioner for the Environment, 2007)
——, *Rivers: New Zealand's shared legacy* (Auckland: Random House, 2013)
Young, Kanalu G. Terry, *Re-thinking the Native Hawaiian Past* (Oxfordshire: Routledge, 1998)
Yunkaporta, Tyson, *Sand Talk: How indigenous thinking can save the world* (Melbourne: Text Publishing, 2019)

REPORTS AND ESSAYS

Awhitu, Nancy, evidence, Te Rohe Potae Inquiry District, *Environmental and Wahi Tapu Report*, Professor Belgrave et al, 2011, p. 57
Hale`ole, S.N., *The Hawaiian Romance of Lā`ieikawai*, trans. Martha Beckwith, reprinted from the *Thirty-third Annual Report of the Bureau of American Ethnology* (Washington: Government Printing Office, 1918)
Ngāti Apa v Attorney-General, New Zealand Court of Appeal, vol. 117, 3, *New Zealand Law Reports*, 2003, p. 643
'Our Atmosphere and Climate 2017: Data to 2016', New Zealand's Environmental Reporting Series, New Zealand Government
Rio Convention on Environment and Development, Samoa's Fifth National Report 2014 to the Convention on Biological Biodiversity, para 2.6.
Samoa's Fifth National Report to the Convention on Biological Biodiversity, Ministry of Natural Resources and Environment, Apia, 2014, introduction, para 2.6, para 3.5
Smith, Huhana, 'Hei Wai Ora: A photo essay', in Betsan Martin & Gary Williams (eds), 'Watersheds and responsible governance of water and the waterways of life, global issues in local contexts', Horowhenua NZ Symposium, November 2009
Tamasese Ta`isi Efi, Tui Atua Tupua, 'An Ethic of Responsibility in Samoan Customary Law', keynote address, Law, Ethics & Responsibility Symposium, Hopuhopu, Ngāruawāhia, Waikato–Tainui College for Research and Development, 25 November 2014
Waitangi Tribunal, *Ko Aotearoa Tenei: Report on the Wai 262 claim* (Wellington: Waitangi Tribunal, Legislation Direct, 2011)
——, *Te Kāhui Maunga Report: The national park district inquiry* (Wellington: Legislation Direct, 2013), p. 1126
——, *The Whanganui River Report*, GP Publications, Wellington, 1999
Wright, Jan, *Change in the High Country: Ecological stewardship and tenure review* (Wellington: Parliamentary Commissioner for the Environment, 2009)

MAGAZINE AND JOURNAL ARTICLES

Bayly, I.A.E., 'Review of how indigenous people managed for water in desert regions of Australia', *Journal of the Royal Society of Western Australia*, 1999, vol. 82, pp. 17–23
Best, Elsdon, 'Maori agriculture: The cultivated food-plants of the natives of New Zealand, with some account of native methods of agriculture, its ritual and origin myths', (Wellington: Government Printer, 1976)
Bonnett, M.I., R.M. McDowall and J.R.E. Sykes, 'Critical habitats for conservation of giant kokopu, *Galaxias argenteus* (Gmelin, 1789)', *Science for Conservation 206*, Department of Conservation, 2002
Brenstrum, Erick, 'Looking back on El Nino', *New Zealand Geographic*, April–June, 1998
Broad, Harry, 'Farming in a fragile landscape', *Terra Nova*, July 1992, pp. 42–43
Burtenshaw, Michael K., 'Maori gourds: An American connection?', *Journal of the Polynesian Society*, Dec. 1999, vol. 108, p. 427
Colenso, William, 'On the vegetable food of the ancient New Zealanders before Cook's visit,' *Transactions of the New Zealand Institute*, vol. 13, 1880, pp. 14–15
Cox, Paul Alan, 'Forests, flying foxes and Samoans: An ecological triangle', *Plant Talks*, vol. 31, Dec 1991, p. 21
Cumberland, Kenneth B., 'A century's change: Natural to cultural vegetation in New Zealand', *Geographical Review 4*, vol. 31, Oct. 1941
Diamond, J., 'Rats as agents of extermination', *Nature*, vol. 318, 1985, pp. 602–03
Downes, T.W., 'Maori mentality regarding the lizard and taniwha in the Whanganui River area', *Journal of the Polynesian Society*, vol. 46, 1937, pp. 206–19
——, 'A tuahu on the Whanganui River', *Journal of the Polynesian Society*, vol. 37, 1928, pp. 165–67

Earle, Sylvia, 'Oceans: The well of life', *Time*, 28 Oct. 1996, pp. 52–53

Flores, M., P. Lazo, G. Campbell and A. Simeone, 'Breeding status of the red-tailed tropicbird and threats to its conservation', *Pacific Science*, Apr. 2017

Gon, Samuel M. et al., 'Āina Momona, Honua Au Loli – Productive Lands, Changing World: using the Hawaiian footprint to inform biocultural restoration and future sustainability in Hawai`i', *Sustainability*, vol. 10, 2018, p. 2

Graeme, Ann, 'Linking land to sea', *Forest & Bird,* Dec. 2019, pp. 46–48

Graves, Don, 'A comparison of methods to apply biochar to temperate soils', *International Journal of Agronomy*, vol. 2018, 2013, pp. 202–61

Grimble, Arthur, 'The migrations of a Pandanus people as traced from a preliminary study of food, food-traditions, and food-rituals in the Gilbert Islands', *Journal of the Polynesian Society*, vol. 42, 1933, pp. 51–84

Hamer, Paul, '"A tangled skein": Lake Horowhenua, Muaūpoko and the Crown 1898–2000', *Waitangi Tribunal Report,* 2015, pp. 8–10

——, 'One in six? The rapid growth of Māori population in Australia', *New Zealand Population Review,* 2008

Handy, E.S.C. and Mary Kawena Pukui, 'The Polynesian family system in Ka-`u, Hawai`i', *Journal of the Polynesian Society*, vol. 59, no. 3, 1950, pp. 234–36

Hau`ofa, Epeli, 'The ocean in us,' *The Contemporary Pacific*, vol. 10, no. 2, 1998, pp. 390–412

Holdsworth, D.K. and A.F. Mark, 'Water and nutrient input:output budgets: Effects of plant cover at seven sites in upland snow tussock grasslands in Eastern and Central Otago, New Zealand', *Journal of the Royal Society of New Zealand*, vol. 20, no. 1, 1990, pp. 1–24

Horrocks, Mark, Ian Smith, Scott Nicol and Rod Wallace, 'Sediment, soil and plant microfossil analysis of Māori gardens at Anaura Bay, eastern North Island, New Zealand: Comparison with descriptions made in 1769 by Captain Cook's expedition', *Journal of Archaeological Science*, vol. 35, September 2008, pp. 2446–64

Ingraham, Neil L., Alan F. Mark and Russell D. Frew, 'Fog deposition by snow tussock grassland on the Otago uplands: Response to recent review of evidence', *NZ Journal of Hydrology*, vol. 47, no. 2, 2008, pp. 107–22

Johns, Dilys, Geoffrey Irwin and Yun Sung, 'An early sophisticated East Polynesian voyaging canoe discovered on New Zealand's coast', *Proceedings of the National Academy of Sciences*, vol. 111, no. 41, 2014

Johnston, Anna, 'The strange career of William Ellis', *Victorian Studies*, vol. 49, no. 3, Spring 2007, pp. 491–501

Jupiter, Stacey, Kostantinos A. Stamoulis, Robert Whitter, Alan Friedlander, Kawika Winter et al., 'Place-based management can reduce human impacts on coral reefs in a changing climate', *Bulletin of the Ecological Society of America*, no. 100, 3 July 2019

Kear, D. and B.L. Wood, 'The hydrology of Upolu: The geology and hydrology of Western Samoa', *New Zealand Geological Survey Bulletin 63,* 1959, pp. 21–23

Kurashima, Natalie, 'The potential of indigenous food production under climate change in Hawai`i', *Nature Sustainability*, vol. 2, no. 3, Feb. 2019

Lichatowich, Jim and Williams, Richard, 'Faith in nature: The missing element in salmon management and mitigation programmes', *Osprey Journal*, May 2015

Lucas, Di, 'Tussock planting time again', *Terra Nova*, Sep. 1991, pp. 53–54

Luomala, Katherine, 'Porpoise and taro in Gilbert Islands: Myths and customs', *Fabula* (Berlin), vol. 18, nos 3–4, 1977

Macfarlane, Robert, 'Should this valley have rights?' *Guardian Weekly*, 8 November 2019, pp. 41–44

——, 'Where the wild things are', *Guardian*, 30 Oct. 2004

Mair, G., 'Notes on fishes in upper Whanganui River', *Transactions of the New Zealand Institute*, 12 May 1879, pp. 315–16

Mark, Alan F. and Katharine J.M. Dickinson, 'Maximising water yield with indigenous non-forest vegetation: A New Zealand perspective', *Frontiers in Ecology and Environment*, vol. 6, no. 1, February 2008

Mathews, R.H., 'The Aboriginal fisheries at Brewarrina', *Journal of the Royal Society of New South Wales*, vol. 37, 1903

McFadgen, Bruce G. and Peter Adds, 'Tectonic activity and the history of Wairau Bar, New Zealand's iconic site of earlier settlement', *Journal of the Royal Society New Zealand*, 2018, pp. 4–6

McNiven, Ian James, 'Dating Aboriginal stone-walled fish traps at Lake Condah, southeast Australia', *Journal of Archaeological Science*, vol. 39, no. 2, 2012, pp. 268–86

Métraux, Alfred, 'The ethnology of Easter Island', Bishop Museum Bulletin no. 160, 1940; reprinted Honolulu, 1971

Murray, P., R. Ford and S. Owen, 'Patterns of island change and persistence offer alternate adaptation pathways for atoll nations', *Nature Communications*, no. 9, article 605, 2018

Noble, M., P. Duncan, D. Perry, K. Prosper, D. Rose, S. Schinierer, G. Tipa, E. Williams, R. Woods and J. Pittock, 'Culturally significant fisheries: Keystones for management of freshwater social-ecological systems', *Ecology and Society*, vol. 21, no. 2, p. 22

Olson, M.D., 'Reconstructing landscapes: The social forest, nature and spirit-world in Samoa', *Journal of the Polynesian Society*, vol. 106, 1997

Palmer, J. Linton, 'A visit to Easter Island, or Rapa Nui, in 1868', *Journal of the Royal Geographic Society of London*, vol. 40, 1870

Pearce, Fred, 'Rivers in the sky: How deforestation is affecting global water cycles', *Yale Environment 360*, July 2018

Poirier, Robert and Doris Schartmueller, 'Indigenous water rights in Australia', *Social Science Journal*, vol. 49, 2012, pp. 317–24

Rigg, T. and J. Bruce, 'The Māori gravel soil of Waimea West, Nelson, New Zealand', *Journal of the Polynesian Society*, vol. 32, 1923, pp. 85–90

Rull, V., N. Canelias-Bolta, O. Margalef, S. Pla-Rabes, A. Saez and S. Giralt, 'Three millennia of climatic, ecological and cultural change on Easter Island', *Frontiers in Ecological Evolution*, 29 March 2016, p. 4

Simpson, Philip G., 'Pohutukawa and biodiversity', *Conservation Advisory Science Notes No. 100*, Department of Conservation, Wellington, 1994

Skinner, W.H., 'Ancient Maori canals: Marlborough', *Journal of the Polynesian Society*, vol. 21, no. 3, 1912, pp. 105–08

Smith, S. Percy, 'Lore of the Whare-Wananga, or teachings of the Maori college on religion, cosmology and history, written down by H.T. Whatahoro from the teachings of Te Matorohanga and Nepia Pohuhu, priests of the whare wananga, East Coast, New Zealand, *Polynesian Society Journal*, Memoir 4, 1915

Stevenson, Phillipa, 'The scandal of our PCP dumps', *Terra Nova*, June, July, August 2002

Summer, Sylva, 'Indigenising water law in the 21st century: Na moku aupuni o ko`olau hui. A native Hawaiian case study, *Cornel Journal of Law and Public Policy*, vol. 16, Summer 2007, p. 579

'The Hawaiian Moon Calendar and the Ways of the Practitioners (Kahuna)', rewritten from articles Lyons Kai`ioho Naone III authored and published in *Na Po`e Hawai`i Magazine*, Spring 1995–Winter 1996

Theroux, Paul, 'Some misconceptions about RLS', *Journal of Pacific History*, vol. 16, no. 3, July 1981, pp. 164–66

Titcomb, Mary with Mary Kawena Pukui, 'Native use of fish in Hawai`i', *Journal of the Polynesian Society*, vol. 60, no. 195, p. 105

Wells, John W., 'Coral growth and geochronometry', *Nature*, vol. 197, 1963, pp. 948–50

Williams, Alan N., Sean Lem, Ian D. Goodwin and Mike Smith, 'Hunter-gatherer response to late Holocene climatic variability in northern and central Australia', *Journal of Quaternary Science*, vol. 5, 2010, pp. 831–38

Young, David, 'Curated cross-roads heritage', *Phanzine*, Aug. 2018

——, 'Faces of the River', *NZ Listener*, 1986, pp. 11–14

——, 'The Old Man and the River', NZ Listener, 31 Mar. 1984, pp. 19–21

MISCELLANEOUS

'Aboriginal Water Rights', Mitchell Library 87170-AGG

Buck, Peter, 'Lectures on traditional Hawaiian history and culture', 1932, MS SC Buck 1301, Bishop Museum, Honolulu

Carbon News, 7 February 2019

Cromarty, Pam, 'A Directory of Wetlands in New Zealand', compiled for Department of Conservation, Wellington, 1996: www.doc.govt.nz/globalassets/documents/science-and-technical/nzwetlands00.pdf

Ellis, William, Ms-Papers-2862, ATL

Burrows, Edwin G., book review of Martha Warren Beckwith, *The Kumulipo: A Hawaiian creation chant, American Anthropologist*, vol. 3, no. 4, Oct.–Dec. 1951, pp. 552–53

Handy, E.S.C. and E.G. Handy, *Native Planters in Old Hawaii: Their life, lore and environment*, Bishop Museum Bulletin 233 (Honolulu: Bishop Museum, 1972)
Joy, Mike, 'Paradise squandered: New Zealand's environmental asset stripping', The 2014 Bruce Jesson Memorial Lecture: www.brucejesson.com/the-2014-bruce-jesson-lecture-mike-joy-paradise-squandered-new-zealands-environmental-asset-stripping/
Kennard, Rachel, 'The potential for Maori customary claims to freshwater' (LLB (Hons) thesis, 2009, University of Otago, Dunedin)
Maude, Harry, 'History of Banaba', in Alaima Talu and Max Quanchi (eds), *Messy Entanglements: The papers of the 10th Pacific History Association Conference*, Tarawa, Kiribati Pacific History Association, Brisbane, 1995
Meadows, Donella H., Dennis L. Meadows, Jorgen Randers and William W. Behrens III, *The Limits to Growth: A report for the Club of Rome's project on the predicament of mankind* (Pan Macmillan, 1974)
Sidey, Richard, 'Forest Spirits: Encounters with ghosts in the Great Bear Rainforest': www.richardsidey.com/spirit-bears
Star-Bulletin, Honolulu, 31 Mar. 2006 & 30 Apr. 2006
Stevenson, C.M., Ramirez, J.M., Morin, F.J. and Barbacci, N., *The Reñaca Papers, VI International Conference on Rapa Nui and the Pacific*, (Los Osos, CA: Easter Island Foundation and University of Valparaíso, 2004)

Acknowledgements

The Fulbright-Creative New Zealand Pacific Writer's Residency in Hawai`i was the platform that gave flight to this project. While it has taken time to complete, a work of this scope could not have been contemplated without the opportunity afforded by the institution founded by Senator J. William Fulbright's post-war idealism.

Vilsoni Heroniko, then director of the Institute of Pacific Studies, was a gracious host who introduced me to a number of occasions, classes, theatres and markets, to beaches and headlands and to folk and folkways in Honolulu. Jerry Finan, former director, Pacific Islands Development Program, East–West Center, and Pacific scholar the late Bob Kiste, were also naturally inclusive. Kawika Winter has been inspirational, living as he does by the abiding aloha principle of his people, of which this work has been a significant beneficiary. Kae`o Duarte, Carlos Andrante, Haunani-Kay Trask and Michelle Matson also contributed. Kiwi-expat couple Robert Sutherland and Anne Kennedy were welcoming and helpful, as were Alan Howard, Jan Rensel, Gary and Soloia Kamauoha and Kim, Takiora and Bryna.

James Lichatowich and Johnny Sundstrom generously inducted me into the ways and politics of salmon; Christopher Rempel and the front-staff at Portland State University Native American Student and Community Center offered First Nation perspectives. Tim Kurtz and Marion Dresner gave willingly of their knowledge. Tchanan Ross and Martin Jacobs played a supporting role in kindness at Siushlaw, while in San Francisco (and sometimes Awaroa), Tom Gaman was a willing touchstone. Amanda Johnson and Rosalie Hastwell in Australia came forward with helpful leads, as did John and Mary Cretney, Margaret Pointer and Ken Ross in New Zealand.

The long-standing Green Drinks Group (Wellington), especially members Wren Green, Clive Anstey, Paul Blaschke and Ralph

Chapman, have been invaluable in offering ecological reading, contacts and perspectives. Roger Lawrence was a vital source on Tamana. Rod Hay and the late David Butler assisted with the Sāmoa forest science; and Dale Smith's advice on photosynthesis was invaluable.

Beginning a book like this is always the easy part; finishing is a different matter. The on-going guidance from old friend Peter Horsley and support from Charles Dawson have been sustaining, the capacious knowledge of Ngārangi-Mata-Tauira-Tataiaro-Rangi-Te-Rangiuia vital. Robyn and Brian Bargh, as founders of Huia Press and publishers of the forerunner of this work, *Woven by Water: Histories of the Whanganui River*, need mention too. To all of them, my deep gratitude – thank you. I also acknowledge all those Māori, threaded through my previous work, who over nearly 40 years have patiently brought me along the pathway that this book has taken. So many of them have passed away. Moe mai rā i roto i te rangimarie.

The staff of the Hamilton and Sinclair Libraries, University of Hawai`i; the Bernice Pauahi Bishop Museum, Hawai`i; the Mitchell Library, Sydney; the Alexander Turnbull Library, Wellington; and the Tasman District and Elma Turner Library, Nelson, have been essential to the task.

Rhys Richards, Roger Lawrence, Eric Brenstrum, James Beattie, Harry Broad, Barry Rigby, Tony Brunt, Bruce Clarkson, Sally Asker and Cheryl Cameron have all provided encouragement, support and guidance, much of it of a historical or ecological nature. Old Pacific hand John O'Grady generously guided the trip to Fiji and to Tarawa, where Farran Redfern made much happen in a short, special time. Blair Styra, Karen White and Aboriginal elder and healer the late Gerry Bostock, all played their special part in the generation of this book.

My family, inevitably, have also contributed. Aliscia Young and Richard Sidey's more than generous role as professional photographers (galaxiid.com) is wonderfully obvious from the cover forward. Miriama and Reuben's support has been most helpful. Scilla's assistance in the finals stages of review has been essential. I thank them all, but from her, there is so much more given than can be said.

Finally, I wish to thank Otago University Press, their publications advisory committee, the two anonymous reviewers and especially publisher Rachel Scott for their immediate enthusiasm for the text and the book project. And to editors Gillian Tewsley, Imogen Coxhead and designer Fiona Moffat for the care they gave to the task in bringing it to publication. After a long exercise in solitude like this, there is nothing quite like the warm enfolding support of brave, capable publishers.

I dedicate this book to the late Howard Ellis: friend, devoted student of nature, conservationist and environmental leader, who brought a rare measure of strategic determination and brilliance, laughter and open-mindedness to his task of a lifetime.

Index

Bold denotes illustrations

About the photographers

RICHARD SIDEY is an award-winning conservation, adventure and wildlife photographer and filmmaker based in Wānaka, New Zealand. Richard has worked for nearly 20 years documenting the wilderness and its inhabitants from Antarctica to the High Arctic. A keen adventurer, bike-packer, pack-rafter, sea kayaker and paragliding pilot, he is seldom far from the wilderness.

ALISCIA YOUNG is an international photographer and award-winning filmmaker based in the Southern Alps of New Zealand. Over the last decade her creativity has taken her to some of the wildest and remotest locations on the planet. Aliscia has always had an affinity with water. She is a freediver and certified PADI open water diver. Nature and the underwater world are where she gathers true inspiration for her work.

Aliscia and Richard are co-founders of Galaxiid, a New Zealand production company creating immersive experiences through visual storytelling.